Optimal and Robust Scheduling for Networked Control Systems

AUTOMATION AND CONTROL ENGINEERING

A Series of Reference Books and Textbooks

Series Editors

FRANK L. LEWIS, Ph.D.,
Fellow IEEE, Fellow IFAC
Professor
Automation and Robotics Research Institute
The University of Texas at Arlington

SHUZHI SAM GE, Ph.D.,
Fellow IEEE
Professor
Interactive Digital Media Institute
The National University of Singapore

Optimal and Robust Scheduling for Networked Control Systems, *Stefano Longo, Tingli Su, Guido Herrmann, and Phil Barber*

Advances in Missile Guidance, Control, and Estimation, *edited by S.N. Balakrishnan, Antonios Tsourdos, and B.A. White*

End to End Adaptive Congestion Control in TCP/IP Networks, *Christos N. Houmkozlis and George A. Rovithakis*

Quantitative Process Control Theory, *Weidong Zhang*

Classical Feedback Control: With MATLAB® and Simulink®, Second Edition, *Boris J. Lurie and Paul J. Enright*

Intelligent Diagnosis and Prognosis of Industrial Networked Systems, *Chee Khiang Pang, Frank L. Lewis, Tong Heng Lee, and Zhao Yang Dong*

Synchronization and Control of Multiagent Systems, *Dong Sun*

Subspace Learning of Neural Networks, *Jian Cheng Lv, Zhang Yi, and Jiliu Zhou*

Reliable Control and Filtering of Linear Systems with Adaptive Mechanisms, *Guang-Hong Yang and Dan Ye*

Reinforcement Learning and Dynamic Programming Using Function Approximators, *Lucian Buşoniu, Robert Babuška, Bart De Schutter, and Damien Ernst*

Modeling and Control of Vibration in Mechanical Systems, *Chunling Du and Lihua Xie*

Analysis and Synthesis of Fuzzy Control Systems: A Model-Based Approach, *Gang Feng*

Lyapunov-Based Control of Robotic Systems, *Aman Behal, Warren Dixon, Darren M. Dawson, and Bin Xian*

System Modeling and Control with Resource-Oriented Petri Nets, *Naiqi Wu and Meng Chu Zhou*

Deterministic Learning Theory for Identification, Recognition, and Control, *Cong Wang and David J. Hill*

Sliding Mode Control in Electro-Mechanical Systems, Second Edition, *Vadim Utkin, Jürgen Guldner, and Jingxin Shi*

Linear Control Theory: Structure, Robustness, and Optimization, *Shankar P. Bhattacharyya, Aniruddha Datta, and Lee H. Keel*

Intelligent Systems: Modeling, Optimization, and Control, *Yung C. Shin and Chengying Xu*

Optimal Control: Weakly Coupled Systems and Applications, *Zoran Gajić, Myo-Taeg Lim, Dobrila Skatarić, Wu-Chung Su, and Vojislav Kecman*

Intelligent Freight Transportation, *edited by Petros A. Ioannou*

Modeling and Control of Complex Systems, *edited by Petros A. Ioannou and Andreas Pitsillides*

Optimal and Robust Estimation: With an Introduction to Stochastic Control Theory, Second Edition, *Frank L. Lewis, Lihua Xie, and Dan Popa*

Feedback Control of Dynamic Bipedal Robot Locomotion, *Eric R. Westervelt, Jessy W. Grizzle, Christine Chevallereau, Jun Ho Choi, and Benjamin Morris*

Wireless Ad Hoc and Sensor Networks: Protocols, Performance, and Control, *Jagannathan Sarangapani*

Stochastic Hybrid Systems, *edited by Christos G. Cassandras and John Lygeros*

Hard Disk Drive: Mechatronics and Control, *Abdullah Al Mamun, Guo Xiao Guo, and Chao Bi*

Autonomous Mobile Robots: Sensing, Control, Decision Making and Applications, *edited by Shuzhi Sam Ge and Frank L. Lewis*

Neural Network Control of Nonlinear Discrete-Time Systems, *Jagannathan Sarangapani*

Fuzzy Controller Design: Theory and Applications, *Zdenko Kovacic and Stjepan Bogdan*

Quantitative Feedback Theory: Fundamentals and Applications, Second Edition, *Constantine H. Houpis, Steven J. Rasmussen, and Mario Garcia-Sanz*

Chaos in Automatic Control, *Wilfrid Perruquetti and Jean-Pierre Barbot*

Differentially Flat Systems, *Hebertt Sira-Ramírez and Sunil K. Agrawal*

Robot Manipulator Control: Theory and Practice, *Frank L. Lewis, Darren M. Dawson, and Chaouki T. Abdallah*

Robust Control System Design: Advanced State Space Techniques, *Chia-Chi Tsui*

Linear Control System Analysis and Design: Fifth Edition, Revised and Expanded, *Constantine H. Houpis, Stuart N. Sheldon, John J. D'Azzo, Constantine H. Houpis, and Stuart N. Sheldon*

Nonlinear Control Systems, *Zoran Vukic*

Actuator Saturation Control, *Vikram Kapila and Karolos Grigoriadis*

Sliding Mode Control In Engineering, *Wilfrid Perruquetti and Jean-Pierre Barbot*

Modern Control Engineering, *P.N. Paraskevopoulos*

Advanced Process Identification and Control, *Enso Ikonen and Kaddour Najim*

Optimal Control of Singularly Perturbed Linear Systems and Applications, *Zoran Gajic*

Robust Control and Filtering for Time-Delay Systems, *Magdi S. Mahmoud*

Self-Learning Control of Finite Markov Chains, *A.S. Poznyak, Kaddour Najim, and E. Gomez-Ramirez*

Optimal and Robust Scheduling for Networked Control Systems

Stefano Longo • Tingli Su
Guido Herrmann • Phil Barber

CRC Press
Taylor & Francis Group
Boca Raton London New York

CRC Press is an imprint of the
Taylor & Francis Group, an **informa** business

CRC Press
Taylor & Francis Group
6000 Broken Sound Parkway NW, Suite 300
Boca Raton, FL 33487-2742

First issued in paperback 2017

ISBN-13: 978-1-4665-6954-6 (hbk)
ISBN-13: 978-1-138-07482-8 (pbk)

Library of Congress Cataloging-in-Publication Data

Longo, Stefano.
Optimal and robust scheduling for networked control systems / authors, Stefano Longo [and three others].
pages cm -- (Control series)
Summary: "This book offers a tool for optimal/robust control system integration via communication networks. This work is shaped around examples and relevant standards of the automotive industry for those examples but the concepts are readily extendable to any application that uses deterministic communication protocols between system components (avionic systems, robots, etc.). The underlying idea is to use the rigorous tools from optimal and robust multivariable control theory to solve the industrial scheduling problem in a transparent manner"-- Provided by publisher.
Includes bibliographical references and index.
ISBN 978-1-4665-6954-6 (hardback)
1. Robust control. 2. Automatic control. 3. Computer networks. I. Title.

TJ217.2.L64 2013
629.8'95--dc23 2012050720

Visit the Taylor & Francis Web site at
http://www.taylorandfrancis.com

and the CRC Press Web site at
http://www.crcpress.com

Dedicated to our families

Contents

List of Figures

List of Tables

List of Acronyms

ACC	Adaptive Cruise Controller
ACCS	Adaptive Cruise-Control System
A/D	Analogue to Digital
AFDX	Avionics Full-Duplex switched ethernet
ARM	Advanced Reduced instruction set computer Machine
CAN	Controller Area Network
CARE	Continuous Algebraic Riccati Equation
CSMA	Carrier Sense Multiple Access
CSMA/BA	CSMA with Bitwise Arbitration
CSMA/CA	CSMA with Collision Avoidance
CSMA/CD	CSMA with Collision Detection
D/A	Digital to Analogue
DARE	Discrete Algebraic Riccati Equation
DOF	Degrees Of Freedom
ECU	Electronic Control Unit
GA	Genetic Algorithm
HEV	Hybrid Electric Vehicle
HIL	Hardware-In-the-Loop
IEEE	Institute of Electrical and Electronics Engineers
iid	independent and identically distributed
LEF	Large Error First
LHS	Left Hand Side
LIN	Local Interconnect Network
LMI	Linear Matrix Inequality
LQ	Linear Quadratic
LQG	Linear Quadratic Gaussian
LQR	Linear Quadratic Regulator
LTI	Linear Time-Invariant
LTV	Linear Time-Varying
MAC	Medium Access Control
MATI	Maximum Allowable Transfer Interval
MEF-TOD	Maximum Error First with Try Once Discard
MIMO	Multiple Input Multiple Output
MJLS	Markovian Jump Linear System
MPC	Model Predictive Control
MUF	Maximum Urgency First

NCS	Networked Control System
NP	Non-deterministic Polynomial-time
ODE	Ordinary Differential Equation
OPP	Optimal Pointer Placement
OSI	Open System Interconnection
PBH	Popov-Belevitch-Hautus
PDU	Protocol Data Unit
PID	Proportional-Integral-Derivative
PSO	Particle Swarm Optimization
QoP	Quality of Performance
QoS	Quality of Service
RHS	Right Hand Side
RR	Round Robin
RTCS	Real-Time Control Systems
SMS	Second-Moment Stability
SOS	Sum-Of-Squares
s.t.	subject to
TDMA	Time Division Multiple Access
TSP	Traveling Salesman Problem
TTCAN	Time Triggered CAN
TTC-SC	Time Triggered CAN with Shared Clock
ZOH	Zero Order Hold

Notation and Symbols

$\mathbb{N}$	Set of natural numbers, *i.e.* $\{1, 2, \ldots\}$.
$\mathbb{N}_0$	$\mathbb{N} \cup \{0\}$.
$\mathbb{R}$	Set of real numbers.
$\mathbb{R}^m$	m-dimensional space with real coordinates.
$\mathbb{Z}$	Set of integers.
$\lvert \cdot \rvert$	Absolute value.
$\lVert \cdot \rVert$	Euclidian norm.
$\mathcal{L}_2$	Lebesgue space with finite $\mathcal{L}_2$-norm.
ℓ_2	Sequence space with finite ℓ_2-norm.
$\lVert \cdot \rVert_{\mathcal{L}_2}$	Induced $\mathcal{L}_2$-gain.
$\lVert \cdot \rVert_{\ell_2}$	Induced ℓ_2-gain.
$\varnothing$	Empty set.
$\in$	Belong to.
$\cup\ \bigcup$	Union.
$\subset, \subseteq$	Subset.
$\max\{\mathcal{A}\}, \min\{\mathcal{A}\}$	Maximum, minimum value of set $\mathcal{A}$.
$\lvert\mathcal{A}\rvert$	Cardinality of set $\boldsymbol{A}$.
$\boldsymbol{A}^T = (\boldsymbol{A})^T, \boldsymbol{a}^T$	Transpose of matrix $\boldsymbol{A}$, transpose of vector $\boldsymbol{a}$.
$\boldsymbol{A} > 0, \boldsymbol{A} \geq 0$	Matrix $\boldsymbol{A}$ positive definite, positive semi-definite (similarly for negative definiteness).
$\operatorname{rank}(\boldsymbol{A})$	Rank of matrix $\boldsymbol{A}$.
$\operatorname{tr}(\boldsymbol{A})$	Trace of a matrix $\boldsymbol{A}$, *i.e.* $\sum_{i=1}^{n} a_{ii}$.
$\operatorname{diag}(a_1, a_2, \ldots, a_n)$	Matrix with vector $\boldsymbol{a} = [a_1\ a_2,\ \ldots,\ a_n]$ in the main diagonal and zeros elsewhere.
$\prod_{i=1}^{n} \boldsymbol{A}(i)$	Matrix product *i.e.* $\boldsymbol{A}(1)\boldsymbol{A}(2)\cdots\boldsymbol{A}(n)$.
$\boldsymbol{I}_m$	Identity matrix of dimensions $(m \times m)$ (or simply $\boldsymbol{I}$ when obvious).
$\mathbf{0}_{m\times n}$	Zero matrix of dimensions $(m \times n)$ (or simply $\mathbf{0}$ when obvious).
$\boldsymbol{e}_k, 1 \leq k \leq p$	Column p-dimensional standard basis vector with a 1 in the kth coordinate and 0 elsewhere (unless otherwise stated).
$\lambda_{max}(\boldsymbol{A}), \lambda_{min}(\boldsymbol{A})$	Maximum, minimum eigenvalue of matrix A.
$\lambda(\boldsymbol{A})$	Set of all eigenvalues of matrix A.
$\lvert\lambda\rvert$	Complex modulus of λ.
$\mathbb{E}\{X\}$	Expected value of the random variable X.

$\Pr(\cdot)$	Probability.
$\lceil \cdot \rceil$	Ceiling function.
$\lfloor \cdot \rfloor$	Floor function.
$\stackrel{\text{def}}{=}$	Equal by definition.
sup	Supremum operator.
$\binom{n}{k}$	Binomial coefficient (the choose function of n and k).
$O(g(x))$	Big O notation. If $f(x) = O(g(x))$, then there are values c and k such that $0 \leq \lvert f(x)\rvert \leq c\lvert g(x)\rvert, \forall x \geq k$.
$\circ$	Hadamard product.
$\otimes$	Kronecker product.
$\mathrm{GCD}(a, b, \ldots)$	Greatest Common Divisor of scalars $a, b, \ldots$.
$\mathrm{LCM}(a, b, \ldots)$	Least Common Multiple of scalars $a, b, \ldots$.
$\mathrm{mod}(a, b)$	Modulo operator.
h	Sampling period.
$j \in \mathbb{N}_0$	Sampling instant.
$\tau(j)$	Time delay at sampling instant j.
J	Quadratic cost.
γ	$\mathcal{H}_\infty$ cost.
σ	Communication sequence (Definition 3.2, p. 51 and Definition 3.6, p. 64).
$p \in \mathbb{N}$	Period of the communication sequence.
Δ	Set of nodes.
$\Delta(i)$	Node i where i is the identifier (Definition 3.1, p. 50).
$d = \lvert\Delta\rvert$	Number of nodes.
$N = (\Delta, \sigma)$	Communication network, a time-varying star (Definition 3.3, p. 51).
ζ	Tuple of sampling times (Definition 5.6, p. 125).
$\mathbb{F}(\mathcal{G}, \mathcal{H})$	Closed-loop mapping from the exogenous input to the regulated output of the generalized plant $\mathcal{G}$ in feedback with controller $\mathcal{H}$.
$\mathbb{L}_D\{\cdot, \cdot\}$	Discrete-time lifting operator (Definition 3.7, p. 69).
$\mathbb{L}_C\{\cdot, \cdot\}$	Continuous-time lifting operator (Definition 8.1, p. 170).
▲	End of Definition.
▼	End of Observation.
♦	End of Example.
•	End of Remark.
■	End of Proof.
$\boldsymbol{A}^c, \boldsymbol{B}^c, \boldsymbol{C}, \boldsymbol{D}$	Matrices of the continuous LTI plant $\mathcal{P}^c$.
$\boldsymbol{A}, \boldsymbol{B}, \boldsymbol{C}, \boldsymbol{D}$	Matrices of the discrete LTI plant $\mathcal{P}$.
$\boldsymbol{A}_K^c, \boldsymbol{B}_K^c, \boldsymbol{C}_K, \boldsymbol{D}_K$	Matrices of the continuous LTI controller $\mathcal{K}^c$.

$\boldsymbol{A}_K, \boldsymbol{B}_K, \boldsymbol{C}_K, \boldsymbol{D}_K$	Matrices of the discrete LTI controller $\mathcal{K}$.
$\boldsymbol{S}_A(\cdot), \bar{\boldsymbol{S}}_A(\cdot)$	Matrices of the actuator scheduler $\mathcal{S}_A$.
$\boldsymbol{S}_S(\cdot), \bar{\boldsymbol{S}}_S(\cdot)$	Matrices of the sensors scheduler $\mathcal{S}_S$.
$\boldsymbol{S}_D(\cdot), \bar{\boldsymbol{S}}_D(\cdot)$	Matrices of the demand scheduler $\mathcal{S}_D$.
$\hat{\boldsymbol{A}}(\cdot), \hat{\boldsymbol{B}}(\cdot), \hat{\boldsymbol{C}}(\cdot), \hat{\boldsymbol{D}}(\cdot)$	Matrices of the LTV augmented plant $\hat{\mathcal{P}}$.
$\hat{\boldsymbol{A}}_K(\cdot), \hat{\boldsymbol{B}}_K(\cdot), \hat{\boldsymbol{C}}_K(\cdot), \hat{\boldsymbol{D}}_K(\cdot)$	Matrices of the LTV augmented controller $\hat{\mathcal{K}}$.
$\hat{\boldsymbol{A}}_{cl}(\cdot), \hat{\boldsymbol{B}}_{cl}(\cdot), \hat{\boldsymbol{C}}_{cl}(\cdot), \hat{\boldsymbol{D}}_{cl}(\cdot)$	Matrices of the LTV augmented closed-loop system $\hat{\mathcal{P}}_{cl}$.
$\bar{\boldsymbol{A}}, \bar{\boldsymbol{B}}, \bar{\boldsymbol{C}}, \bar{\boldsymbol{D}}$	Matrices of the discrete-time lifted plant (with ZOH) $\bar{\mathcal{P}}$.
$\breve{\boldsymbol{A}}, \breve{\boldsymbol{B}}, \breve{\boldsymbol{C}}, \breve{\boldsymbol{D}}$	Matrices of the discrete-time lifted plant (without ZOH) $\breve{\mathcal{P}}$.
$\tilde{\boldsymbol{A}}_k, \tilde{\boldsymbol{B}}_k$	Matrices of the k-lifted plant.
$\boldsymbol{x}(t), \boldsymbol{x}(j) \in \mathbb{R}^{n_x}$	State vector of $\mathcal{P}^c$, $\mathcal{P}$.
$\hat{\boldsymbol{x}}(j) \in \mathbb{R}^{n_x+n_u}$	State vector of $\hat{\mathcal{P}}$.
$\boldsymbol{x}_K(t), \boldsymbol{x}_K(j) \in \mathbb{R}^{n_k}$	State vector of $\mathcal{K}^c$, $\mathcal{K}$.
$\hat{\boldsymbol{x}}_K(j) \in \mathbb{R}^{n_u+n_d+n_k+n_y}$	State vector of $\hat{\mathcal{K}}$.
$\boldsymbol{x}_A(j) \in \mathbb{R}^{n_u}$	State vector of $\mathcal{S}_A$.
$\boldsymbol{x}_S(j) \in \mathbb{R}^{n_y}$	State vector of $\mathcal{S}_S$.
$\boldsymbol{x}_D(j) \in \mathbb{R}^{n_d}$	State vector of $\mathcal{S}_D$.
$\hat{\boldsymbol{x}}_{cl}(j) \in \mathbb{R}^{n_x+n_u+n_d+n_k+n_y}$	State vector of $\hat{\mathcal{P}}_{cl}$.
$\boldsymbol{u}(t), \boldsymbol{u}(j) \in \mathbb{R}^{n_u}$	Output vector of $\mathcal{K}^c$, $\mathcal{K}$.
$\hat{\boldsymbol{u}}(j) \in \mathbb{R}^{n_u}$	Scheduled input vector of $\mathcal{P}$ (unless otherwise stated).
$\bar{\boldsymbol{u}}(pl) \in \mathbb{R}^{pn_u}$	Discrete-time lifted version of $\boldsymbol{u}(j)$.
$\boldsymbol{y}(t), \boldsymbol{y}(j) \in \mathbb{R}^{n_y}$	Measured output vector of $\mathcal{P}^c$, $\mathcal{P}$.
$\hat{\boldsymbol{y}}(j) \in \mathbb{R}^{n_y}$	Scheduled output vector of $\mathcal{P}$ or input vector of $\mathcal{K}$ (unless otherwise stated).
$\bar{\boldsymbol{y}}(pl) \in \mathbb{R}^{pn_y}$	Discrete-time lifted version of $\boldsymbol{y}(j)$.
$\boldsymbol{d}(j) \in \mathbb{R}^{n_d}$	Demand input vector of $\mathcal{S}_D$.
$\hat{\boldsymbol{d}}(j) \in \mathbb{R}^{n_d}$	Scheduled demand input vector of $\mathcal{K}$.
$\bar{\boldsymbol{d}}(pl) \in \mathbb{R}^{pn_d}$	Discrete-time lifted version of $\boldsymbol{d}(j)$.
$\boldsymbol{w}(j) \in \mathbb{R}^{n_w}$	Exogenous input vector of $\mathcal{P}$ (unless otherwise stated).
$\bar{\boldsymbol{w}}(pl) \in \mathbb{R}^{pn_w}$	Discrete-time lifted version of $\boldsymbol{w}(j)$.
$\boldsymbol{z}(t), \boldsymbol{z}(j) \in \mathbb{R}^{n_z}$	Performance output vector of $\mathcal{P}^c$, $\mathcal{P}$ (unless otherwise stated).

Preface

The art and issues of using one medium for conveying more than one message can be considered to have its roots in the cacophony of the prehistoric jungles. Mankind may have brought order to the debate, and the invention of the telephone and later radio meant that, given enough time, an almost limitless amount of information could be conveyed over unprecedented distances, but along with this ability came the issue of who could transmit, when. An early example of an agreed protocol was when maritime users of the 500kHz radio band would maintain silence and listen for distress signals for two minutes every quarter of an hour. An outcome of the Titanic disaster.

Remote control also flourished with the electrical age. At first, telegraph signals were sent to human operators to control the movement of trains. More recently standard 2-20mA current loop signals are wired from central control rooms to remote sensors and actuators on large chemical and industrial plants.

Automatic control had been long established before electronics; the machines of the industrial revolution had mechanical governors performing proportional, integral and derivative control. The mathematics used to describe, analyze and solve the stability issues of feedback control was also well advanced.

It was, however, late in the last century, that the development of digital electronics facilitated a communication and control revolution. Many signals could be sent through the same pair of wires to and from many different sensors and actuators. The viability of networked control systems was established.

In the automotive field, the cost, weight and reliability benefits of digitally multiplexed wiring were being recognized and bulky, heavy, expensive copper wiring looms and connectors were replaced with new microprocessor control units. By the turn of the millennium, high-speed serial communication systems, such as the Control Area Network (CAN), were becoming prevalent.

Time division multiplexed control becomes a networked control when several systems or control loops want to use the same communication mechanism at the same time. The CAN protocol mentioned above includes contention prioritizing so that it is quite clear which message has priority and which must wait their opportunity. As the network traffic increases, this becomes more and more of an issue. Fortunately the development of the capability of the individual control units and the transport layers also accelerated. Data transfer rates of hundreds of thousands of bits per second became normal, permitting the replacement of protocols such as CAN with time division multiplexed protocols such as FlexRay.

The growth of computer science as a profession also brought with it ideas of data abstraction and functional decomposition where, in order to cope with the complexity of the emerging systems, information on the source and nature of data was deliberately obscured and the requirements of signals to be acted on was devolved to smart actuators whose internal operation became irrelevant to the higher level functions. No longer did a car window control have to worry about the current in its motors, it had to worry about whether to allow the driver to open and close the window with the ignition on, or with the engine running, or whether to allow the children in the back to open and close their windows.

Developments of automatic control and the reliability of electronics creates opportunities for much more critical closed loop systems. Electronic control of the engine power of cars; the throttle, the fuel injection and the ignition timing has become universal. Systems such as anti-lock braking (ABS) and stability control (ESP) have the potential for computers to apply or release the brakes automatically. Power stations, railway systems and aircraft rely on electronic control for their safe operation, particularly in situations too complex for humans to understand and correctly respond to.

It is this unique combination of the two critical and rapidly developing fields of control theory and network analysis that is the subject of this book: the analysis and robust integration of control systems across a time deterministic bus such as FlexRay.

Tools known from robust and optimal control theory are framed in a networked control context to permit the reliable, robust and performance-oriented off-line design of control systems across control networks with time deterministic behavior. The designer's certainty of the selection of the right network configuration is given through well-known measures for robust performance, while the speed of the offline optimization is guaranteed through using fast stochastic algorithms. This is an additional unique point of this book, as it also develops an engineering tool with a focus on robustness and performance in networked control integration.

How to read this book

This book covers the topic at different levels, from technical results to practical applications. It has not been designed to be read from first to last page. This section provides a road map for the book (see Figure 1 for a graphical representation).

Chapter 1 is a short motivational introduction. **Chapter 2** is a comprehensive review of the techniques commonly used to handle Networked Control Systems (NCSs) and it should give a feel of the problems and challenges involved. **Chapter 3** describes our framework for the modelling of NCSs with

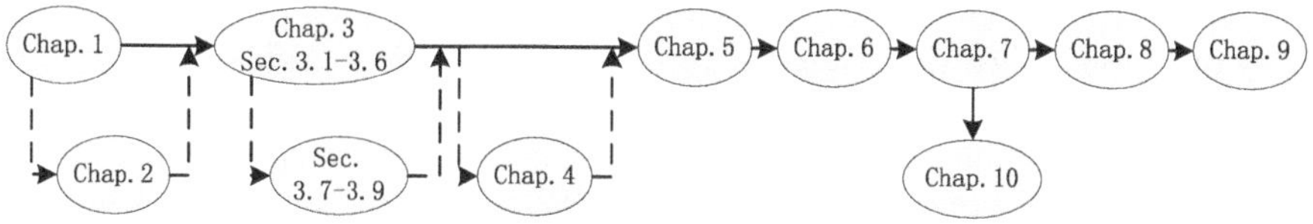

FIGURE 1: Book road map.

time-triggered communication. This framework is meant to unify existing techniques under a unique method which can handle a large class of NCSs. The modeling techniques described in Sections 3.2-3.6 are used throughout the book and they are fundamental for the understanding of the following chapters. Sections 3.7-3.9 are useful extensions to the method but not necessary for the understanding of the other chapters (with exception of some concepts from Section 3.7 that will be used in Section 9.5). **Chapter 4** provides technical results needed to prove the preservation of the structural properties of NCSs. The interested reader can learn how these results were obtained; however, they are not necessary for the understanding of the following chapters. **Chapter 5** discusses the procedures adopted for the communication sequence optimization. This chapter is important to understand the results of the following chapters. **Chapter 6** presents the techniques for the codesign of controller and communication sequences. In **Chapter 7**, we present a technique where we assume that a controller is given and therefore only the communication sequence needs to be designed. **Chapter 8** extends the idea of the previous chapter for the design of robust communication sequences. In **Chapter 9**, the design techniques developed in the previous chapters are applied to hardware in the loop automotive control systems. **Chapter 10** is an extension to the nonlinear case.

Acknowledgments

The authors would like to acknowledge the department of Automotive Engineering at Cranfield University, the department of Mechanical Engineering at the University of Bristol, and Beijing Institute of Technology for the support during the preparation of this book. Thanks to the Royal Society, UK, for an equipment grant (Research Grant/Round 2007/R2), and to the National Natural Science Foundation of China and the Royal Society, UK, for a highly useful and successful international collaboration grant (Grant No. 61011130163/JP090823). Also, thanks to Jaguar and Land Rover for the useful discussions, to TTE®-Systems for help with the hardware used in Chapter 9, and to Vector Informatik GmbH for providing the hardware and software tools used in Section 9.5.

Some of the material in this book (text, figures, and tables) has been previously published, by the same authors, in scientific journals.

Chapter 4 is based on the publication [S. Longo, G. Herrmann and P. Barber, Stabilizability and detectability in networked control, *IET Control Theory & Applications*, vol. 4, no. 9, pp. 1612-1626, 2010] and the material is reproduced by permission of the Institution of Engineering and Technology.

Chapters 5 and 6 are based on the publications [S. Longo, G. Herrmann and P. Barber, Optimization approaches for controller and schedule codesign in networked control, in 6th IFAC symposium on Robust Control Design (ROCOND09), vol. 6-1, 2009], [S. Longo, G. Herrmann and P. Barber, Controllability, observability in networked control, in 6th IFAC symposium on Robust Control Design (ROCOND09), vol. 6-1, 2009] and the material is reproduced by permission of the International Federation of Automatic Control.

Chapter 8 and part of Chapter 9 are based on the publications [S. Longo, G. Herrmann and P. Barber, Robust scheduling of sampled data networked control systems, *IEEE Control System Technology*, vol. PP, no. 99, pp. 1-9, 2011], [S. Longo, T. Su, G. Herrmann, P. Barber and U. Gerlinger, Scheduling of the FlexRay static segment for robust controller integration, 2011 IEEE Multi-Conference on Systems and Control, 2011] and the material is reproduced by permission of the Institute of Electrical and Electronic Engineers.

Chapter 10 is based on the publication [T. Su, S. Longo, G. Herrmann and P. Barber, Computation of an optimal communication schedule in a nonlinear networked control system using Sum-of-Squares, *Systems & Control Letters*, vol. 61, no. 3, pp. 387396, 2012] and the material is reproduced by permission of Elsevier.

Computer simulations and graphs were produced using MATLAB®. MATLAB and Simulink are registered trademarks of The Mathworks, Inc. For product information, please contact:
The MathWorks, Inc.
3 Apple Hill Drive
Natick, MA 01760-2098 USA
Tel: 508-647-7000
Fax: 508-647-7001
E-mail: info@mathworks.com
Web: www.mathworks.com

Author Biographies

Dr. Stefano Longo was born in Syracuse, Italy. He received his M.Sc. and Ph.D., both in control systems, respectively from the University of Sheffield (UK) in 2007 and the University of Bristol (UK) in 2011. His Ph.D. thesis was awarded the prestigious Institution of Engineering and Technology (IET) Control and Automation Prize.

In November 2010 he was appointed for the position of Research Associate in the Department of Electrical and Electronic Engineering at Imperial College London (UK) where he worked in the intersection of the Control & Power and Circuit & Systems Research Groups.

Since August 2012 he has been a Lecturer in Vehicles Electrical and Electronic Systems in the Department of Automotive Engineering at Cranfield University (UK).

His work and his research interests gravitate around the problem of implementing advanced control algorithms in hardware, where the controller design and the hardware implementation are not seen as two separate and decoupled problems, but as a single interconnected one.

Dr. Tingli Su was born in Liaoyang, China. She received her Bachelor degree in Mechatronic Engineering (major in control theory and application) from Beijing Institute of Technology in 2007, and directly started her Ph.D. program (at the same university). In 2009 she was invited by the University of Bristol as an exchange Ph.D. student, sponsored by the Chinese Scholarship Council during the first year and Jaguar & Land Rover Research the following year.

During the exchange period, she explored the field of nonlinear Network Control Systems (NCS) with communication constraints. Optimization problems for this type of systems, subject to performance and robustness requirements, were considered. Moreover, the application of the novel time-triggered communication protocol FlexRay to nonlinear NCS was also investigated.

Dr. Guido Herrmann received the German degree Diplom-Ingenieur der Elektrotechnik (with highest honours) from the Technische Universität zu Berlin and was awarded for this the Erwin-Stephan Prize. He spent a year of his undergraduate studies at Heriot-Watt-University (UK) funded by a grant from the German Academic Exchange Service. In 2001, he received a Ph.D. from the University of Leicester. The Ph.D. was sponsored first by the

Daimler-Benz-Foundation (Germany) and later by a Marie-Curie Fellowship (European Commission).

From 2001 to 2003, he worked in the mechatronics and micro-systems group of the A*Star Data Storage Institute (Singapore) doing research and consultancy for the data storage industry. From 2003 until February 2007, he was a Research Associate, Research Fellow and a Lecturer in the Department of Engineering at Leicester University. In March 2007, he took up a permanent lecturing position at the Department of Mechanical Engineering of the University of Bristol. In August 2009, he was promoted to Senior Lecturer. Since August 2012, he has been a Reader in Control and Dynamics. He was at several occasions a Visiting Lecturer and a Visiting Professor in Malaysia and Singapore; in September 2005, he was a Visiting Professor at the Data Storage Institute.

His research considers the development and application of novel, robust, and nonlinear control systems. He has been collaborating with several institutions in Australia, China, Malaysia, Singapore and the USA, e.g. the A*Star Data-Storage Institute, Singapore, the Beijing Institute of Technology, the National University of Singapore, and the University of Texas, Arlington. He has been working with companies such as Western Digital on hard disk drive servo problems and Jaguar Land Rover on networked control systems and vehicle parameter estimation. He is a Senior Member of the IEEE, a Technical Editor of the IEEE/ASME Transactions on Mechatronics journal, and an Associate Editor of the International Journal on Social Robotics.

Dr. Phil Barber, after gaining his degree in Electrical and Control Engineeing at Birmingham University, including a year out at GEC Hirst's Research Centre working on fibre optic sensors and fluxgate magnetometers, started his career in the auto industry in 1985 at the then Lucas Research Centre in Birmingham working on spark advance angle perturbation methods for driveline shunt mitigation and other real time engine control algorithms such as the use of floating point arithmetic to derive measures for engine roughness.

He gained his Ph.D. in 1992 from the University of Warwick after researching the cycle-by-cycle transient fuelling characteristics of fuel injected engines for 3 years as a research fellow.

Joining Jaguar in 1990, and after working on drivetrain simulation, he went on to be part of the team working on steer by wire; lane keeping and adaptive cruise control (ACC) demonstrators for the PROMETHEUS project. The Adaptive Cruise Control won the Henry Ford Technical Award, the highest award for technical achievement in the Ford Motor Corporation in 2000.

From 2000, working in mainstream engineering he oversaw the changeover of ACC radar manufacture from a military background to automotive production standards. He went on to leading the ACC delivery team in mainstream, then the teams for Brake Electronics and Software Delivery.

He returned to research in 2008 as the Technical Specialist for Chassis Systems and Vehicle Capability. Current interests include vehicle dynamics, state estimation, distributed and networked systems for real time vehicle control, and regenerative braking.

He is trained in lean manufacturing methods from the Halewood Lean Learning Academy (2007). As member of the Institution of Engineering and Technology, (Membership number 24295008), he serves on the executive team of their Automotive and Road Transport Systems Network. He is also registered with the Engineering Council as a Chartered Engineer.

1

Introduction

CONTENTS

A control system is a device or a collection of devices used to guide and regulate the behavior and operation of other devices or dynamical systems. In mathematics and engineering, control theory is the discipline that studies the behavior of dynamical systems.

"According to [35], we may call the period from 1868 to the early 1900s the primitive period of automatic control. It is standard to call the period from then until 1960 the classical period, and the period from 1960 through present times the modern period [72]. During those periods, our society went from agricultural, mechanical or steam engine, electrical, electronic, to computer ages, while our control theory and applications experienced the changes from *ad hoc* applied mathematical problems, frequency-domain design, state space approach, and discrete event and hybrid dynamical systems, to intelligent control algorithms.

With no exception, every breakthrough in technology has caused a milestone change or paradigm shift in automatic control. Now we have entered the age of a networked society, people are expecting and experiencing a new connected lifestyle in a connected world where you can 'compute anywhere, connect anything'. What would be the corresponding milestone change or paradigm shift for automatic control? To be specific, what will we design and use to control networked devices and systems in the new age?" [130, p. 26].

1.1 Overview

1.1.1 Real-time control systems

In a real-time system, the value of a task depends not only on the correctness of its computation but also on the time at which results are available. Control systems constitute an important subclass of real-time systems. Rapid changes in technology have increased the number of cheap and powerful embedded devices where almost all control algorithms are nowadays implemented. In order to cut down costs, Real-Time Control Systems (RTCS) often have computing and communication resource limitations. Such limitations are the main cause of time intervals between a sampling instant and the corresponding actuating instant. This time interval is known as delay and its detrimental effect comes from the fact that controller and/or actuators are forced to use 'old' information about the plant/controller. Delays may vary from one sample to another giving rise to delay jitter.

1.1.2 Networked control systems

Multiple Input Multiple Output (MIMO) RTCS are becoming complex and the traditional point-to-point communication has been replaced by an increasing number of networked architectures. Large amounts of data are now transmitted over the same medium.

In classical sampled-data control theory, the control engineer often makes the assumption that the sampling time is equidistant and periodic and the delays are negligible or at least constant. Furthermore, control systems are idealized in a way that information exchange is perfect and happens instantaneously. This is rarely true and the problem acquires new dimensions if the control system is a Networked Control System (NCS) (sometimes called a Network-Based Control Systems). In an NCS, sensors, actuators and controllers are spatially distributed and interconnected via a shared communication medium. More generally, in an NCS all control loops are closed through a real-time network as shown in Figure 1.1.

The term NCS is being used to describe a large range of systems. The following example should give a taste of the type of situations where NCSs arise.

Example 1.1 *Consider a platoon of unmanned vehicles as represented in Figure 1.2. Each vehicle has a radar that measures the distance to the vehicle in front and a controller that acts on the throttle and brake actuators to keep such distance constant. The controller in each vehicle has been carefully designed to guarantee stability and some level of performance which, for this example, is achieved by a slightly underdamped transient behavior (oscillations around the settling point). If vehicle i encounters a disturbance it will oscillate before*

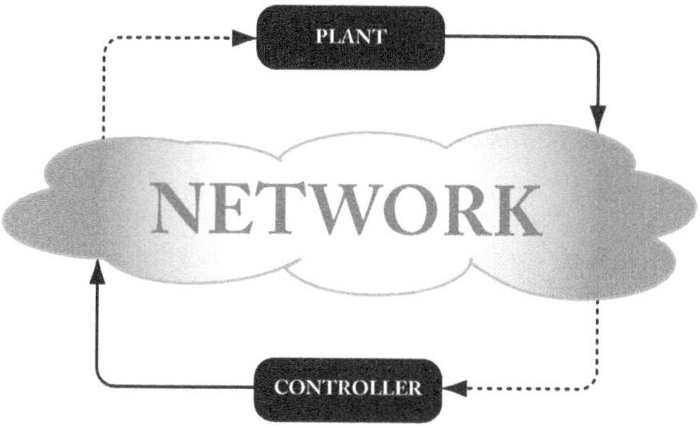

FIGURE 1.1: Schematic diagram of a networked control system (NCS): the traditional control loops are now closed through a network.

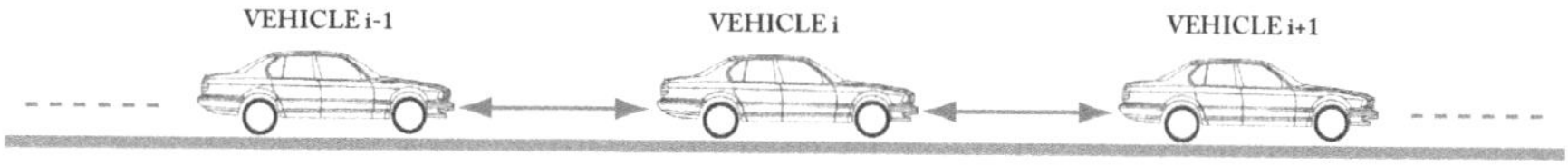

FIGURE 1.2: A platoon of unmanned vehicles: an example of an NCS.

reaching steady-state again. Vehicle $i-1$ (the one behind) will oscillate to track its (oscillating) reference point (the vehicle in front) and will add some more oscillations due to its own dynamics. The oscillation will increasingly propagate through the platoon until a vehicle will crash. How is it possible that stable control systems ended up in such an unstable scenario? The reason is because the unmanned vehicle platoon is an NCS (a system of interconnected subsystems) where the interconnections between subsystems (vehicles) have been ignored at the controller design stage. This NCS can be represented by Figure 1.3. Now that the problem has been identified, controllers can be

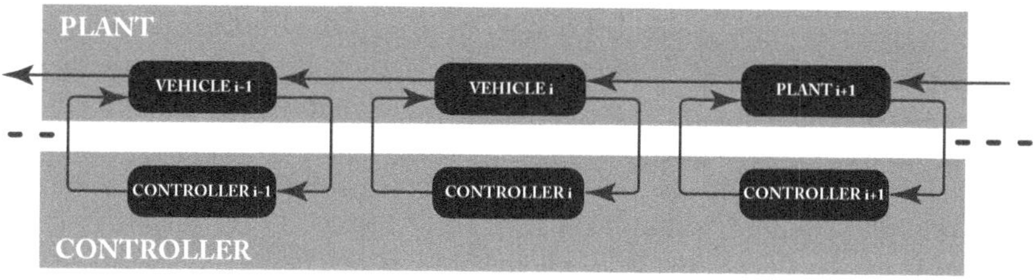

FIGURE 1.3: Interconnected systems: new problems like string stability arise.

designed to ensure overall stability (or more precisely, string stability). But, since the subsystems are interconnected, is it sufficient for a local controller, to achieve good performance, only to have information about the subsystem's states? It is likely that performance can be increased if controller i can have *information about vehicles* $i+1$, $i+2$, *and so on. But how many exactly? This and others are all interesting problems related to NCS, however the focus here is on a more specific aspect of NCSs.*

Let us assume that it has been decided that controller i *achieves satisfactory performance if it has information about three vehicles in front,* i.e. *from subsystem* $i+1$, $i+2$ *and* $i+3$. *Such information can only be exchanged by wireless communication, and it is unlikely that there will be a dedicated channel (frequency carrier) for each signal. Most probably, all vehicles will have to share their communication medium,* i.e. *they will have to schedule their wireless communication. Now, the next question is: what is the 'best' way of scheduling communication between subsystems? And since all vehicles can be seen as a whole large plant and all controllers as a whole large controller, what is the 'optimal' communication schedule between plant and controller?* ♦

If pre-planning of a particular communication policy at the design stage is possible, then the last question of Example 1.1 is one of the problems this book is trying to address. However, Example 1.1 is just a special case where only sensor signals need to be scheduled. A more generic situation would be the one of a system with many sensor, demand, and actuator signals that share the same communication medium and therefore need to be scheduled. The following is another example.

Example 1.2 *A plant with three outputs (sensors) and two inputs (actuators) needs to be controlled. The controller is digital and it is connected to the plant via a network. Due to the limited bandwidth, the controller can only take some of the sensor measurements and/or send a control signal to the actuators at each sampling instant. Let us assume that, for that particular NCS configuration, at time instant* $j=1$ *only data from sensors 1 and 3 are available, at* $j=2$ *only actuator 1 can be controlled, at* $j=3$ *only actuator 2 can be controlled and at* $j=4$ *only data from sensor 2 are available. This communication pattern repeats periodically such that at* $j=5$ *data from sensors 1 and 3 are available and so on. This scenario is depicted in Figure 1.4. If we have the freedom (as it is often the case) to design the communication pattern, what will be the one that is closer, according to some performance measure, to the ideal situation where all sensors and actuators can communicate to the controller simultaneously? In other words, what is the optimal communication sequence between the control system components? Furthermore, if we also have the freedom to design a controller together with a communication sequence, what will the 'optimal' overall design be?* ♦

As Example 1.2 suggests, control over a network, which is synonymous with limited communication in the feedback loop(s), introduces new challenges for the control engineer.

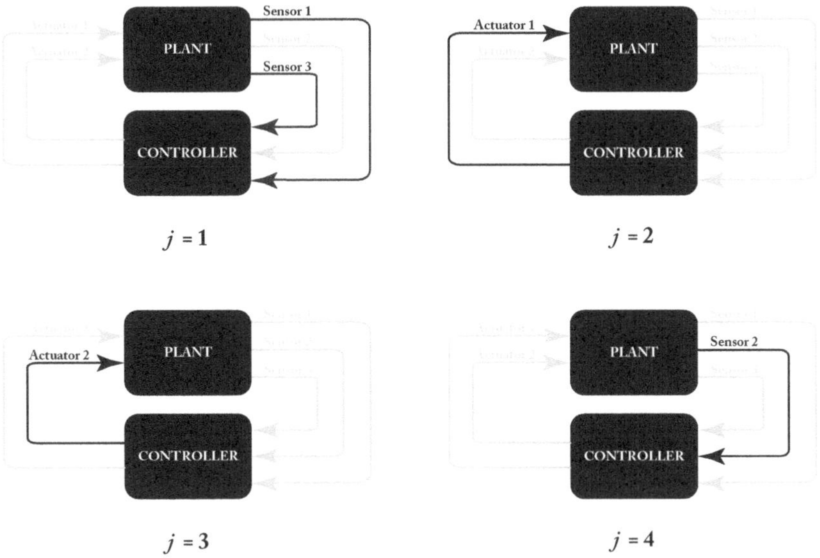

FIGURE 1.4: NCS with scheduling: only some signals can be sent at each time instance.

1.1.3 Limited communication systems

This section formalizes the concepts that arose in the examples in the previous section. An NCS, as defined here, is a control system with limited demand-to-controller, sensor-to-controller and controller-to-actuator communication. The NCS components are the following.

- A shared, finite bandwidth *communication medium* (also called *communication channel* or *bus*) which is the communication network between the NCS's components.

- The *nodes*, which are connection points attached to the communication medium. A node is an active electronic device capable of sending and receiving information over the communications channel. A node may be a source of information if it relays sensor and/or demand signals, a means of action if it is connected to one or more actuator(s), a controller if it performs control algorithms or any combination of those. If a node contains sensor(s), actuator(s) and a local controller it may be referred to as a subsystem.

In such systems, information is sent and received in a serial fashion in the form of packets; information cannot be exchanged all at the same time (there is a limit to the amount of bandwidth supported by the network) as there is no guarantee of non-corrupted delivery and often only one node can transmit at a time. Nodes attached to the shared network have to perform a number of tasks including analogue-to-digital (A/D) and/or digital-to-analogue (D/A) conversion and signal encoding/decoding. If a node implements a control algorithm, it will be called a *controller node.*

The input and output of the block denoted as 'PLANT' in Figure 1.1 are discrete signals. The implicit assumption is that the plant includes a means of D/A and A/D conversion. The block denoted as 'CONTROLLER' is a digital controller. Also, in that diagram, the solid lines represent a connection with up-to-date information (ideal signal) while the dashed line is the connection with limited information (due to the network). These assumptions and convention are used throughout this book.

The type of NCSs considered here can be represented by the diagram in Figure 1.5. The shared communication medium is a communication bus and

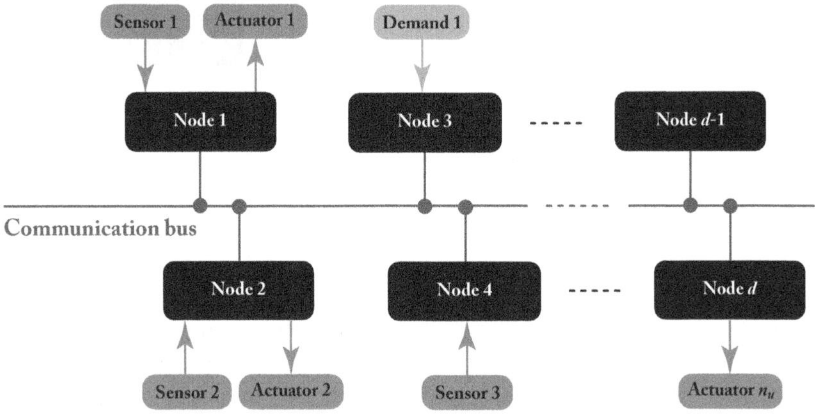

FIGURE 1.5: A distributed system in an NCS configuration: nodes' access to the communication medium is limited.

the nodes can be subsystems (*e.g.* nodes 1 and 2), actuator nodes (*e.g.* node d), sensor nodes (*e.g.* node 4) or controller nodes (*e.g.* node $d-1$) spatially distributed and connected to each other. The plant is a (usually continuous-time) process where sensors and actuators are attached. The controller is the set of the digital control algorithms that may run in a single node or in more than one node in a decentralized fashion.

1.2 Motivation

NCSs enable the integration of an increasing number of complex control systems implemented on distributed control units. The problem is increased as nowadays the manufacturers tend to produce their own control algorithms which need to be integrated into existing NCSs. Therefore, NCSs increase modularity, and flexibility and allow quick and easy maintenance at low cost.

NCSs are essential and have become the standard in the automotive industry [70], avionic systems [44] (see Figure 1.6), robots [93, 39] and automated

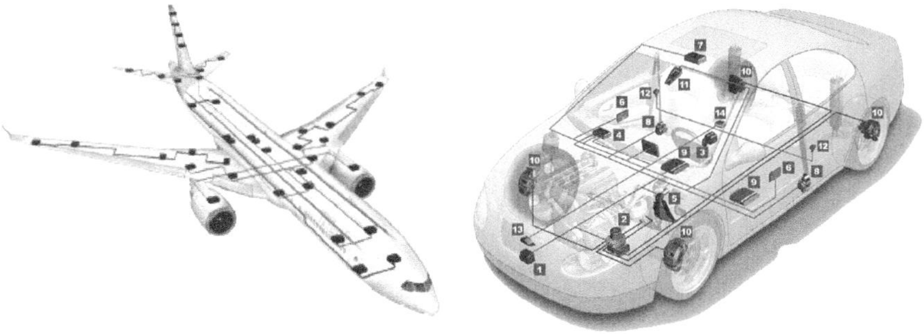

FIGURE 1.6: On the left the NCS of a passenger airplane using a time-deterministic protocol, Avionics Full-Duplex Switched Ethernet (AFDX): the significant weight reduction results in 20% higher fuel efficiency. On the right the NCS of a modern car: a dedicated line for each Electronic Control Unit (ECU) would be impractical.

manufacturing systems [73]. They reduce hardwiring and costs of installation and implementation. For instance, a typical modern car is an NCS with a high-speed Controller Area Network (CAN) in front of the firewall for engine, transmission and traction control, and a low-speed CAN for locks, windows and other devices [128].

In the process of development of real-time systems, there is often a gap between control theory and computer science [122]. Traditionally, there is a separation between the controller design and its implementation [6]. The control engineers design a (optimal) controller and afterwards the software engineers implement this controller on some hardware. The limitation on the practical implementation (the limited communication) leads to an overall suboptimal design. This is equivalent to designing a controller for a plant model with important unmodeled dynamics.

This book aims at analyzing in more detail the properties of NCSs from the control aspect but taking into consideration the issues related to the practical implementation. The aim is to contribute to the bridging of the gap between control theory and computer science since we believe that, if the bandwidth constraints of the communication medium are considered at the controller design stage, the performance of the controller significantly increases.

2

Control of plants with limited communication

CONTENTS

Imagine the situation where an air traffic controller has to provide separation between aircrafts in an area of intense air traffic. The air traffic controller communicates with the aircraft pilots using a push-to-talk radio system, which has many issues such as the fact that only one transmission can be made on a frequency at a time or transmissions will merge together and be unreadable. Due to the poor visibility conditions, pilots are obliged to follow the air controller instruction. The air controller, at his desk, gathers information from the ground instruments and from the pilots to decide what instruction should be transmitted to each pilot.

The National Air Traffic Services is in charge of providing a policy for the pilots and for the air controller that minimizes the risk of mid-air collision under these conditions. Leaving the freedom to transmit whenever information is available or is requested would increase the chances of simultaneous unreadable transmission, controller overload, delayed communication and in general non-determinism. If the transmission is somehow scheduled, then some of the previous problems are solved but a critical piece of information might have to dangerously wait before it is granted access while non-critical communication is occupying the radio frequency. What would an optimal policy be? Should the air traffic controller, in making decisions, also consider the communication policy itself? Should the pilots be 'smart' in a sense that they should also make some basic local decisions?

2.1 Introduction

An NCS is a limited communication system. Limited communication systems arise because of the use of a finite bandwidth shared network to close the feedback loops of some (normally distributed) systems. The presence of the network inevitably introduces new challenges for the control engineering community that lie in the intersection of control and communication theory. Since information exchange has to be somehow time multiplexed, the inevitable effect is the so called network-induced delay. Delays are a well known cause of performance degradation and destabilization of control loops. Network-induced delays can be infinitely long (packet dropout) and non-deterministic (queues and bandwidth contention). The typical non-deterministic behavior of control networks becomes difficult to model and it is not uncommon that probabilistic methods (Markov chains and Poisson processes) merged with standard tools from control and information theory are used.

Probably, the first attempt to summarize the issues of networking of control systems was made by [103]. More recently in [22], techniques used to handle network-induced delays based on state augmentation, queuing and probability theory, nonlinear control and perturbation theory and scheduling are presented. In [49, 144] a large amount of original work is reviewed (emphasizing the high research interest on the area) in a more philosophical and conceptual manner. In [49] several models (sampled-data, model-based and hybrid) are examined. One of the latest surveys on recent results on NCSs is presented in [48] where several results on estimation, analysis, and controller synthesis for NCSs are gathered together and compared. Many of the results presented rely on Lyapunov-based techniques and only provide sufficient conditions for stability of the NCSs. For linear systems, Linear Matrix Inequalities (LMIs) are often used. The conclusion is that the majority of work only addresses the stability issues of NCSs and performance has been neglected. Also, the use of Lyapunov techniques leads to conservative results.

Tools from information theory have often been adopted to explore the issues of control under limited communication. For example, the work of [18] defines for the first time the concept of 'minimum attention' in control. It is pointed out that control algorithms with small values of $\|\frac{\partial \boldsymbol{u}}{\partial \boldsymbol{x}}\|$ and $\|\frac{\partial \boldsymbol{u}}{\partial t}\|$ requires less frequent updating. This concept can be cast into an 'attention functional', that integrates over all time and space, of the form

$$attention = \int_{\Omega} \phi \left(\boldsymbol{x}, t, \frac{\partial \boldsymbol{u}}{\partial \boldsymbol{x}}, \frac{\partial \boldsymbol{u}}{\partial t} \right) d\boldsymbol{x} dt. \tag{2.1}$$

Another example is the work of [120] where the network in an NCS problem was thought of as a channel with optimal encoding/decoding and maximum available bit-rate. Quantization schemes and chaotic behavior analysis have also been used by other authors to extend the knowledge of control under limited communication (*e.g.* [40]).

In the context of this study, the techniques used to control NCSs have been grouped into three classes (see Figure 2.1).

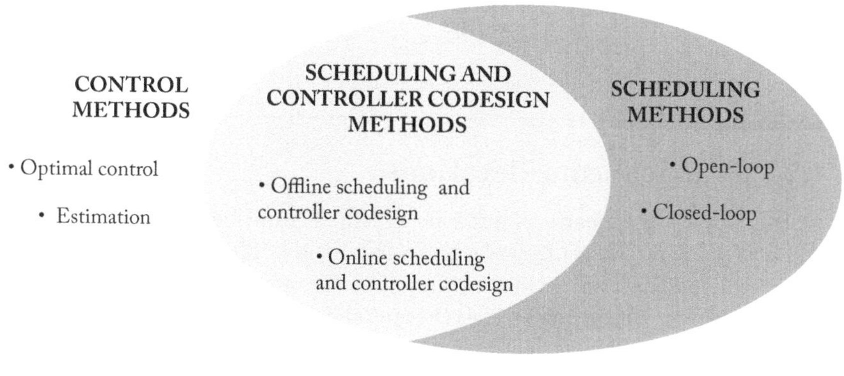

FIGURE 2.1: Representation of classification of control and scheduling techniques for NCSs.

- *Control methods.* Given a network, a controller is designed to cater for the effect (uncertainty, non-determinism) of the network to guarantee system Quality of Performance (QoP).
- *Scheduling methods.* Given a controller, a (usually dynamic) scheduling algorithm is designed to minimize the effects of the network implementation and guarantee network Quality of Service (QoS).
- *Scheduling and controller codesign methods.* Given a plant, an optimal schedule and controller are designed simultaneously treating the communication as a system constraint and guaranteeing both QoP and QoS.

The above methods (each substantially different from each other) have their advantages and disadvantages related to the practical implementation. What is common to these three methods is that an effort is made to design a controller (and/or a communication scheduler) that considers the practical networking implementation. With the inclusion of the limited communication paradigm into the controller design, new issues for structural properties (controllability and observability) and stabilizability arise.

The remainder of this chapter is organized as follows. Section 2.2 is an introduction to practical issues of general networks closely related to the type of networked control considered in this book. Section 2.3 illustrates some techniques used to model different types of NCSs. Section 2.4 discusses the control design approaches where the controller takes into account the network but there is no control over the network. Section 2.5 discusses the design approaches where a controller is given and there is direct control over the network. The codesign of a communication schedule together with a controller is analyzed in Section 2.6. Finally, in Section 2.7, structural and stability issues are discussed.

2.2 Practical considerations

At this point, the general definition of NCSs should be reduced to only those types of NCS architectures related to this study. There are two major types of control systems that utilize communication networks and these are *remote control systems* (also known as tele-operation control, *e.g.* [55]) and *shared-network control systems* [130]. Remote control systems involve remote data acquisition and/or remote control and have been generally used for convenience and safety. Shared-network control systems consist of many relatively close demand, sensor, actuator and controller nodes that share a common communication medium. The advantages of these architectures were discussed in Section 1.2. Here, NCSs that are shared-networked control systems are considered.

2.2.1 Real-time networks

In a network, messages are exchanged among the various nodes in the form of packets. In *data networks* (*e.g.* Ethernet), data are sent occasionally and in large packets. It is essential that all packets reach their destination and, although short transmission delays are desirable, longer delays are not critical. On this type of network, a great effort is made to maximize the amount of information in a packet, hence, large packets are used to reduce the ratio between actual data and overheads. Contrary to data networks, data in *control networks* (*e.g.* Controller Area Network) are continuously transmitted in smaller packets. It is essential that packets meet their critical time requirements. Packet loss, delays and non-determinism seriously affect the stability and performance of the networked system and may not be acceptable for safety-related control systems. Often only 'fresh' data are considered useful.

A large number of networks and communication protocols were standardized a long time before control over real-time networks was considered as an issue. It is common that networks and communication protocols originally designed for non-real-time applications are now used for real-time control (*e.g.* Ethernet-like networks are considered in [102] for control applications). The consequence is that control methods need to adapt to the standard off-the-shelf technology.

A practical point to consider here is the communication paradigm adopted by the networked system. Real-time communication is managed by the Medium Access Control (MAC) sub-layer of layer 2 of the Open System Interconnection (OSI) model. There are two main, fundamentally different, communication paradigms: contention-based and contention-free MAC protocols. Their features will be briefly reviewed.

2.2.1.1 Contention-based paradigms

Contention-based paradigms (*e.g.* event-triggered protocols) are random access architectures. The communication medium access is requested by events occurring in a node at arbitrary times. They must be able to handle collision via arbitration mechanisms (by assigning priority to messages) or by detecting collisions. Hence, they cannot be deterministic since events usually occur in a random fashion. Typical examples are the family of Carrier Sense Multiple Access (CSMA) protocols. Carrier sense means that each node is aware of communication in the network and will wait until the network is idle before transmitting. Multiple access means that many nodes can potentially begin transmitting as soon as the network enters the idle state. Modifications of the pure CSMA are the following.

- *CSMA/CD.* This is a CSMA with a Collision Detection protocol. Collision detection means that, if more than one node writes into the network at the same time, nodes are able to detect that their message has been corrupted (because of the collision). This is possible since nodes, while transmitting, are listening to the network. If a collision is detected, an algorithm (the standard binary exponential back off) will determine the random time interval until transmission is retried. No message prioritization is supported, making it difficult to bound delays.

- *CSMA/CA*. This is a CSMA with a Collision Avoidance protocol. Collision avoidance means that an effort is made by the nodes to reduce the probability of collision between multiple messages sent to the network. This is achieved by algorithms that handle random back off time and contention.
- *CSMA/BA*. This is a CSMA with a Bitwise Arbitration protocol. Bitwise arbitration means that each message is assigned a priority during the node pre-programming. This priority is coded into the message identifier which are the first bits that are written into the network. Messages are broadcasted and each node, while transmitting, is listening to the network. If, for instance, a node writes a 1 but reads a 0 it means that some other node is transmitting a 0 and therefore it stops transmission. On the other hand, the node that transmitted the 0 is unaware of other node transmissions and continues sending its message. For this reason, 0 and 1 are known as dominant and recessive bits respectively. Since identifiers should be unique for each node, at the end of the transmission of the identifier only one node is granted access and will continue sending its message unaware of contention and, most importantly, uninterrupted. CSMA/BA behaves like a logical AND and this type of arbitration is non-destructive. The determinism of CSMA/BA is used in the Controller Area Network (CAN) since it is attractive for use in RTCS (see Figure 2.2 for an example). To allow

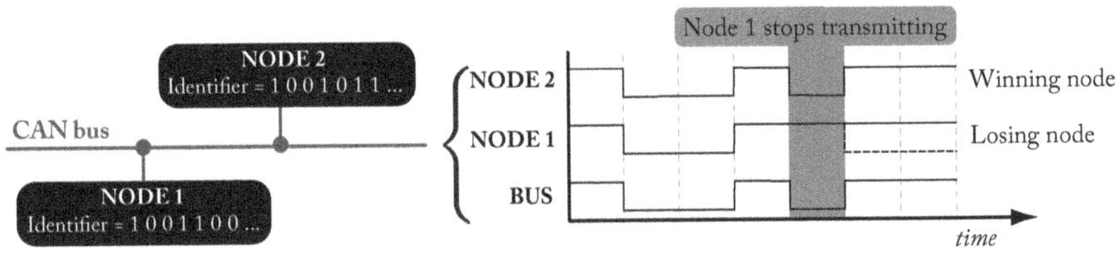

FIGURE 2.2: A two nodes NCS under CAN communication (contention-base with bitwise arbitration paradigm): nodes transmit their identifier and the one with highest priority will win the contention.

bitwise arbitration, the data rate must be relatively slow and typically this is inversely proportional to the physical distance between nodes.

Among the most recent results on event-triggered NCSs is the work of [34, 132, 133].

2.2.1.2 Contention-free paradigms

Contention-free (*e.g.* time-triggered protocols) are controlled access architectures. The medium access is solely dependent on the progression in time and it therefore allows the pre-planning of a particular communication sequence at the design stage controlled by specific algorithms. See Figure 2.3 for an example of a two node NCS in a contention-free paradigm. It is clear that

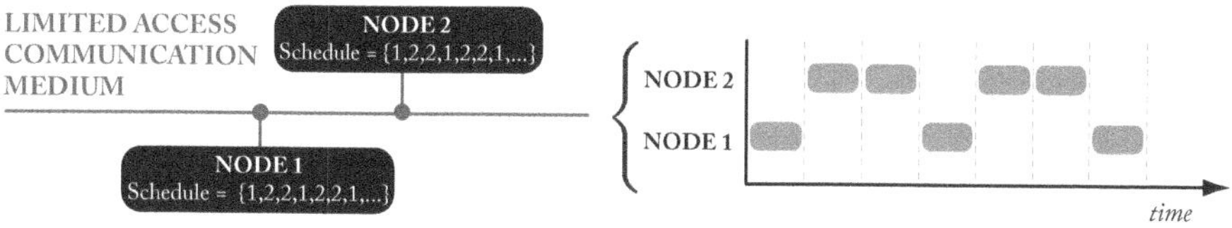

FIGURE 2.3: A two nodes NCS under a contention-free paradigm: nodes transmit in their allocated time slots only.

contention-free architectures are more suited for hard real-time applications [68] and they will result in a more highly-predictable system behavior [79, 68] at the expense of flexibility. Master/slave configurations can be used to give contention-free characteristics to protocols originally designed as contention-based (*e.g.* TTCAN) and some modern protocols support contention-free and contention-based types of communication simultaneously (*e.g.* FlexRay) to get the benefit from both. Typical examples are the Time Division Multiple Access (TDMA) and token bus protocols.

- *TDMA.* In TDMA, all nodes attached to the network must share a global clock. This allows division of time into time slots and assignment of each slot to one node only. In this way, nodes are able to share the network in a deterministic manner (no collisions). Time slot allocation is often performed offline by defining a periodic sequence of node access called TDMA rounds (see Figure 2.3). TDMA protocols can be centralized or distributed. In a centralized architecture, a master node coordinates the medium access by periodically writing into the network with the double scope of generating a global clock and assigning slots. Of course, less bandwidth is available to the slave nodes because some is used by the master node. In an ideal decentralized architecture, all node clocks are synchronized when the system is started, and every node writes to the network when it is its turn. In practice, because of clock drift, clock synchronization algorithms must be implemented.

- *Token bus.* In the token bus protocol, nodes are arranged in a ring topology and each node knows the address of its neighbors. The access is regulated by the nodes passing the 'token' whenever they have finished transmission or a maximum time has elapsed. A special algorithm is needed to notify nodes of connection to and disconnection from the ring.

2.3 Models for NCSs

The modeling and analysis of NCSs will be limited to linear time-invariant (LTI) systems. The assumption is that the distributed plant is described by the equations

$$\begin{aligned} \dot{\boldsymbol{x}}(t) &= \boldsymbol{A}^c\boldsymbol{x}(t) + \boldsymbol{B}^c\hat{\boldsymbol{u}}(t) + \boldsymbol{w}(t), \quad \boldsymbol{x}(0) = \boldsymbol{x}_0, \\ \boldsymbol{y}(t) &= \boldsymbol{C}\boldsymbol{x}(t) + \boldsymbol{D}\hat{\boldsymbol{u}}(t) + \boldsymbol{v}(t), \end{aligned} \tag{2.2}$$

where $\boldsymbol{x}(t) \in \mathbb{R}^{n_x}$ is the state, $\hat{\boldsymbol{u}}(t) \in \mathbb{R}^{n_u}$ is the controlled input, $\boldsymbol{y}(t) \in \mathbb{R}^{n_y}$ is the measurement vector and $\boldsymbol{w}(t) \in \mathbb{R}^{n_x}$ and $\boldsymbol{v}(t) \in \mathbb{R}^{n_y}$ are uncorrelated zero-mean Gaussian white noise processes with covariance matrices $\boldsymbol{R}_w \geq \boldsymbol{0}$ and $\boldsymbol{R}_v > \boldsymbol{0}$ respectively. $\boldsymbol{A}^c$, $\boldsymbol{B}^c$, $\boldsymbol{C}$ and $\boldsymbol{D}$ are matrices of appropriate sizes. Without loss of generality, assume that $\boldsymbol{D} = \boldsymbol{0}_{n_y \times n_u}$. Since an NCS is a sampled-data system, the control input $\hat{\boldsymbol{u}}(t)$ is a discrete input signal created by a zero-order-hold element for a constant sampling period h, *i.e.*

$$\hat{\boldsymbol{u}}(t) = \hat{\boldsymbol{u}}(jh) = constant \quad \text{for} \quad t \in [jh, (j+1)h) \quad \forall j \in \mathbb{N}_0, \tag{2.3}$$

where $\mathbb{N}_0 = \mathbb{N} \cup \{0\}$. The model in (2.2) has to be sampled with a periodic sampling interval h where, for a signal $\boldsymbol{f}(t)$

$$\boldsymbol{f}(jh) = \lim_{t \to jh^-} \boldsymbol{f}(t) = \boldsymbol{f}(jh^-). \tag{2.4}$$

The sampled-data time-invariant system for constant input $\hat{\boldsymbol{u}}(jh)$ (defined in (2.3)) is

$$\begin{aligned} \boldsymbol{x}(jh+h) &= \boldsymbol{A}\boldsymbol{x}(jh) + \boldsymbol{B}\hat{\boldsymbol{u}}(jh) + \boldsymbol{w}(jh), \quad \boldsymbol{x}(0) = \boldsymbol{x}_0, \\ \boldsymbol{y}(jh) &= \boldsymbol{C}\boldsymbol{x}(jh) + \boldsymbol{v}(jh), \end{aligned} \tag{2.5}$$

where

$$\begin{aligned} &\boldsymbol{A} = \acute{\boldsymbol{A}}(h), \quad \boldsymbol{B} = \acute{\boldsymbol{B}}(h), \\ &\acute{\boldsymbol{A}}(\tau) = e^{\boldsymbol{A}^c\tau}, \quad \acute{\boldsymbol{B}}(\tau) = \int_0^\tau e^{\boldsymbol{A}^c s} ds \boldsymbol{B}^c, \end{aligned} \tag{2.6}$$

and j is the sampling instant. (Although the sampling of the noise process may seem unusual, the discrete model considered here, as the literature suggests, is a common starting point for NCS analysis.)

2.3.1 Modeling the contention-based paradigm

As discussed in Section 2.2.1.1, contention-based paradigms are characterized by randomness in the medium access with consequent randomness in information exchange. Since the dominant effect of this type of networks is to introduce delays into the control loop, the natural way to tackle the problem is by finding suitable models for the delays. An endless amount of work has been done on network-induced delays (*e.g.*, among the most recent, [36, 27, 25, 134]). This work is mainly based on $\mathcal{H}_\infty$ performance measures and solutions of LMIs. On a less conservative note is the work of [92].

Denote $\tau^{sc}(j)$ the sensor-to-controller and $\tau^{ca}(j)$ controller-to-actuator delay (any controller computation delay can be included in one of these). The simplest model assumes that delays are constant and smaller than the sampling time (see Figure 2.4(a)). If delays are not constant then constant delays can be achieved by introducing a buffer longer than the worst case delay [81]. This is shown in Figure 2.4(b) where the non-constant dotted line represents a signal with variable delay. The obvious drawback is that all delays are now

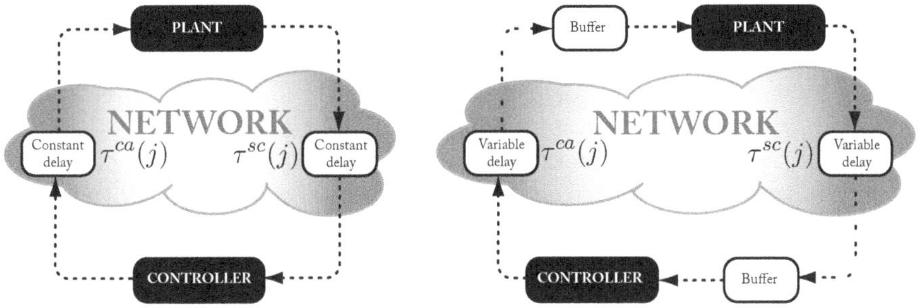

(a) (b)

FIGURE 2.4: An NCS model as a delayed feedback loop. In (a) the delays are assumed to be constant. In (b) varying delays (illustrated by a non-constant dotted line) are rendered constant by introducing two buffers.

equal to the worst case. The assumption of constant delays produces a good model only when the process time constant is much larger than the introduced delays. If messages are time-stamped, then the delays are known and the standard result of systems with delays is presented here [1]. Assume that $\tau^{sc}(j) + \tau^{ca}(j) < h$ and by integrating (2.5) over a sampling interval obtain

$$\boldsymbol{x}(jh+h) = \boldsymbol{A}\boldsymbol{x}(jh) + \boldsymbol{B}_0(\tau^{sc}(j), \tau^{ca}(j))\hat{\boldsymbol{u}}(jh) \tag{2.7}$$

$$+ \boldsymbol{B}_1(\tau^{sc}(j), \tau^{ca}(j))\hat{\boldsymbol{u}}(jh-h) + \boldsymbol{w}(j), \tag{2.8}$$

where

$$\boldsymbol{A} = e^{\boldsymbol{A}^c h},$$

$$\boldsymbol{B}_0(\tau^{sc}(j), \tau^{ca}(j)) = \int_0^{h-\tau^{sc}(j)-\tau^{ca}(j)} e^{\boldsymbol{A}^c s} ds \boldsymbol{B}^c,$$

$$\boldsymbol{B}_1(\tau^{sc}(j), \tau^{ca}(j)) = \int_{h-\tau^{sc}(j)-\tau^{ca}(j)}^{h} e^{\boldsymbol{A}^c s} ds \boldsymbol{B}^c. \tag{2.9}$$

If the network-induced delays are random, they can be modeled by letting the distribution of the network delays be governed by the states of an underlying Markov chain. This is shown in [92] and the transition probability is described by

$$q_{ik} = \Pr(r(j+1) = k | r(j) = i), \quad i, k \in \{\textit{states of the chain}\}, \tag{2.10}$$

where $r(j)$ is the current state of the Markov chain. However, according to [22], finding the Markov relation of a delay such as q_{ik} is a challenging task. (For simplicity, the sampling period h will be omitted when obvious.)

2.3.2 Modeling the contention-free paradigm

Under the contention-free paradigm, communication is deterministic. Delays are known since the medium access is controlled by a global scheduler and no arbitration takes place (clearly, the assumption is that of an ideal channel where no signal corruption or packet loss occurs). This paradigm is usually modeled by time-varying system equations. The time dependency is introduced by the scheduled communication. Only a limited number of control and/or sensor signals can access the communication channel at any time tick. Signals that did not grant access can be treated in the following ways:

- assign a value equal to zero (this will be called the *non-zero-order-hold* case)
- assign a value equal to the previous value (this will be called the *zero-order-hold* case)
- assign a model-predicted value (this will be called the *model-based* case).

The above cases are listed in order of implementation complexity and performance optimality. Each case will be examined in more detail.

2.3.2.1 The non-zero-order-hold case

Consider the LTI system equations in (2.5) without the white noise processes. To account for the limited communication, let $\boldsymbol{S}_A(j) \in \mathbb{R}^{n_u \times n_u}$ and $\boldsymbol{S}_S(j) \in \mathbb{R}^{n_y \times n_y}$ be binary matrices[1] with zeros everywhere apart from some entries in the diagonal which are equal to 1. $\boldsymbol{S}_A(j)$ and $\boldsymbol{S}_S(j)$ will be called the actuator and sensor scheduling matrices respectively since their effect (as it will be shown in the next chapter) is to schedule signal access to the network. The limited communication system for the non-Zero-Order-Hold (non-ZOH) case can be represented by the following Linear Time-Variant (LTV) system equations

$$\begin{aligned} \boldsymbol{x}(j+1) &= \boldsymbol{A}\boldsymbol{x}(j) + \boldsymbol{B}\boldsymbol{S}_A(j)\boldsymbol{u}(j), \quad \boldsymbol{x}(0) = \boldsymbol{x}_0, \\ \hat{\boldsymbol{y}}(j) &= \boldsymbol{S}_S(j)\boldsymbol{C}\boldsymbol{x}(j). \end{aligned} \tag{2.11}$$

[1] A formal definition of these matrices will be given later.

Let $\boldsymbol{u}(j)$ and $\boldsymbol{y}(j) = \boldsymbol{C}\boldsymbol{x}(j)$ be the up-to-date actuator and sensor signal vectors respectively (*i.e.* if there was no network), then, the effective vectors are $\hat{\boldsymbol{u}}(j) = \boldsymbol{S}_A(j)\boldsymbol{u}(j)$ and $\hat{\boldsymbol{y}}(j) = \boldsymbol{S}_S(j)\boldsymbol{y}(j)$ respectively. In [149, 150, 56, 51, 52], this model is utilized and it is called the 'extended plant'. It can be represented diagrammatically by Figure 2.5(a) with the convention that less dense dotted lines are signals with limited information.

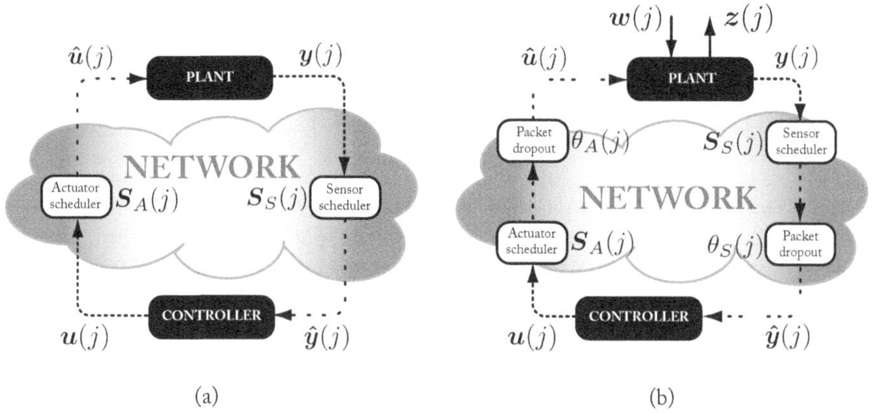

FIGURE 2.5: An NCS modeled as a limited communication system. In (a) only a limited number of signals can access the network (illustrated by a less dense dotted line). In (b) a stochastic packet dropout (illustrated by missing dots in the dotted line) as well as scheduled communication is considered.

It is argued in [149, 150, 56, 51] that choosing to 'ignore' the actuators and sensors that are not actively communicating (*i.e.* reset their signal values to zero) may benefit the system performance but a formal proof is not provided. Indeed, its simplicity in analysis may be regarded as an advantage.

In [58], a different model is presented to characterize the effect of the network in a non-ZOH fashion. The network is modeled as a communication channel where only some messages can be transmitted at any time (as before) but with the addition that messages can be also lost in a random fashion. This approach suits an $\mathcal{H}_\infty$ design. Consider the NCS in Figure 2.5(b) where the plant is described by the discrete-time equations

$$\begin{bmatrix} \boldsymbol{x}(j+1) \\ \boldsymbol{z}(j) \\ \boldsymbol{y}(j) \end{bmatrix} = \begin{bmatrix} \boldsymbol{A} & \boldsymbol{B}_1 & \boldsymbol{B}_2 \\ \boldsymbol{C}_1 & \boldsymbol{D}_{11} & \boldsymbol{D}_{12} \\ \boldsymbol{C}_2 & \boldsymbol{D}_{21} & \boldsymbol{0} \end{bmatrix} \begin{bmatrix} \boldsymbol{x}(j) \\ \boldsymbol{w}(j) \\ \hat{\boldsymbol{u}}(j) \end{bmatrix}, \tag{2.12}$$

where $\boldsymbol{x}(j) \in \mathbb{R}^{n_x}$ is the state, $\hat{\boldsymbol{u}}(j) \in \mathbb{R}^{n_u}$ is the controlled input, $\boldsymbol{y}(j) \in \mathbb{R}^{n_y}$ is the measurement vector, $\boldsymbol{w}(j) \in \mathbb{R}^{n_w}$ is the disturbance vector and $\boldsymbol{z}(j) \in \mathbb{R}^{n_z}$ is the performance output vector. Also, the pairs $(\boldsymbol{A}, \boldsymbol{B}_2)$ and $(\boldsymbol{C}_2, \boldsymbol{A})$ are respectively controllable and observable. It is assumed that the multiple sensor and actuator signals have to share the communication medium in a way that only one signal can be transmitted at any sampling time. To account for the limited communication, let (as before) $\boldsymbol{S}_A(j) \in \mathbb{R}^{n_u \times n_u}$ and $\boldsymbol{S}_S(j) \in \mathbb{R}^{n_y \times n_y}$ be binary matrices with zeros everywhere apart from some entries in the diagonal which are equal to 1. To account for the stochastic packet dropout, let $\theta_A(j), \theta_S(j) \in \{0,1\}$ be independent and identically distributed (iid) Bernoulli processes. Then, the limited communication system with packet dropout can be represented by the following LTV system equations

$$\begin{bmatrix} \boldsymbol{x}(j+1) \\ \boldsymbol{z}(j) \\ \boldsymbol{y}(j) \end{bmatrix} = \begin{bmatrix} \boldsymbol{A} & \boldsymbol{B}_1 & \theta_A(j)\boldsymbol{B}_2\boldsymbol{S}_A(j) \\ \boldsymbol{C}_1 & \boldsymbol{D}_{11} & \theta_A(j)\boldsymbol{D}_{12}\boldsymbol{B}_2\boldsymbol{S}_A(j) \\ \theta_S(j)\boldsymbol{S}_S(j)\boldsymbol{C}_2 & \theta_S(j)\boldsymbol{S}_S(j)\boldsymbol{D}_{21} & \boldsymbol{0} \end{bmatrix} \begin{bmatrix} \boldsymbol{x}(j) \\ \boldsymbol{w}(j) \\ \hat{\boldsymbol{u}}(j) \end{bmatrix}. \tag{2.13}$$

Notice that the system in (2.13) is a more generalized version of the system in (2.11) since it includes a model for packet dropout and considers the exogenous input and the performance output vectors. Furthermore, only periodic scheduling matrices with period p are normally considered (*i.e.* $\boldsymbol{S}_A(j) = \boldsymbol{S}_A(j+p), \boldsymbol{S}_S(j) = \boldsymbol{S}_S(j+p)$).

2.3.2.2 The zero-order-hold case

Consider the LTI system in (2.5) without the white noise processes. Let $\boldsymbol{S}_A(j)$ and $\boldsymbol{S}_S(j)$ be defined in the same way as before. For simplicity, consider the case where limited communication only applies between controller and actuators (the sensor-to-controller communication is not constrained, *i.e.* $\boldsymbol{S}_S(j) = \boldsymbol{I}$ for all j). Since the signals that are not currently being sent are holding their previous value, this will be called the Zero-Order-Hold (ZOH) case. The scheduler can be represented by

$$\hat{\boldsymbol{u}}(j) = \bar{\boldsymbol{S}}_A(j)\hat{\boldsymbol{u}}(j-1) + \boldsymbol{S}_A(j)\boldsymbol{u}(j), \quad \hat{\boldsymbol{u}}(0) = \hat{\boldsymbol{u}}_0, \tag{2.14}$$

where $\hat{\boldsymbol{u}}(j)$ is the limited control vector, $\boldsymbol{u}(j)$ is the up-to-date control vector, $\bar{\boldsymbol{S}}_A(j) = \boldsymbol{I} - \boldsymbol{S}_A(j)$ and $\boldsymbol{I}$ is the identity matrix. The scheduler dynamics in (2.14) can be understood as follows: the signal vector $\hat{\boldsymbol{u}}(j)$ has all its entries equal to $\hat{\boldsymbol{u}}(j-1)$ (the previous vector) apart from some entries (given by the position of the unit elements of the diagonal of $\boldsymbol{S}_A(j)$) which will be equal to the ones of $\boldsymbol{u}(j)$ (the up-to-date control signal). The plant model, including the limited control communication, can now be modified as

$$\begin{aligned} \boldsymbol{x}(j+1) &= \boldsymbol{A}\boldsymbol{x}(j) + \boldsymbol{B}\hat{\boldsymbol{u}}(j) \\ &= \boldsymbol{A}\boldsymbol{x}(j) + \boldsymbol{B}\bar{\boldsymbol{S}}_A(j)\hat{\boldsymbol{u}}(j-1) + \boldsymbol{B}\boldsymbol{S}_A(j)\boldsymbol{u}(j). \end{aligned} \tag{2.15}$$

Merge the two system equations (2.14)–(2.15) to form the system with augmented states

$$\begin{aligned} \hat{\boldsymbol{x}}(j+1) &= \hat{\boldsymbol{A}}(j)\hat{\boldsymbol{x}}(j) + \hat{\boldsymbol{B}}(j)\boldsymbol{u}(j), \quad \hat{\boldsymbol{x}}(0) = \begin{bmatrix} \boldsymbol{x}_0^T & \boldsymbol{u}_0^T \end{bmatrix}^T, \\ \boldsymbol{y}(j) &= \hat{\boldsymbol{C}}\hat{\boldsymbol{x}}(j). \end{aligned} \tag{2.16}$$

where

$$\hat{\boldsymbol{x}}(j) = \begin{bmatrix} \boldsymbol{x}(j) \\ \hat{\boldsymbol{u}}(j-1) \end{bmatrix},$$

$$\hat{\boldsymbol{A}}(j) = \begin{bmatrix} \boldsymbol{A} & \boldsymbol{B}\bar{\boldsymbol{S}}_A(j) \\ \boldsymbol{0} & \bar{\boldsymbol{S}}_A(j) \end{bmatrix}, \quad \hat{\boldsymbol{B}}(j) = \begin{bmatrix} \boldsymbol{B}\boldsymbol{S}_A(j) \\ \boldsymbol{S}_A(j) \end{bmatrix}, \quad \hat{\boldsymbol{C}} = \begin{bmatrix} \boldsymbol{C} & \boldsymbol{0} \end{bmatrix}. \tag{2.17}$$

Note that, because of the matrices $\boldsymbol{S}_A(j)$ and $\bar{\boldsymbol{S}}_A(j)$, the equation in (2.17) describes the dynamics of an LTV system. This model, for periodic $\boldsymbol{S}_A(j)$ and $\bar{\boldsymbol{S}}_A(j)$, is utilized in the work of [105, 80, 14, 75] and can be represented diagrammatically by Figure 2.6(a) with the usual dotted line convention.

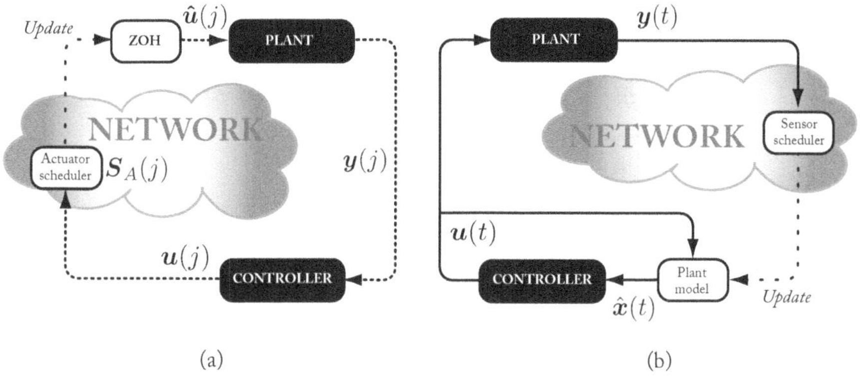

FIGURE 2.6: An NCS modeled as a 'smart' limited communication system. In (a) only a limited number of signals can access the network to update the values of the zero-order-hold. In (b) the update value is used to estimate the plant states through a plant model.

The usual way to proceed from (2.16) is to assume that the scheduling policy is periodic with period p. This allows down-sampling of the system with the aim to create a higher dimensional system in order to eliminate the time dependance from the system matrices (discrete-time lifting technique). This system (used in [105, 80, 14, 75]) will be called the *augmented lifted plant*.

The same ZOH strategy can be adopted to model an NCS with stochastic packet loss. In fact, the scheduling strategies discussed above (where only a limited number of signals can access the network) can be thought of as a closed-loop system where, at every sampling time, only some known packets are not lost. This model is used in [107] for an $\mathcal{H}_\infty$ design. Consider the LTI system in (2.5) without the white noise processes. This time, consider a network placed between sensors and controller only (controller and actuators are collocated). Let $\theta_S(j)$ be a Bernoulli process given by

$$\Pr(\theta_S(j) = 0) = q, \quad \Pr(\theta_S(j) = 1) = 1 - q, \tag{2.18}$$

representing the probability that a packet is lost. Then, consider the following packet loss model for the sensor reading

$$\hat{\boldsymbol{y}}(j) = \theta_S(j)\boldsymbol{y}(j) + (1 - \theta_S(j))\hat{\boldsymbol{y}}(j-1). \tag{2.19}$$

Notice the similarities between (2.19) and (2.14) and appreciate the difference between $\boldsymbol{S}_S(j)$ and $\theta_S(j)$. Hence, proceed in a similar way by considering the augmented system

$$\begin{aligned} \hat{\boldsymbol{x}}(j+1) &= \hat{\boldsymbol{A}}(j)\hat{\boldsymbol{x}}(j) + \hat{\boldsymbol{B}}\boldsymbol{u}(j), \quad \hat{\boldsymbol{x}}(0) = \begin{bmatrix}\boldsymbol{x}_0^T & \hat{\boldsymbol{y}}_0^T\end{bmatrix}^T, \\ \hat{\boldsymbol{y}}(j) &= \hat{\boldsymbol{C}}(j)\hat{\boldsymbol{x}}(j). \end{aligned} \tag{2.20}$$

where

$$\begin{aligned} \hat{\boldsymbol{x}}(j) &= \begin{bmatrix}\boldsymbol{x}(j) \\ \hat{\boldsymbol{y}}(j-1)\end{bmatrix}, \quad \hat{\boldsymbol{A}}(j) = \begin{bmatrix}\boldsymbol{A} & \boldsymbol{0} \\ \theta_S(j)\boldsymbol{C} & (1-\theta_S(j))I\end{bmatrix}, \\ \hat{\boldsymbol{B}}(j) &= \begin{bmatrix}\boldsymbol{B} \\ \boldsymbol{0}\end{bmatrix}, \quad \hat{\boldsymbol{C}} = \begin{bmatrix}\theta_S(j)\boldsymbol{C} & (1-\theta_S(j))I\end{bmatrix}. \end{aligned} \tag{2.21}$$

This model is a Markovian Jump Linear System (MJLS) and the reason why it has been presented in this section is for the similarities with the previous model using the ZOH strategy. In reality, a MJLS model would be more suited for a contention-based paradigm where the probability of packet collision (which is effectively a packet loss) could be naturally described by (2.18).

In summary, in the ZOH case, the plant dynamics are augmented to include the communication medium dynamics. This increases complexity but it is regarded as a more realistic representation. From the practical point of view, it should be remembered that each actuator (or more precisely, each actuator node) must be able to recognize and decode a message sent to it. This implies a minimum of embedded electronics in the actuator node able to easily store the old messages until a new one arrives.

2.3.2.3 The model-based case

So far two models were considered: one where the non-updated signal is reset to zero (non-ZOH case) and one where it is held constant (ZOH case). The next step is to consider the case where the non-updated signal is estimated using a model of the plant's dynamics. Generally speaking, this is the 'best' a node can do when the required information is not available. Inspired by the idea of 'minimum attention control' of [18], in [84] a model is proposed where the data coming from the network is used to instantaneously update the state of the controller (see Figure 2.6(b)). Although the model-based control in [84] was proposed by the authors as a method to handle limited communication in a general NCS framework, it is introduced here as an NCS model under the contention-free paradigm because it seems a natural extension of the previously presented models. Only the case where limited communication applies between plant sensors and controller is considered in [84] (the controller-to-actuator communication is not constrained).

Consider the continuous LTI plant in (2.2) with $\boldsymbol{C} = \boldsymbol{I}$. Let the plant model be

$$\dot{\boldsymbol{x}}_m(t) = \boldsymbol{A}_m \boldsymbol{x}_m(t) + \boldsymbol{B}_m \boldsymbol{u}(t), \tag{2.22}$$

where the input $\boldsymbol{u}(t) = -\boldsymbol{K}\boldsymbol{x}_m(t)$ is generated by the estimator-based controller (see Figure 2.6(b), where $\hat{\boldsymbol{x}}(t) \stackrel{\text{def}}{=} \boldsymbol{x}_m(t)$). Here $\boldsymbol{x}_m(t)$ is the state of the plant model and will be used to generate an estimation error as shown in (2.23). Assume that the full state vector is available (if not, an observer based controller can be used [84]). The state estimation error is defined as

$$\boldsymbol{e}(t) = \boldsymbol{x}(t) - \boldsymbol{x}_m(t), \tag{2.23}$$

and since it is reset to zero at the sampling time t_j (because at time t_j sensor communication is assumed to be granted) the error dynamics are

$$\begin{aligned} &\dot{\boldsymbol{e}}(t) = (\boldsymbol{A}^c - \boldsymbol{A}_m + \boldsymbol{B}^c\boldsymbol{K} - \boldsymbol{B}_m\boldsymbol{K})\boldsymbol{x}(t) + (\boldsymbol{A}^c - \boldsymbol{A}_m - \boldsymbol{B}^c\boldsymbol{K} - \boldsymbol{B}_m\boldsymbol{K})\boldsymbol{e}(t), \\ &\boldsymbol{e}(t_j) = \boldsymbol{0}. \end{aligned} \tag{2.24}$$

By defining $\boldsymbol{z}(t) = \left[\boldsymbol{x}(t)^T \ \boldsymbol{e}(t)^T\right]^T$, the overall closed-loop evolves according to

$$\begin{aligned} &\dot{\boldsymbol{z}}(t) = \Lambda \boldsymbol{z}(t), \\ &\boldsymbol{z}(t_j) = \left[\boldsymbol{x}(t^-)^T \ \boldsymbol{0}^T\right]^T, \quad \forall t \in [t_j, t_{j+1}), \quad t_{j+1} - t_j = h(j), \end{aligned} \tag{2.25}$$

where $\boldsymbol{x}(t_j^-)$ is the limit from below of $\boldsymbol{x}(\tau)$ as $\tau \uparrow t_j$ and

$$\boldsymbol{\Lambda} = \begin{bmatrix} \boldsymbol{A}^c + \boldsymbol{B}^c\boldsymbol{K} & -\boldsymbol{B}^c\boldsymbol{K} \\ \boldsymbol{A}^c - \boldsymbol{A}_m + \boldsymbol{B}^c\boldsymbol{K} - \boldsymbol{B}_m\boldsymbol{K} & \boldsymbol{A}^c - \boldsymbol{A}_m - \boldsymbol{B}^c\boldsymbol{K} - \boldsymbol{B}_m\boldsymbol{K} \end{bmatrix}. \tag{2.26}$$

Note that $h(j)$ is the update time. The system in (2.25) with initial conditions $\boldsymbol{z}(t_0) = [\boldsymbol{x}(t_0)^T \ \boldsymbol{0}^T]^T = z_0$ has the response

$$\boldsymbol{z}(t) = e^{\boldsymbol{\Lambda}(t-t_j)} \left(\prod_{k=1}^{j} \boldsymbol{M}(k) \right) z_0, \quad t \in [t_j, t_{j+1}), \quad t_{j+1} - t_j = h(j), \tag{2.27}$$

where

$$\boldsymbol{M}(k) = \begin{bmatrix} \boldsymbol{I} & 0 \\ 0 & 0 \end{bmatrix} e^{\boldsymbol{\Lambda} h(k)} \begin{bmatrix} \boldsymbol{I} & 0 \\ 0 & 0 \end{bmatrix}. \tag{2.28}$$

It should be noted that in this model-based representation, the state estimation error, $\boldsymbol{e}(t)$, is reset to zero at every update time $h(j)$. If, as for the previous cases, only a limited number of states can be updated every sampling time, then, only those entries of the vector $\boldsymbol{e}(t)$ are reset to zero at every time tick. The result would be an LTV system that, for a periodic schedule, will also be periodic. Hence, it can be down-sampled and transformed into an LTI system (discrete-time lifting technique).

From the practical point of view, the complexity of computation at the controller node has increased from simply holding the old value for the ZOH case to continuously computing an estimation. Considering the increasing power of embedded systems, microprocessor speed compared to the relatively long network-induced delays, a model-based approach might be considered reasonable.

2.4 Control methods

Control methods for NCSs refer to those controller design techniques where the network limitations (delays, packet loss and jitter) are taken into account but no control over the network takes place. In other words, it is assumed that a network (with a communication protocol) is given, and a controller is designed to minimize its detrimental effect. Control methods aim at guaranteeing QoP [130, Ch. 4].

2.4.1 Stochastic and robust control

Let us begin with the optimal stochastic control method of [92]. Consider the LTI system in (2.5) with $\boldsymbol{C} = \boldsymbol{I}$. Then, the optimal control $\hat{\boldsymbol{u}}(j)$ that minimizes the cost function

$$J = \mathbb{E}\left\{\boldsymbol{x}(M)^T\boldsymbol{Q}_M\boldsymbol{x}(M) + \sum_{j=0}^{M-1}\begin{bmatrix}\boldsymbol{x}(j)\\ \hat{\boldsymbol{u}}(j)\end{bmatrix}^T \boldsymbol{Q}\begin{bmatrix}\boldsymbol{x}(j)\\ \hat{\boldsymbol{u}}(j)\end{bmatrix}\right\}, \tag{2.29}$$

where

$$\boldsymbol{Q}_M > \boldsymbol{0}, \quad \boldsymbol{Q} = \begin{bmatrix}\boldsymbol{Q}_{11}\boldsymbol{Q}_{12}\\ \boldsymbol{Q}_{12}^T\boldsymbol{Q}_{22}\end{bmatrix} \geq \boldsymbol{0}, \quad \boldsymbol{Q}_{22} > \boldsymbol{0} \tag{2.30}$$

is given by

$$\hat{\boldsymbol{u}}(j) = -\boldsymbol{L}(j, \tau^{sc}(j))\begin{bmatrix}\boldsymbol{x}(j)\\ \hat{\boldsymbol{u}}(j-1)\end{bmatrix}, \quad \forall j \in \mathbb{N}. \tag{2.31}$$

This optimal controller is linear but depends on the sensor-to-controller delay. The computation of the matrix $\boldsymbol{L}(j, \tau^{sc}(j))$ (shown in [92]) assumes that $\tau^{sc}(j)$ and $\tau^{ca}(j)$ are random variables. This is an unrealistic assumption since both $\tau^{sc}(j)$ and $\tau^{ca}(j)$ depend on the network load. If the delays are modeled by letting their distribution be governed by the states of the underlying Markov chain of (2.10), then the optimal control strategy becomes

$$\hat{\boldsymbol{u}}(j) = -\boldsymbol{L}(j, \tau^{sc}(j), r(j)) \begin{bmatrix} \boldsymbol{x}(j) \\ \hat{\boldsymbol{u}}(j-1) \end{bmatrix}, \quad \forall j \in \mathbb{N}, \tag{2.32}$$

where $r(j)$ is the current state of the Markov chain. Since $\boldsymbol{L}(j, \tau^{sc}(j), r(j))$ now depends on $r(j)$, the controller must know the current state of the Markov chain. This method has been extended to the output feedback case and it has been shown in [92] that a time-varying Kalman filter can be used to compute an estimate of the plant states (the separation principle holds).

A stochastic Markovian model has been used in [108] to solve an $\mathcal{H}_\infty$ problem. A plant of the form

$$\begin{bmatrix} \boldsymbol{x}(j+1) \\ \boldsymbol{z}(j) \\ \boldsymbol{y}(j) \end{bmatrix} = \begin{bmatrix} \boldsymbol{A}(\theta(j)) & \boldsymbol{B}_1(\theta(j)) & \boldsymbol{B}_2(\theta(j)) \\ \boldsymbol{C}_1(\theta(j)) & \boldsymbol{D}_{11}(\theta(j)) & \boldsymbol{D}_{12}(\theta(j)) \\ \boldsymbol{C}_2(\theta(j)) & \boldsymbol{D}_{21}(\theta(j)) & \boldsymbol{0} \end{bmatrix} \begin{bmatrix} \boldsymbol{x}(j) \\ \boldsymbol{w}(j) \\ \boldsymbol{u}(j) \end{bmatrix}, \tag{2.33}$$

is considered where $\boldsymbol{x}(j) \in \mathbb{R}^{n_x}$ is the state, $\boldsymbol{u}(j) \in \mathbb{R}^{n_u}$ is the controlled input, $\boldsymbol{y}(j) \in \mathbb{R}^{n_y}$ is the measurement vector, $\boldsymbol{w}(j) \in \mathbb{R}^{n_w}$ is the disturbance vector and $\boldsymbol{z}(j) \in \mathbb{R}^{n_z}$ is the performance output vector. The system matrices are of appropriate size and they are functions of a discrete-time Markov chain taking values in a finite set $\mathcal{N} = \{1, 2, \ldots, M\}$ and with transition probabilities given by

$$q_{ik} = \Pr\left(\theta(j+1) = j | \theta(j) = i\right), \quad q_{ik} \geq 0, \quad \sum_{k=1}^{M} q_{ik} = 1. \tag{2.34}$$

The dependence on $\theta(j)$ indicates that the plant is in mode $i \in \mathcal{N}$ (*i.e.* $\theta(j) = i$). This is in fact a generalization of the dropout model in (2.20) because, apart from including error and disturbance vectors, (2.33) assumes a Markov chain taking values in any finite set of states $\mathcal{N} = \{1, 2, \ldots, M\}$ instead of $\mathcal{N} = \{1, 2\}$ as in (2.20). The aim is to design a controller of the form

$$\begin{aligned} \boldsymbol{x}_K(j+1) &= \boldsymbol{A}_K(\theta(j))\boldsymbol{x}_K(j) + \boldsymbol{B}_K(\theta(j))\boldsymbol{y}(j) \\ \boldsymbol{u}(j) &= \boldsymbol{C}_K(\theta(j))\boldsymbol{x}_K(j), \end{aligned} \tag{2.35}$$

(where $\boldsymbol{x}_K(j) \in \mathbb{R}^{n_k}$ is the controller state) that minimizes the closed-loop gain from $\boldsymbol{w}(j)$ to $\boldsymbol{z}(j)$ using the $\mathcal{H}_\infty$-norm as a measure. The main result of [108] is the derivation of necessary and sufficient LMI conditions for the synthesis of the $\mathcal{H}_\infty$ optimal controller.

In [58], a method for designing an $\mathcal{H}_\infty$ controller for the model in (2.13) is considered. Since the limited communication system state matrices are p-periodic, the controller will also be p-periodic. The controller takes the form of

$$\begin{aligned} \boldsymbol{x}_K(j+1) &= \boldsymbol{A}_K(j-1, \theta_S(j-1), \theta_A(j-1))\boldsymbol{x}_K(j) \\ &\quad + \boldsymbol{B}_K(j-1, \theta_A(j-1))\hat{\boldsymbol{y}}(j-1) \\ \boldsymbol{u}(j) &= \boldsymbol{C}_K(j, \theta_A(j))\boldsymbol{x}_K(j) + \boldsymbol{D}_K(j)\hat{\boldsymbol{y}}(j). \end{aligned} \tag{2.36}$$

Notice that the controller state matrix $\boldsymbol{A}_K(j-1, \theta_S(j-1), \theta_A(j-1))$ depends on $\theta_A(j-1)$ which is the information regarding the arrival of the control input $\boldsymbol{u}(j)$ (indeed, this is a strong assumption). Hence, the controller must receive acknowledgment of $\theta_A(j-1)$ at time j. In [58], an exact characterization for the controller synthesis is obtained and a necessary and sufficient condition is stated in terms of LMIs. The bounded real lemma-type proof given is based on the results of [108]. Furthermore, the controller is designed around the given matrices $\boldsymbol{S}_A(j), \boldsymbol{S}_S(j)$ (which are p-periodic) but their optimality is not treated. Thus, the controller is optimal only for that particular communication schedule.

The $\mathcal{H}_\infty$ design approach proposed in [147] is for a plant model which includes parameter uncertainty, network-induced delay and data dropout. The analysis is based on a Lyapunov-Krasovskii functional method and a new technique using a slack matrix variable is provided. A memoryless type controller was designed solving a set of LMIs. Other $\mathcal{H}_\infty$ control methods for robustness in the presence of packet loss are presented in [135, 60].

2.4.2 Estimation

Estimation over 'lossy networks' and model-based control techniques are especially important in applications like remote sensing, space exploration and sensor networks [48]. The idea is to minimize the network bandwidth usage by using local open-loop observers to reduce the required communication at the expense of computational load (this is known as controlled communication). Recent work on observers for NCSs has been carried out in [12, 101] and in the references therein. This is motivated by the growing number of smart sensor and actuator nodes with enough processor power to perform local computation [141].

The model-based control for networked system of [84] uses a plant model at the controller node to estimate the plant's states (with and without full state feedback) that are not available at the sampling time (see Figure 2.6(b)). This was discussed in Section 2.3.2.3 as a method for dealing with missing information in a contention-free communication paradigm. In reality, this method is more general and also applies to contention-based networks or whenever there are communication constraints between sensors and controller. Another model-based approach is used in [146] to artificially centralize a physically decentralized NCS. The approach implements local estimators that use global plant states. Although the work of [146] is an estimation-based control method, it will be discussed later (in Section 2.5) together with other scheduling methods since it offers an interesting dynamic scheduling technique for NCSs.

The problem of estimation over lossy networks is widely discussed in [48]. The LTI system in (2.5) with $\boldsymbol{B} = \boldsymbol{0}$ is considered. Consider a lossy channel modeled by the stochastic process $\theta(j) \in \{0, 1\}$ meaning that if a message is delivered, then $\theta(j) = 1$ and *vice versa*. The optimal estimate is given by

$$\tilde{\tilde{\boldsymbol{x}}}(j|j-1) = \mathbb{E}\left\{\boldsymbol{x}(j)|\theta(l), \forall l \leq j-1; \boldsymbol{y}(l), \forall l \leq j-1 \text{ s.t. } \theta(l) = 1\right\}, \quad (2.37)$$

given the information $\{\theta(l), \forall l \leq l-1\} \cup \{\boldsymbol{y}(l) : \theta(l) = 1, \forall j-1\}$ to the remote estimator at time j. The notation $\mathbb{E}\{X\}$ denotes the expected value of the random variable X. The time-varying Kalman filter equations used to recursively compute this estimate are given in [48]. A variation to the above is made by assuming that smart sensors have enough computational power. In that case, an optimal state estimate can be computed locally (before transmission through the network to the remote estimator) using the stationary Kalman filter

$$\begin{aligned} &\tilde{\boldsymbol{x}}(j+1|j+1) = \boldsymbol{A}\tilde{\boldsymbol{x}}(j|j) + \boldsymbol{F}\left(\boldsymbol{y}(j+1) - \boldsymbol{C}\boldsymbol{A}\tilde{\boldsymbol{x}}(j|j)\right), \\ &\forall j \in \mathbb{N}, \quad \tilde{x}(0|0) = \mathbf{0}, \end{aligned} \quad (2.38)$$

where

$$\boldsymbol{F} = \boldsymbol{P}\boldsymbol{C}^T(\boldsymbol{C}\boldsymbol{P}\boldsymbol{C}^T + \boldsymbol{R}_v)^{-1}, \quad (2.39)$$

and

$$\boldsymbol{P} = \boldsymbol{A}\boldsymbol{P}\boldsymbol{A}^T + \boldsymbol{R}_w - \boldsymbol{A}\boldsymbol{P}\boldsymbol{C}^T(\boldsymbol{C}\boldsymbol{P}\boldsymbol{C}^T + \boldsymbol{R}_v)^{-1}\boldsymbol{C}\boldsymbol{P}\boldsymbol{A}^T > \mathbf{0}. \quad (2.40)$$

The remote estimator will now receive the local estimate $\tilde{\boldsymbol{x}}(j|j)$ to compute the optimal estimate

$$\tilde{\tilde{\boldsymbol{x}}}(j|j-1) = \mathbb{E}\left\{\boldsymbol{x}(j)|\theta(l), \forall l \leq j-1; \tilde{\boldsymbol{x}}(l), \forall l \leq j-1 \text{ s.t. } \theta(l) = 1\right\}, \quad (2.41)$$

given the information $\{\theta(l), \forall l \leq l-1\} \cup \{\tilde{\boldsymbol{x}}(l) : \theta(l) = 1, \forall j-1\}$. The advantage of local estimation is that stability is preserved for larger drop rates than if the raw measurement were sent through the network [142] however, local estimation may be computationally expensive.

In [142], in order to reduce bandwidth usage, the case is considered where, at the sensor node, a decision is made whether to send the measurement or not (controlled communication). The decision depends on a local generated random variable $\theta(j)$ with probability distribution $\Lambda(j, \tilde{\boldsymbol{e}}(j)) \in [0, 1]$ where $\tilde{\boldsymbol{e}}(j) = \tilde{\boldsymbol{x}}(j|j) - \tilde{\tilde{\boldsymbol{x}}}(j|j-1)$ is an error between the local estimate and a copy of the remote estimate. By taking into account the network dropouts with probability $\pi \in [0, 1)$ then the stochastic decision becomes

$$\begin{aligned} &\Pr(\theta(j) = 1) = (1-\pi)\Lambda(j, \tilde{\boldsymbol{e}}(j)) \quad \Rightarrow \quad \text{local estimate delivered} \\ &\Pr(\theta(j) = 0) = 1 - (1-\pi)\Lambda(j, \tilde{\boldsymbol{e}}(j)) \quad \Rightarrow \quad \text{local estimate not delivered.} \end{aligned} \quad (2.42)$$

The problem of optimal controlled communication has been considered in [141] but without packet dropout and with the simplification that the smart sensor can measure all states (no need of local estimation). The optimal communication policy is given by finding the optimal distribution $\Lambda(j, \tilde{e}(j))$ that minimizes the cost

$$J = \lim_{M \to \infty} \frac{1}{M} \mathbb{E} \left\{ \sum_{j=0}^{M} \|\boldsymbol{e}(j+1)\|^2 + \lambda \theta(j) \right\}, \tag{2.43}$$

where $\boldsymbol{e}(j) = \boldsymbol{x}(j) - \tilde{\boldsymbol{x}}(j|j-1)$, $\lambda > 0$ and $\| \cdot \|$ is the Euclidean norm. The optimal policy is stationary and can be calculated using dynamic programming and value iteration.

2.5 Scheduling methods

Scheduling methods for NCSs refer to those communication scheduling techniques that take into account the dynamics of the plant and its associated controller. In other words, the assumption is that a controller designed for a network-free system is given and a communication scheduling policy must be provided to minimize the effect of the network. Scheduling methods aim at guaranteeing QoS [130, Chapter 4].

In NCSs, not only is the schedulability of tasks in each node important, but also the schedulability of the information exchange between nodes through the network. The correct execution of tasks in a node depends on the information received by other nodes which is also subject to scheduling. This is referred to as multiprocessor scheduling [109] and the problem of optimal scheduling of tasks among processors is NP-complete [37]. Since RTCS are very sensitive to delays, especially time-varying and non deterministic delays, it is essential that communication scheduling not only guarantees schedulability but also guarantees stability and ultimately performance of the NCS. Scheduling of communication in NCSs is more complex than the traditional task scheduling as it requires knowledge of the dynamics of the plant.

In RTCSs, resources can be categorized into *computing resources* and *communication resources*. The processor is the computing resource shared among the various tasks. The network is the communication resource shared among different nodes' messages. Both tasks and messages can be periodic, aperiodic or sporadic, have a execution/transmission time and deadlines. It is easy to see an 'homomorphism' between computing and communication resources sharing in the sense that both could be analyzed and treated in a similar way. Although this is an accepted argument (*e.g.* [138, p. 52]) the following fundamental differences should be considered.

- Task scheduling is preemptive (the execution of a task can be interrupted and resumed some time later) while message scheduling cannot allow preemption.

- Tasks running on a processor are centralized in the sense that the processor has full knowledge of the task states and variables. Node messages are decentralized in the sense that the node knowledge of states and variables of other nodes is limited and depends on the scheduling policy itself.

- Typical issues like message collision, network arbitration, and packet loss in communication scheduling do not have a counterpart in tasks processor sharing (this is a consequence of the decentralized nature of NCSs).

It is the second point above that imposes the largest limitation on communication scheduling and the structure of the NCS determines the size of this limitation. For instance, consider three cases:

(i) sensors and controller are collocated but actuators are distributed

(ii) controller and actuators are collocated but sensors are distributed

(iii) sensors, controller and actuators are distributed.

In case (i), there is global knowledge of the plant outputs and the scheduler can decide which actuator to control accordingly. In case (ii), the sensor nodes should adopt a (distributed) scheduling policy that gives network access to the sensor with the most important information for the controller (*e.g.* the largest error from previously sent message). Since there is no global knowledge of the sensor states (or error), it is difficult to establish which sensor should have priority. Case (iii) is the most complex case since now, both sensor readings and control commands should be scheduled in some optimal way.

A communication scheduler will be called 'open-loop' if the scheduling does not depend on the plant and/or network states. It will be called 'closed-loop' if the scheduling depends on the plant and/or network states.

2.5.1 Open-loop scheduling

The simplest open-loop communication policy is the Round Robin (RR) scheduling. RR assigns time slots to each node equally, in order and without priority. RR scheduling can be easily implemented in contention-free paradigms like TDMA or token bus. A variation to the simple RR algorithm is to allocate more time slots to nodes that require more 'attention' or nodes from which information is more important. This is achieved by establishing a periodic schedule *a priori* according to some optimality criteria. In contention-based paradigms open-loop scheduling consists of assigning each node a communication period and, by allowing the network to be sufficiently underloaded, some performance, under the worst case scenario, is guaranteed. In non-destructive bitwise arbitration protocols like CSMA/BA the optimal choice of priorities, for example, is deadline monotonic *i.e.* the nodes with the shortest deadline are assigned the highest priority. Although the above mentioned scheduling policy is simpler to implement than closed-loop methods (discussed in the next section), it is often the preferred choice for safety-critical control systems [68, 76]. The reasons are: open-loop scheduling methods are easy to understand and their behavior is easily predictable; off-the-shelf technology can be used without the need of implementing new protocols, or model-based techniques.

2.5.2 Closed-loop scheduling

Closed-loop scheduling methods use information about the plant and/or network to dynamically schedule the communication among nodes. The key idea, as for feedback control, is to generate an error and then use the communication scheduling as the manipulated variable to bring this error to zero. The problem is to find a suitable error measure and a way to deal with it.

To solve this problem, a method is proposed in [128, 129], called MEF-TOD (maximum-error-first with try-once-discard), for communication protocols like CAN that allow bit-wise arbitration (CSMA/BA). Consider the LTI plant in (2.2) without the white noise processes and the associated (stabilizing) controller

$$\begin{aligned}\dot{\boldsymbol{x}}_K(t) &= \boldsymbol{A}_K^c \boldsymbol{x}_K(t) + \boldsymbol{B}_K^c \boldsymbol{y}(t)\\ \boldsymbol{u}(t) &= \boldsymbol{C}_K \boldsymbol{x}_K(t) + \boldsymbol{D}_K \boldsymbol{y}(t),\end{aligned} \tag{2.44}$$

where $\boldsymbol{x}_K(t) \in \mathbb{R}^{n_k}$ is the controller state. Let $\hat{\boldsymbol{y}}(t)$ and $\hat{\boldsymbol{u}}(t)$ be the last transmitted measurement and control vectors respectively (since they are broadcasted, they are globally known). The autonomous dynamics of the NCS are given by

$$\dot{\boldsymbol{z}}(t) = \begin{bmatrix}\dot{\boldsymbol{x}}_{cl}(t)\\ \dot{\boldsymbol{e}}(t)\end{bmatrix} = \boldsymbol{A}_{cl}^c \begin{bmatrix}\boldsymbol{x}_{cl}(t)\\ \boldsymbol{e}(t)\end{bmatrix} = \begin{bmatrix}\boldsymbol{A}_{cl,11}^c & \boldsymbol{A}_{cl,12}^c\\ \boldsymbol{A}_{cl,21}^c & \boldsymbol{A}_{cl,22}^c\end{bmatrix}\begin{bmatrix}\boldsymbol{x}_{cl}(t)\\ \boldsymbol{e}(t)\end{bmatrix}, \tag{2.45}$$

where

$$\boldsymbol{x}_{cl}(t) = \begin{bmatrix}\boldsymbol{x}(t)\\ \boldsymbol{x}_K(t)\end{bmatrix}, \quad \boldsymbol{e}(t) = \begin{bmatrix}\hat{\boldsymbol{y}}(t) - \boldsymbol{y}(t)\\ \hat{\boldsymbol{u}}(t) - \boldsymbol{u}(t)\end{bmatrix}, \tag{2.46}$$

and

$$\begin{aligned}\boldsymbol{A}_{cl,11}^c &= \begin{bmatrix}\boldsymbol{A}^c + \boldsymbol{B}^c\boldsymbol{D}_K\boldsymbol{C} & \boldsymbol{B}^c\boldsymbol{C}_K\\ \boldsymbol{B}_K^c\boldsymbol{C} & \boldsymbol{A}_K^c\end{bmatrix}, \quad \boldsymbol{A}_{cl,12}^c = \begin{bmatrix}\boldsymbol{B}^c\boldsymbol{D}_K & \boldsymbol{B}^c\\ \boldsymbol{B}_K^c & \boldsymbol{0}\end{bmatrix},\\ \boldsymbol{A}_{cl,21}^c &= -\begin{bmatrix}\boldsymbol{C} & \boldsymbol{0}\\ \boldsymbol{0} & \boldsymbol{C}_K\end{bmatrix}\boldsymbol{A}_{cl,11}^c, \quad \boldsymbol{A}_{cl,22}^c = -\begin{bmatrix}\boldsymbol{C} & \boldsymbol{0}\\ \boldsymbol{0} & \boldsymbol{C}_K\end{bmatrix}\boldsymbol{A}_{cl,12}^c.\end{aligned} \tag{2.47}$$

The entries of the error vector $\boldsymbol{e}_i(t), i = 1, 2, \ldots, n_u + n_y$ (which can also be vectors) are reset to zero every time an update is sent through the network (without network, $\boldsymbol{e}(t) = \boldsymbol{0}$), and what entry should be reset to zero at any time tick depends on the scheduling strategy. In the MEF-TOD protocol, the priority of each node's message is proportional to the Euclidian norm of $\boldsymbol{e}_i(t)$. This information is not globally known, therefore there is no guarantee that another message, in another node, has been assigned the same priority. With the bitwise arbitration technology of CAN, priorities can be encoded into the most significant bits of the message identifier. The least significant bits are fixed and assigned to each node uniquely so that, if more than one node tries to send a message, only one will be granted access due to its fixed higher priority. The nodes that fail to win the competition will discard their message and generate a new one for the next attempt. The key feature of this protocol is that priority is assigned dynamically but, because of the decentralized structure of the NCS, arbitration indeterminism cannot be avoided. This protocol has been also considered in [28] for the codesign of an observer that asymptotically reconstructs the plant states. Sufficient conditions in terms of matrix inequalities for the existence of an observer-protocol pair are given.

The estimation-based closed-loop scheduling proposed in [146] can be regarded as an improvement (at the expense of computational burden) of the MEF-TOD. In [146], the discrete case is considered, hence, all the variables can be defined as before but for discrete-time steps. The scheduling policy is similar. The distributed sensor measurements are scheduled by creating global knowledge of the node states using identical estimators at each node. The estimator of node i has sufficient computational power to estimate every node (scalar) output $\hat{y}_k(j)$, $k = 1, 2, \dots d$ (where d is the number of nodes) including its own, $\hat{y}_i(j)$. The estimation is based on the information broadcasted to the network. However, node i knows its true output value $y_i(j)$ and therefore an absolute value of the error, $e_i(j) = |y_i(j) - \hat{y}_i(j)|$ can be calculated. If $e_i(j) > g_i$ where g_i is a predefined threshold value, then the node 'realizes' that every other node is using an incorrect measurement and broadcasts the true measurement $y_i(j)$. In [141] the same architecture as in [146] is used but the authors also propose a stochastic communication logic based on doubly stochastic Poisson processes. A message is broadcasted according to a Poisson process whose rate depends on the error $e_i(j)$.

A remark should be made on the fact that the above proposed scheduling policies (and many others) rely on the common belief that sending the most recent observation is optimal (this is the core of the MEF-TOD method). Although it may sound counterintuitive, the results in [106] show that sending the most recent observation, in systems where observations are occasionally lost, is not optimal. Conditions for the existence of a linear combination of past and present measurements, which minimizes the state estimation error covariance, are derived.

In [126], it is assumed that each node (which is a subsystem with full state availability) has global knowledge of the network utilization (currently used bandwidth). Its local information (state variables and local bandwidth) and the current network bandwidth utilization are used to dynamically assign the sampling time to each node. Let each subsystem in a node be an LTI system described by (2.5) with $\boldsymbol{C} = \boldsymbol{I}$. Let h_i be the sampling time of node $i = 1, 2, \dots d$ where d is the number of nodes, h_l and h_s the longest and shortest possible sampling times and c a parameter that represents the criticalness of the sampling. Then, the current sampling time is calculated using the following nonlinear heuristic function

$$h_i(j+1) = (h_l - h_s)e^{-c\|\boldsymbol{x}(j)\|} + h_s, \quad \forall i = 1, 2, \ldots, d, \tag{2.48}$$

where h_s is a function of the available bandwidth calculated from the measures of the global and local bandwidth utilization [126]. The variables h_l and c are used to heuristically adjust the shape of the exponential function. Note that for small $\|\boldsymbol{x}(j)\|$ (where $\|\cdot\|$ is the Euclidian norm) the sampling time will be close to the longest possible (h_l), and for large $\|\boldsymbol{x}(j)\|$ it will be close to the shortest (h_s).

In [145], the proposed scheduling algorithm, which the authors call LEF (large error first), schedules node communication according to their state distances from the equilibrium point. What is fundamentally different in [145] is the assumption of a master-slave node configuration that allows a centralized scheduling decision. The master node scans all the states of the slave nodes and decides which node should have priority. In theory, since priorities are assigned globally, message collisions can be avoided. For instance, in a TTCAN protocol, the master node sends a message between node messages. The master node message will contain the identifier of the nodes that should access the bus next. All nodes receive this message but only the one with that identifier will acknowledge it. This mechanism has the double purpose of synchronizing communication, centralizing scheduling and improving determinism at the expense of higher bus utilization.

Inspired by the MEF-TOD and LEF scheduling methods, in [137] another feedback-based network scheduler is proposed which is called MUF (maximum urgency first). The scheduling is based on a weighted measure of the processess states but the scheduler is directly connected to each node by a separate communication medium. In this configuration, the scheduler works with up-to-date information of node states that is continuously available. This, in a sense, defeats the original purpose of scheduling distributed nodes with limited communication. More recently, an LMI-based scheduling method to maximize the allowable delay bound for NCS scheduling was proposed by [66].

2.6 Scheduling and controller codesign methods

Scheduling and controller codesign methods refer to those techniques used to jointly solve the problem of designing an optimal controller together with an optimal scheduling policy. The benefits, from an optimality point of view, are obvious. The traditional separation between control engineer and software engineer (and the consequent overall suboptimal design) are eliminated. The discussion that follows will focus on methods where an offline (static) scheduler is designed with its associated optimal controller and where scheduling policy and controller are dynamically calculated (online scheduling). Recent results on codesign methods for dependable NCSs can be found in [8].

2.6.1 Offline scheduling and controller codesign

Offline scheduling and control is effectively an open-loop scheduling method but the controller is designed and optimized around that particular scheduling policy. For obvious practical reasons, all offline schedules must be periodic. Furthermore, it is reasonable to assume that such schedules are implemented in contention-free protocols, otherwise the original purpose of having an optimal controller with an optimal schedule is defeated. Section 2.3 reviews some of the methods where the scheduling mechanism is taken into account by merging the limited communication channel dynamics into the controlled plant dynamics. The result is a time-varying model that, for a p-periodic sequence, will also be p-periodic.

Consider the non-ZOH description of the NCS given in (2.11). The work of [149] proposes a method for finding p-periodic sequences of scheduling matrices $\{\boldsymbol{S}_A(0), \boldsymbol{S}_A(1), \ldots, \boldsymbol{S}_A(p) = \boldsymbol{S}_A(0), \ldots\}$, $\{\boldsymbol{S}_S(0), \boldsymbol{S}_S(1), \ldots, \boldsymbol{S}_S(p-1), \boldsymbol{S}_S(p) = \boldsymbol{S}_S(0), \ldots\}$ that preserve the structural properties (controllability and observability) of the original system when it is augmented with the network dynamics (the extended plant). This is obtained by recursively constructing a controllability matrix formed by a sufficient number of selected linearly independent vectors. Such a sequence always exists, for controllable plants, and will be discussed in more detail in Section 2.7.2. Once suitable sequences for $\boldsymbol{S}_A(j)$ and $\boldsymbol{S}_S(j)$ are selected, an optimal controller is designed for (2.11) (with the addition of Gaussian noises) such that the quadratic cost function

$$J = \mathbb{E}\left\{ \boldsymbol{x}^T(M)\boldsymbol{Q}\boldsymbol{x}(M) + \sum_{j=0}^{M-1} \left(\boldsymbol{x}^T(j)\boldsymbol{Q}\boldsymbol{x}(j) + \boldsymbol{u}^T(j)\boldsymbol{u}(j)\right) \right\} \tag{2.49}$$

is minimized. This is the standard Linear Quadratic Gaussian (LQG) problem that is solved by designing a Kalman filter (that gives the optimal state estimate $\tilde{\boldsymbol{x}}$) and a Linear Quadratic (LQ) optimal feedback gain $\boldsymbol{L}(j)$. The resulting optimal controller that minimizes J is given by the feedback law

$$\boldsymbol{u}(j) = -\boldsymbol{L}(j)\tilde{\boldsymbol{x}}(j). \tag{2.50}$$

The method of [149] is an LQG controller design for a given schedule. Although this schedule guarantees controllability and observability, it is not optimal. In [54], the same analysis is extended to the case where communication is subject to known delays. The delay compensator method proposed in [81] is adopted.

The problem of designing an optimal controller for an optimal sequence is analyzed in [105]. Consider the LTI system equations in (2.2) without the white noise processes, with $\boldsymbol{D} = \boldsymbol{0}$, $\boldsymbol{C} = \boldsymbol{I}$ and the associated LQ problem

$$\min_{\hat{\boldsymbol{u}}} \boldsymbol{x}(M)^T \boldsymbol{Q}^{c0} \boldsymbol{x}(M) + \int_0^M \left(\boldsymbol{x}(t)^T \boldsymbol{Q}^{c1} \boldsymbol{x}(t) + \hat{\boldsymbol{u}}(t)^T \boldsymbol{Q}^{c2} \hat{\boldsymbol{u}}(t)\right) dt, \quad (2.51)$$

where $\boldsymbol{Q}_{c1} \geq \boldsymbol{0}$, $\boldsymbol{Q}_{c2} > \boldsymbol{0}$ and $\boldsymbol{Q}_{c0} \geq \boldsymbol{0}$ have been chosen for a desired closed-loop response. Obtain the sampled-data model of (2.2) given in (2.5) and its augmented representation (2.16). Recall that (2.16) is the augmented plant model which includes the limited communication dynamics. Such a model can only be constructed for a given sequence of scheduling matrices $\boldsymbol{S}_A(j)$. Once again, only consider p-periodic sequences ($\boldsymbol{S}_A(j) = \boldsymbol{S}_A(j+p), \forall j \geq 0$). The equivalent LQ problem for the augmented system can be written as

$$\min_{\boldsymbol{u}} \hat{\boldsymbol{x}}(M)^T \hat{\boldsymbol{Q}}_0 \hat{\boldsymbol{x}}(M) + \sum_{j=0}^{M} \begin{bmatrix} \hat{\boldsymbol{x}}(j) \\ \boldsymbol{u}(j) \end{bmatrix}^T \begin{bmatrix} \hat{\boldsymbol{Q}}_1(j) & \hat{\boldsymbol{Q}}_{12}(j) \\ \hat{\boldsymbol{Q}}_{12}(j)^T & \hat{\boldsymbol{Q}}_2(j) \end{bmatrix} \begin{bmatrix} \hat{\boldsymbol{x}}(j) \\ \boldsymbol{u}(j) \end{bmatrix}, \quad (2.52)$$

where the matrices $\hat{\boldsymbol{Q}}_0, \hat{\boldsymbol{Q}}_1(j), \hat{\boldsymbol{Q}}_2(j), \hat{\boldsymbol{Q}}_{12}(j)$ can be found in [105][2]. Notice that these matrices are time-varying (periodic) because of $\boldsymbol{S}_A(j)$. The discrete-time lifting technique (mentioned in Section 2.3.2.2 and discussed in more detail in the next chapter) is used to eliminate the periodicity of both the augmented plant and the correspondent LQ cost. At this point, the problem is in the form of a standard LQ design problem but only for a given sequence of matrices, $\boldsymbol{S}_A(j)$. The solution to what turns out to be a complex combinatorial optimization problem is a heuristic algorithm based on a partition of the problem into three subproblems. This partitioning allows a local search approach to be effective for small single-channel problems but strong assumptions are made. Only the single-channel case is considered *i.e.* only one control signal can be updated at any time tick. The problem of finding sequences that preserve controllability of the lifted augmented system is not addressed but, contrary to [149], a more realistic model (with a ZOH) is considered and a method for designing (sub)optimal sequences is given. However, this method leads to search problems over large trees [75].

What is proposed in [75] is a method which determines, more efficiently, (sub)optimal sequences and avoids the combinatorial explosion. This is a branch and bound based algorithm that, at every step, expands a set of candidate sequences by backward iteration. The branches of the sequences that do not perform better for all the states (compared to the others) are pruned. The measure of 'better' is given by a pruning parameter R that needs to be chosen *a priori*. A proof of optimality for sufficiently large R is given. This is in fact a compromise between an almost exhaustive search for a near-to-optimal solution and a faster search for a suboptimal solution. The result depends on the

[2]Note that there is a mistake in [105] in evaluating the matrix $\hat{\boldsymbol{Q}}_2(j)$. More precisely, their matrix is not time-dependent and it is non-singular (while in fact it is indeed time-dependent and singular). This mistake allows the authors to continue their analysis by using the inverse of $\hat{\boldsymbol{Q}}_2(j)$ that, of course, is not defined. This, and other issues found in [105] are discussed in Chapter 6.

bounding (or pruning) parameter. The measure of optimality is the solution of a set of Riccati equations for the standard quadratic time-varying problem (no lifting is used, resulting in a periodic feedback controller).

Similarly, in [14], the LQ problem is translated into a mixed-integer quadratic programming formulation and solved using a branch and bound based method (the advantage of this formulation is the existence of many efficient academic and commercial solvers). It is dependent on the chosen initial states and developed for finite horizon problems only.

In [80], the augmented plant model in (2.16) is also used but the corresponding controller is designed under an $\mathcal{H}_\infty$ and $\mathcal{H}_2$ performance index. For a given communication sequence, the problem is formulated into a periodic control problem for which a direct LMI design method is developed. The heuristic search method for the (sub)optimal sequence consists of gradually increasing the sequence period and computing a (presumably exhaustive) search algorithm until a pre-defined tolerance is satisfied.

To summarize, the main differences between the previously reviewed work are:

- the performance index: LQ in [149, 105, 75, 14] and $\mathcal{H}_\infty/\mathcal{H}_2$ in [80]
- the adopted model: time-invariant in [105, 14] and periodic in [149, 75, 80]
- the optimality of the scheduling: non-optimal in [149], heuristically optimal in [105, 80] and (sub)optimal in [75, 14].

Also, only [149] provides a method that guarantees controllability (and observability) of a scheduling sequence while, in the others, this problem is not treated (or it is implicitly assumed that, if all nodes are scheduled at least once, then controllability and/or observability is preserved).

2.6.2 Online scheduling and controller codesign

The offline scheduling and control methods presented in the previous section are attractive, especially for practical applications, because they are easy to implement and because they increase the degree of determinism in the NCS. However, the argument against static schedules is that, despite the fact the controller has been optimized for that schedule, disturbances are not accounted for. Furthermore, it becomes more difficult to schedule sporadic messages (that normally share the same medium) and, if a new node is added, the whole optimization problem needs to be solved again.

The idea of online scheduling and control is to design, at every sampling instance, a controller with a communication schedule. A solution to this problem has been presented in [14] using a Model Predictive Control (MPC) approach for the scheduling of the controller-to-actuator communication only. In MPC, an optimal control problem is solved online at each sampling period. Let σ be the communication schedule between controller and actuators, then the MPC problem consists of minimizing the quadratic cost

$$J = \min_{\boldsymbol{u},\sigma} \quad \hat{\boldsymbol{x}}(M)^T \hat{\boldsymbol{Q}}_0 \hat{\boldsymbol{x}}(M) + \sum_{j=0}^{M-1} \begin{bmatrix} \hat{\boldsymbol{x}}(j) \\ \boldsymbol{u}(j) \end{bmatrix}^T \begin{bmatrix} \hat{\boldsymbol{Q}}_1(j) & \hat{\boldsymbol{Q}}_{12}(j) \\ \hat{\boldsymbol{Q}}_{12}(j)^T & \hat{\boldsymbol{Q}}_2(j) \end{bmatrix} \begin{bmatrix} \hat{\boldsymbol{x}}(j) \\ \boldsymbol{u}(j) \end{bmatrix}, \tag{2.53}$$

subject to the augmented system in (2.16) with $\boldsymbol{C} = \boldsymbol{I}$ and the constraints on σ (the communication sequence). The solution to this problem is a control and communication schedule based on the future evolution of the system over the horizon M. As for standard MPC approaches, only the first control action and the first element of the schedule are considered while the rest is disregarded. The obvious drawback of this approach comes from the excessively expensive computational resources required. In view of this, a heuristic scheduling algorithm, called Optimal Pointer Placement (OPP), is proposed in [14]. OPP is a compromise between the advantages of online scheduling and those of offline scheduling. A p-periodic communication schedule with the associated p sets of controller gains are optimized offline. During run time, instead of solving the large MPC optimization problem described above, a smaller problem is solved. The solution of this problem is a 'pointer' that points to the optimal actuator for control with a corresponding controller gain.

The idea of MPC is also considered in [102] but this time buffers are introduced at each actuator node able to store a sequence of future control inputs. Their argument is that networks like Ethernet, originally designed as data networks, are often adopted for low level control technology. As discussed in Section 2.2.1, data networks have the capability of transmitting large packets at the expense of longer delays and higher packet dropout probability. Hence, a predictive controller could send, in a single packet, a set of present and future control inputs that are stored and sequentially used at the actuator node until an updated set of control values is received. On the other hand, assuming that acknowledgments form part of the network protocol, the controller receives information regarding the buffer state and uses this information to solve the MPC problem. This approach is generalized for discrete-time nonlinear multiple-input plant models with state $\boldsymbol{x}(j)$ and input vector $\hat{\boldsymbol{u}}(j)$ of the form

$$\boldsymbol{x}(j+1) = f(\boldsymbol{x}(j), \hat{\boldsymbol{u}}(j)), \quad j \in \mathbb{N}_0, \tag{2.54}$$

and, at each time instant j, the following cost function is minimized

$$J = F(\tilde{\boldsymbol{x}}(j+M)) + \sum_{l=j}^{j+M-1} L(\tilde{\boldsymbol{x}}(l), \tilde{\boldsymbol{u}}(l)), \tag{2.55}$$

where $\tilde{\boldsymbol{x}}(\cdot)$ and $\tilde{\boldsymbol{u}}(\cdot)$ are the predicted states and inputs respectively and $L(\cdot,\cdot)$ and $F(\cdot)$ are penalizing weighting functions. Also in this case, an expensive optimization problem needs to be solved at every sampling instance. Furthermore, the node acknowledgments on receiving a message type of protocol (which is in effect a feedback message that shares the same communication medium) means that the overall bandwidth used for control is halved. An interesting question at this point is: when is it better to have feedback information from the nodes at the expense of bandwidth utilization? And also, what is the optimal trade off between open-loop and closed-loop scheduling for control and communication scheduling?

A different approach for a simultaneous protocol and controller design has been presented in [29]. The difference of this approach, compared to the previous ones, is that only the network induced error is considered and therefore the protocol is implementable on CAN-like NCSs (the previously shown scheduling codesign methods are based on the plant's states). In [29], the MEF-TOD protocol of [128, 129] (presented in Section 2.5.2) is used with the same definition of networked induced error given in (2.46). The scheduling matrix is defined as

$$\boldsymbol{S} = s_{r,c} = \begin{bmatrix} \boldsymbol{S}_S & \boldsymbol{0} \\ \boldsymbol{0} & \boldsymbol{S}_A \end{bmatrix}, \quad \boldsymbol{S} \in \mathcal{S} = \{\boldsymbol{S}(1), \boldsymbol{S}(2), \ldots \boldsymbol{S}(d)\}, \tag{2.56}$$

where d is the number of nodes. $\boldsymbol{S}_A$ and $\boldsymbol{S}_S$ are diagonal matrices[3] with a 1 or a 0 in their diagonals and $\sum_{r,c} s_{r,c} = 1$ (*i.e.* only one node, actuator or sensor, communicates at any time). Assume that a ZOH strategy is used with the continuous LTI plant in (2.2) (without noise) and its discrete representation in (2.5). Then, the model for this NCS is given by

$$\begin{bmatrix} \boldsymbol{x}_T(t_{i+1}, i) \\ \boldsymbol{e}(t_{i+1}, i) \end{bmatrix} = \begin{bmatrix} \boldsymbol{A}_T & \boldsymbol{B}_T \\ \boldsymbol{C}_T(\boldsymbol{I} - \boldsymbol{A}_T) & \boldsymbol{I} - \boldsymbol{C}_T\boldsymbol{B}_T \end{bmatrix} \underbrace{\begin{bmatrix} \boldsymbol{I} & \boldsymbol{0} \\ \boldsymbol{0} & \boldsymbol{S}_T \end{bmatrix}}_{\overset{\text{def}}{=} \tilde{\boldsymbol{S}}} \begin{bmatrix} \boldsymbol{x}_T(t_i, i-1) \\ \boldsymbol{e}(t_i, i-1) \end{bmatrix}, \tag{2.57}$$

where

$$\boldsymbol{x}_T(t_i, i-1) = \begin{bmatrix} \boldsymbol{x}(t_i, i-1) \\ \boldsymbol{x}_K(t, i-1) \end{bmatrix}, \quad \boldsymbol{e}(t_i, i-1) = \begin{bmatrix} \hat{\boldsymbol{y}}(t, i) - \boldsymbol{y}(t, i) \\ \hat{\boldsymbol{u}}(t, i) - \boldsymbol{u}(t, i) \end{bmatrix}, \tag{2.58}$$

and

$$\begin{aligned} \boldsymbol{A}_T &= \begin{bmatrix} \boldsymbol{A} + \boldsymbol{B}\boldsymbol{D}_K\boldsymbol{C} & \boldsymbol{B}\boldsymbol{C}_K \\ \boldsymbol{B}_K\boldsymbol{C} & \boldsymbol{A}_K \end{bmatrix}, \quad \boldsymbol{B}_T = \begin{bmatrix} \boldsymbol{B}\boldsymbol{D}_K & \boldsymbol{B} \\ \boldsymbol{B}_K & \boldsymbol{0} \end{bmatrix}, \\ \boldsymbol{C}_T &= \begin{bmatrix} \boldsymbol{C} & \boldsymbol{0} \\ \boldsymbol{0} & \boldsymbol{C}_K \end{bmatrix}, \quad \boldsymbol{S}_T = \begin{bmatrix} \boldsymbol{S}_S(i) & \boldsymbol{0} \\ \boldsymbol{D}_K(\boldsymbol{I} - \boldsymbol{S}_S(i)) & \boldsymbol{S}_A(i) \end{bmatrix}. \end{aligned} \tag{2.59}$$

Also, $t_i = (i-1)h$, and the use of the two-dimensional time arguments (t, i) denotes continuous time $t \geq 0$ and the total number of jumps $i \in \mathbb{N}$. The matrix $\boldsymbol{S}$ is treated as a control input. $\boldsymbol{x}_K$ is the controller state vector. The

[3]These matrices are similar to the previously introduced matrices $\boldsymbol{S}_A(j), \boldsymbol{S}_A(j)$ but the dependance on j has been dropped for consistency with the notation of [29].

controller matrices $\boldsymbol{A}_K$, $\boldsymbol{B}_K$, $\boldsymbol{C}_K$ and $\boldsymbol{D}_K$ are assumed to be unknown and have to be designed. Having established the necessary and sufficient condition for quadratic stability, [29] present an algorithm to simultaneously design a controller and a protocol. The algorithm solves a sequence of LMIs.

2.7 Structural and stability analysis

Consider the LTI system in (2.2) and its sampled-data representation in (2.5). If $\boldsymbol{A}^c$ does not have two eigenvalues with equal real and imaginary parts that differ by an integral multiple of $\frac{2\pi}{h}$ (*i.e.* the sampling frequency is non-pathological [20]), then

$$\begin{aligned} (\boldsymbol{A}^c, \boldsymbol{B}^c) \ \textit{controllable} \quad &\Rightarrow \quad (\boldsymbol{A}, \boldsymbol{B}) \ \textit{controllable} \\ (\boldsymbol{C}, \boldsymbol{A}^c) \ \textit{observable} \quad &\Rightarrow \quad (\boldsymbol{C}, \boldsymbol{A}) \ \textit{observable}. \end{aligned} \tag{2.60}$$

This is a sufficient condition for preservation of controllability and observability after discretization (necessary and sufficient conditions can be found in [67]). If a system is controllable/observable then it is also stabilizable/detectable. Since NCSs have been modeled here mainly as sampled-data systems, this is a starting point for structural (controllability and observability) analysis. Addressing the stability of feedback loops that are closed over a network is the next problem. Quantization issues are ignored here. The analysis focuses on delays, stochastic packet dropout and scheduling. Structural and stability properties for the codesign problem are discussed separately.

2.7.1 Stability with delays and packet dropout

The usual way to investigate the effect of delays in NCSs is by considering the continuous LTI equations of a combined plant and controller [48],

$$\begin{aligned} \dot{\boldsymbol{x}}(t) &= \boldsymbol{A}^c \boldsymbol{x}(t) + \boldsymbol{B}^c \hat{\boldsymbol{y}}(t) \\ \boldsymbol{y}(t) &= \boldsymbol{C}\boldsymbol{x}(t), \end{aligned} \tag{2.61}$$

where a delay element (the network) is placed between the sampled output $\boldsymbol{y}(t)$ and the plant input $\hat{\boldsymbol{y}}(t)$. This analysis is analogous to the one in (2.2)-(2.6) apart from the fact that here the combined plant and controller model is considered. Because of the delay $\tau(j)$, $0 < \tau(j) < h$ (where j is the sampling instant and h is the sampling period), the system input is

$$\hat{\boldsymbol{y}}(t) = \begin{cases} \hat{\boldsymbol{y}}(j-1), & t \in [j, j+\tau(j)) \\ \hat{\boldsymbol{y}}(j), & t \in [j+\tau(j), j+1) \end{cases}. \tag{2.62}$$

The sampled data representation of (2.61), in the presence of the delay $\tau(j)$, is

$$\boldsymbol{x}(j+1) = e^{\boldsymbol{A}^c h}\boldsymbol{x}(j) + e^{\boldsymbol{A}^c(h-\tau(j))}\left(\int_0^{\tau(j)} e^{\boldsymbol{A}^c s}ds\right)\boldsymbol{B}^c\hat{\boldsymbol{y}}(j-1) + \left(\int_0^{h-\tau(j)} e^{\boldsymbol{A}^c s}ds\right)\boldsymbol{B}^c\hat{\boldsymbol{y}}(j), \quad (2.63)$$

and the autonomous dynamics of the NCS are given by

$$\begin{bmatrix}\boldsymbol{x}(j+1)\\ \hat{\boldsymbol{y}}(j)\end{bmatrix} = \boldsymbol{A}(h,\tau(j))\begin{bmatrix}\boldsymbol{x}(j)\\ \hat{\boldsymbol{y}}(j-1)\end{bmatrix}, \quad (2.64)$$

where

$$\boldsymbol{A}(h,\tau(j)) = \begin{bmatrix} e^{\boldsymbol{A}^c h} + \left(\int_0^{h-\tau(j)} e^{\boldsymbol{A}^c s}ds\right)\boldsymbol{B}^c\boldsymbol{C} & e^{\boldsymbol{A}^c(h-\tau(j))}\left(\int_0^{\tau(j)} e^{\boldsymbol{A}^c s}ds\right)\boldsymbol{B}^c \\ \boldsymbol{C} & \boldsymbol{0}\end{bmatrix}. \quad (2.65)$$

2.7.1.1 Constant delays

When the delay is constant (*i.e.* $\tau(j) = \tau$ for all $j \in \mathbb{N}$) the analysis for stability is straightforward. The NCS in (2.61) is exponentially stable if and only if $\boldsymbol{A}(h,\tau)$ is Schur (*i.e.* all its eigenvalues have magnitude strictly less than 1) [48, 17]. Constant delays can be achieved by using a buffer as discussed in Section 2.3.1 (see Figure 2.4(b)). The analysis for the case where delays are longer than the sampling period is more complicated. This is considered in [24] where a sufficient LMI condition for the stability analysis are proposed. An analysis of an NCS modeled as delayed differential equations is presented in [48]. The authors show the benefit of such models that are valid for any delay (even for delays longer than the sampling time).

2.7.1.2 Variable sampling and delays

For the more general case, where the network delay $\tau(j)$ is not constant and the sampling h is not periodic, [48] adapts the results from [151] to provide a Lyapunov-based argument to prove stability of the NCS under these conditions.

Theorem 2.1 ([48]) *Assume the existence of the constants* h_{min}, h_{max}, τ_{min}, τ_{max} *such that*

$$0 \le h_{min} \le h \le h_{max}, \quad 0 \le \tau_{min} \le \tau(j) \le \tau_{max}, \quad \forall j \in \mathbb{R}. \quad (2.66)$$

The NCS described by (2.61)-(2.62) *is exponentially stable for variable sampling and delay if there exists a matrix* $\boldsymbol{P} = \boldsymbol{P}^T > \boldsymbol{0}$ *such that*

$$\boldsymbol{A}(h,\tau)^T\boldsymbol{P}\boldsymbol{A}(h,\tau) - \boldsymbol{P} < \boldsymbol{0}, \quad \forall h \in [h_{min}, h_{max}], \tau \in [\tau_{min}, \tau_{max}]. \quad (2.67)$$

Such a matrix $\boldsymbol{P}$ is not easy to find for all values of h and τ. A randomized algorithm is proposed in [151] to find the largest h_{max} for which stability is guaranteed when $h_{min} = \tau_{min} = \tau_{max} = 0$.

More results on stability with variable delays are given in [33, 139, 153].

2.7.1.3 Model-based NCS

The stability of the model-based NCS discussed in Section 2.3.2.3 is analyzed in [84, 85]. Consider the model-based system in (2.25)-(2.27).

Theorem 2.2 ([84]) *If there exists a constant h such that $h = t_{j+1} - t_j$, then the system described by (2.27) is globally exponentially stable around the solution $\boldsymbol{z}(t_j) = [\boldsymbol{x}(t_j)^T \boldsymbol{e}(t_j)^T]^T = [\mathbf{0}^T\ \mathbf{0}^T]^T$ if and only if the eigenvalues of $\boldsymbol{M}(j)$ (see (2.28) with $h(k) = h$) are strictly inside the unit circle.*

The same authors, in [85], consider the case where the update time $h(j)$ is not constant but contained within some interval. Stability is proved using Lyapunov's second method. Consider the model-based system in (2.25)-(2.27).

Theorem 2.3 ([85]) *The system described by (2.27) is Lyapunov asymptotically stable for $h \in [h_{min}, h_{max}]$ if there exists a matrix $\boldsymbol{P} > \mathbf{0}$ such that*

$$\boldsymbol{P} - \boldsymbol{M}\boldsymbol{P}\boldsymbol{M}^T = \boldsymbol{Q}, \quad \forall h \in [h_{min}, h_{max}], \tag{2.68}$$

where $\boldsymbol{M}$ is given in (2.28).

Theorem 2.3 may be used to derive an interval $h_{min} \leq h \leq h_{max}$ for which stability is guaranteed (that will depend on the choice of $\boldsymbol{P}$). Proofs of Theorems 2.2 and 2.3 can be found in [84, 85] together with other results that involve stability conditions for iid random $h(j)$ and also for $h(j)$ driven by a finite state Markov chain.

2.7.1.4 Packet dropout

The stability of the NCS with stochastic packet dropout, described in (2.12), has been analyzed by [108] using results from the stability theory of MJLS. The notion of second-moment stability (SMS) is adopted. With the plant model in (2.12) and the controller structure in (2.35), the closed-loop system becomes

$$\begin{bmatrix} \boldsymbol{x}_{cl}(j+1) \\ \boldsymbol{z}(j) \end{bmatrix} = \begin{bmatrix} \boldsymbol{A}_{cl}(\theta(j)) & \boldsymbol{B}_{cl}(\theta(j)) \\ \boldsymbol{C}_{cl}(\theta(j)) & \boldsymbol{D}_{cl}(\theta(j)) \end{bmatrix} \begin{bmatrix} \boldsymbol{x}_{cl}(j) \\ \boldsymbol{w}(j) \end{bmatrix}, \tag{2.69}$$

where

$$
\begin{aligned}
\boldsymbol{x}_{cl}(j+1) &= \begin{bmatrix}\boldsymbol{x}(j+1)^T & \boldsymbol{x}_K(j+1)^T\end{bmatrix}^T,\\
\boldsymbol{A}_{cl}(\theta(j)) &= \begin{bmatrix}\boldsymbol{A}(\theta(j)) & \boldsymbol{B}_2(\theta(j))\boldsymbol{C}_K(\theta(j))\\ \boldsymbol{B}_K(\theta(j))\boldsymbol{C}_2(\theta(j)) & \boldsymbol{A}_K(\theta(j))\end{bmatrix},\\
\boldsymbol{B}_{cl}(\theta(j)) &= \begin{bmatrix}\boldsymbol{B}_1(\theta(j))\\ \boldsymbol{B}_K(\theta(j))\boldsymbol{D}_{21}(\theta(j))\end{bmatrix},\\
\boldsymbol{C}_{cl}(\theta(j)) &= \begin{bmatrix}\boldsymbol{C}_1(\theta(j)) & \boldsymbol{D}_{12}(\theta(j))\boldsymbol{C}_K(\theta(j))\end{bmatrix},\\
\boldsymbol{D}_{cl}(\theta(j)) &= \boldsymbol{D}_{11}(\theta(j)).
\end{aligned}
\tag{2.70}
$$

The conclusion is that the closed-loop system in (2.69) is SMS and has $\mathcal{H}_\infty$-gain less than a scalar $\gamma > 0$ if and only if there exists a matrix $\boldsymbol{Z} > \boldsymbol{0}$, $\boldsymbol{Z} \in \mathbb{R}^{(n_x+n_k)\times(n_x+n_k)}$ such that

$$
\begin{bmatrix}
\begin{bmatrix}\boldsymbol{Z} & \boldsymbol{0}\\ \boldsymbol{0} & \gamma^2\boldsymbol{I}\end{bmatrix} & (\star)^T & \cdots & (\star)^T\\
\sqrt{p_1}\begin{bmatrix}\boldsymbol{A}_{cl}(1)\boldsymbol{Z} & \boldsymbol{B}_{cl}(1)\\ \boldsymbol{C}_{cl}(1) & \boldsymbol{D}_{cl}(1)\end{bmatrix} & \begin{bmatrix}\boldsymbol{Z} & \boldsymbol{0}\\ \boldsymbol{0} & \boldsymbol{I}\end{bmatrix} & \cdots & (\star)^T\\
\vdots & & \ddots & \vdots\\
\sqrt{p_M}\begin{bmatrix}\boldsymbol{A}_{cl}(M)\boldsymbol{Z} & \boldsymbol{B}_{cl}(M)\\ \boldsymbol{C}_{cl}(M) & \boldsymbol{D}_{cl}(M)\end{bmatrix} & \begin{bmatrix}\boldsymbol{0} & \boldsymbol{0}\\ \boldsymbol{0} & \boldsymbol{0}\end{bmatrix} & \cdots & \begin{bmatrix}\boldsymbol{Z} & \boldsymbol{0}\\ \boldsymbol{0} & \boldsymbol{I}\end{bmatrix}
\end{bmatrix} > \boldsymbol{0},
\tag{2.71}
$$

where the matrices $(\star)^T$ are obvious from symmetry and the assumption that the probability matrix for the Markov process satisfies the constraint $q_{ik} = q_k$ for all $i, k \in \mathcal{N}$ is made. Notice that (2.71) is a bilinear matrix inequality since it is linear in the controller parameters (for a fixed scaling matrix $\boldsymbol{Z}$) or in $\boldsymbol{Z}$ (for fixed controller matrices).

Various results for stability with packet dropout are given in [136, 82] while the mixed problem of packet dropout and delays has been treated in [152, 116, 124, 32].

2.7.1.5 Scheduling algorithms

Communication scheduling can be seen as a technique to distribute the unavoidable delays between control loops in an optimal manner. If the plant in (2.61) is open-loop unstable, then (2.64) will become unstable for sufficiently long delays. Therefore, the problem of guaranteeing stability when scheduling algorithms are designed is often equivalent to finding an upper bound τ_{max} on the loop delay, and making sure that messages are scheduled with a period less than τ_{max}. Such upper bound is often called the Maximum Allowable Transfer Interval (MATI). Recall that, in an open-loop (or static) schedule, scheduling must be periodic with a period p.

Lemma 2.1 ([129]) *Assume that a static network scheduler starting at time* t_0 *with integer periodicity* p, *a MATI* τ_{max} *and maximum growth in error* *in τ_{max} seconds strictly bounded by $\beta \in (0, \infty)$ is given. Then, for any time $t \geq t_0 + p\tau_{max}$,*

$$\|\boldsymbol{e}(t)\| < \frac{\beta p(p+1)}{2}, \tag{2.72}$$

where $\boldsymbol{e}(t)$ is given in (2.46) and β is a constant.

It turns out that, if p is let to be the number of nodes competing instead, then Lemma 2.1 is also valid for the MEF-TOD scheduler (Section 2.5.2). In fact, the worst case error bound of the MEF-TOD scheduler is the same as that of the special RR scheduler (Section 2.5.1). The bound is conservative for both scheduling algorithms because τ_{max} is a deadline. The assumption in [129] is that a stabilizing controller has been designed *a priori*. Also, recall that without a network $\boldsymbol{e}(t) = \boldsymbol{0}$, and therefore the dynamics in (2.45) reduce to $\dot{\boldsymbol{x}}_{cl}(t) = \boldsymbol{A}^c_{cl,11}\boldsymbol{x}_{cl}(t)$. Thus, there exists a unique matrix $\boldsymbol{P}$ such that

$$(\boldsymbol{A}^c_{cl,11})^T\boldsymbol{P} + \boldsymbol{P}\boldsymbol{A}^c_{cl,11} = -\boldsymbol{Q}, \tag{2.73}$$

where $\boldsymbol{P}$ and $\boldsymbol{Q}$ are positive definite symmetric matrices (in [129], $\boldsymbol{Q} = \boldsymbol{I}$ is used). At this point, the general condition of stability for an NCS under open-loop or closed-loop scheduling can be stated.

Theorem 2.4 ([129, 151]) *Let an NCS be described by the continuous dynamics of (2.45), with p sensor nodes operating under MEF-TOD, or with integer periodicity p under static scheduling. If the maximum allowable transfer interval satisfies*

$$\tau_{max} < \min \left\{ \frac{ln(2)}{p\|\boldsymbol{A}^c_{cl}\|}, \frac{1}{4\|\boldsymbol{A}^c_{cl}\|\left(\sqrt{\frac{\lambda_{max}(\boldsymbol{P})}{\lambda_{min}(\boldsymbol{P})}}+1\right)p(p+1)}, \right. \\ \left. \frac{\lambda_{min}(\boldsymbol{Q})}{8\lambda_{max}(\boldsymbol{P})\|\boldsymbol{A}^c_{cl}\|^2\left(\sqrt{\frac{\lambda_{max}(\boldsymbol{P})}{\lambda_{min}(\boldsymbol{P})}}+1\right)p(p+1)} \right\}, \tag{2.74}$$

then the NCS is globally exponentially stable.

The proof of Theorem 2.4 is derived based on Lyapunov's second method and treats the network-inducederror term as a vanishing perturbation.

An improvement to the above maximum allowed transfer interval has been presented in [88] and it is shown that $\mathcal{L}_p$ stability of the system is preserved if

$$\tau_{max} < \frac{1}{L}\ln\left(\frac{L+\gamma}{\rho L+\gamma}\right), \tag{2.75}$$

where ρ characterizes the stability properties of the protocol, L characterizes the possible growth of the error between the real values of inputs and outputs and their last transmitted values via the network and γ is the $\mathcal{L}_p$ disturbance gain for robustness properties of the closed-loop system but without the network in the loop. The authors introduce a class of globally asymptotically stable protocols and show that both the RR and the MEF-TOD protocols belong to this class. The analysis is generalized for nonlinear systems and it is proven that, if the closed-loop system without limited (or ideal) communication is $\mathcal{L}_p$ stable, then $\mathcal{L}_p$ stability is preserved with the limited communication for sufficiently small MATIs. It is also shown that, for the same controller, MEF-TOD protocols lead to larger MATI bounds than RR protocols. In general, it is shown that MEF-TOD protocols provide better performance than RR protocols. Input-to-state stability has been studied by the same authors in [89] and Lyapunov-based results on improved bounds on MATI are given in [19].

2.7.2 Structural properties and stability for the codesign problem

In this section the structural properties and stability for the codesign problem, discussed in Section 2.6, are reviewed.

Consider the non-ZOH NCS model of (2.11). This model has been used in [150] where a method is presented for finding periodic scheduling matrices $\boldsymbol{S}_A(j)$ and $\boldsymbol{S}_S(j)$ that preserve observability and reachability of the closed-loop system with limited communication. For the preservation of reachability, let c be the number of times a 1 appears in $\boldsymbol{S}_A(j)$ (*i.e.* the number of actuators that can be controlled at every time tick).

Theorem 2.5 ([150, 51]) *Consider the plant in* (2.11) *and let it evolve from* $j = 0$ *to* $j = j_f$. *Also, let* $\boldsymbol{A} \in \mathbb{R}^{n_x \times n_x}$ *be invertible and the pair* $(\boldsymbol{A}, \boldsymbol{B})$ *be controllable. For any integer* $1 \leq c < n_u$ *there exists an integer* $p > 0$ *and a* p*-periodic sequence of matrices* $\boldsymbol{S}_A(j)$ *with* $p \leq \lceil \frac{n_x}{c} \rceil n_x$ *such that the plant in* (2.11) *is controllable on* $[j, j_f]$ *for all* j, *and therefore controllable* ($\lceil \cdot \rceil$ *is the ceiling function).*

The proof of Theorem 2.5 provides an algorithm which selects linearly independent columns from $[\boldsymbol{B}, \boldsymbol{AB}, \ldots, \boldsymbol{A}^{p-1}\boldsymbol{B}]$ so that

$$\operatorname{rank}\left(\left[\boldsymbol{A}^{j_f-1}\boldsymbol{B}\boldsymbol{S}_A(0) \quad \boldsymbol{A}^{j_f-2}\boldsymbol{B}\boldsymbol{S}_A(1) \quad \ldots \quad \boldsymbol{B}\boldsymbol{S}_A(j_f-1)\right]\right) = n_x. \quad (2.76)$$

This result is in fact an upper bound, $\lceil \frac{n_x}{c} \rceil n_x$, on the period p of the periodic matrices $\boldsymbol{S}_A(j)$. For LQG control, it is sufficient to guarantee the weaker properties of stabilizability and detectability [51]. The problem of finding matrices $\boldsymbol{S}_S(j)$ that preserve observability is the dual problem and an equivalent result is presented. Moreover, [150] considers the case of output feedback stabilization where an observer design is implemented. In [50], the same results of [150] have been used for a model that adds to the model in (2.11) sensor-to-controller and controller-to-actuator delays. The stability analysis is similar.

Consider now the ZOH NCS model of (2.16). This augmented model has been considered by [56] to find a method for constructing scheduling matrices $\boldsymbol{S}_A(j), \boldsymbol{S}_S(j)$ that preserve observability/reachability of the closed-loop system with the limited communication. The results are very similar to the one of [150] (that considered the non-ZOH model). If fact, Theorem 2.5 also applies for the ZOH model if the upper bound on the period p is modified to

$$p \leq \left\lceil \frac{n_x}{c} \right\rceil (n_x + 1). \tag{2.77}$$

In [51], the authors have noticed that the period p is usually shorter than the upper bound of Theorem 2.5. Their results show that a bound of the form $p \leq n_x$ is sufficient. The analysis is extended to include the more complex ZOH strategy of (2.16). Under this model, a sequence of matrices $\boldsymbol{S}_A(j)$ that preserve controllability can have a period $p \leq n_x$ only if the following condition is satisfied.

Corollary 2.1 ([51]) *Consider the plant in* (2.16)*. Let* $\boldsymbol{A} \in \mathbb{R}^{n_x \times n_x}$ *be invertible and the pair* $(\boldsymbol{A}, \boldsymbol{B})$ *be controllable. Let* $n_1, n_2, \ldots, n_{n_u}$ *be the controllability indices of* $(\boldsymbol{A}, \boldsymbol{B})$*. For* $i = 1, \ldots, n_u - 1$*, let* C_i *be the matrix that contains the first* $\sum_{k=1}^{i} n(k)$ *columns of*

$$\begin{aligned}&\big[\boldsymbol{b}_1, \boldsymbol{A}\boldsymbol{b}_1, \ldots, \boldsymbol{A}^{n_1-1}\boldsymbol{b}_1, \boldsymbol{A}^{n_1}\boldsymbol{b}_2, \ldots, \boldsymbol{A}^{n_1+n_2-1}\boldsymbol{b}_2,\\ &\ldots, \boldsymbol{A}^{n_1+\ldots n_{n_u-1}}\boldsymbol{b}_{n_u}, \ldots \boldsymbol{A}^{n-1}\boldsymbol{b}_{n_u}\big],\end{aligned} \tag{2.78}$$

where $\boldsymbol{b}_i$ *is the* i^{th} *column of* $\boldsymbol{B}$*. Let* $q(i) = \sum_{t=1}^{i-1} n(t)$ *and* $\boldsymbol{V}_i \in \mathbb{R}^{n_x}$ *be a unit vector orthogonal to the range of* C_i *with column* $\boldsymbol{A}^{q(i)}\boldsymbol{b}_i$ *deleted. If*

$$\boldsymbol{V}_i^T \sum_{k=0}^{q(i)} \boldsymbol{A}^k \boldsymbol{b}_i \neq 0 \quad \forall i = 2, \ldots, m-1, \tag{2.79}$$

then there exists a p-periodic sequence of matrices $\boldsymbol{S}_A(j)$*, with* $p \leq n_x$*, such that the controllable subspace of* (2.16) *contains the span of the first* n_x *states in* (2.16) *(controllability is preserved).*

If the condition in (2.79) is not satisfied, then controllability can still be guaranteed for longer periods with the bound given in (2.77). Once again, preserving observability is the dual problem. In summary, the analysis of [150, 56, 51, 52] proves that communication sequences that preserve reachability and observability exist, have an upper bound on their period and can be found by a simple algorithm.

In [29], several conditions for quadratic stabilization of the NCS model in (2.57) have been derived. The concept of 'weak partial state control Lyapunov functions' is introduced. This is used for protocol synthesis because such protocols are implementable on CAN (if they do not depend on plant or controller states, but only on the network induced errors). The NCS model in (2.57) is quadratically stable if for a controller in the form of (2.44) and a protocol $S = \Sigma(\boldsymbol{e}(t,i))$, $\Sigma : \mathbb{R}^{n_x+n_y} \to \mathcal{S}$ there exists a matrix $\boldsymbol{P} = \boldsymbol{P}^T > \boldsymbol{0}$ such that the Lyapunov function

$$\boldsymbol{V}(\boldsymbol{x}(t,i), \boldsymbol{e}(t,i)) = \begin{bmatrix}\boldsymbol{x}(t,i) & \boldsymbol{e}(t,i)\end{bmatrix} \boldsymbol{P} \begin{bmatrix}\boldsymbol{x}(t,i) \\ \boldsymbol{e}(t,i)\end{bmatrix}, \tag{2.80}$$

satisfies

$$\Delta \boldsymbol{V}(\boldsymbol{x}, \boldsymbol{e}, \Sigma(\boldsymbol{e})) = \begin{bmatrix}\boldsymbol{x}(t,i) & \boldsymbol{e}(t,i)\end{bmatrix}\left(\tilde{\boldsymbol{S}}(\boldsymbol{e})^T \left(\boldsymbol{A}_{cl}^c\right)^T \boldsymbol{P} \boldsymbol{A}_{cl}^c \tilde{\boldsymbol{S}}(\boldsymbol{e}) - \boldsymbol{P}\right)\begin{bmatrix}\boldsymbol{x}(t,i) \\ \boldsymbol{e}(t,i)\end{bmatrix} < \boldsymbol{0},$$

$$\forall \begin{bmatrix}\boldsymbol{x}(t,i) \\ \boldsymbol{e}(t,i)\end{bmatrix} \neq \begin{bmatrix}0 \\ 0\end{bmatrix}, \tag{2.81}$$

where

$$\tilde{\boldsymbol{S}}(\boldsymbol{e}) \stackrel{\text{def}}{=} \tilde{\boldsymbol{S}}|_{S=\Sigma(\boldsymbol{e}(t,i))}, \tag{2.82}$$

is the matrix resulting from the protocol $S = \Sigma(\boldsymbol{e}(t,i))$ (see (2.57)) which here depends on $\boldsymbol{e}(t,i)$. Furthermore:

- the NCS in (2.57) is said to be 'protocol quadratically stabilizable' if for a given controller (2.44) there exists a function $\Sigma : \mathbb{R}^{n_x+n_y} \to \mathcal{S}$ such that (2.57) with the protocol $S = \Sigma(e(t,i))$ is quadratically stable;
- the NCS in (2.57) is said to be 'protocol and controller quadratically stabilizable' if there exists a controller in the form of (2.44) such that (2.57) is 'protocol quadratically stable'.

The analysis in [29] starts from a natural but not tractable problem and builds up to derive conditions for protocol and controller quadratic stability of NCS that only requires knowledge of local networked induced errors. Such conditions are necessary for real-world NCSs because it cannot be assumed that every node has a measure of the network-induced error of other nodes.

2.8 Nonlinear NCSs

In practice, the plant may have significant nonlinearities, which also may require nonlinear control schemes. This inevitably increases the complexity of the NCS from the scheduling, control and codesign point of view. In Chapter 10, the problem is tackled by using reasonable approximations of the nonlinearities, as well as the discrete and sampled-data representations of the plant and the controller. This will allow the use of a sum-of-squares (SOS) approach.

Although there is still no universal methodology to solve nonlinear NCS problems, researchers have made great effort working in this field: The input-output stability properties of NCS were considered in [88] where a unifying framework for scheduling protocols that preserve stability properties of the system was provided. The input-to-state stability has been studied in [88]. The work in [99] investigates the stabilization of the nonlinear system in an NCS via the approximate discrete-time model. The result of [98] provides a method for robust state feedback control for NCS with linear plants subject to delays and mild nonlinearities, so that linear matrix inequalities can be employed. Another general framework for nonlinear NCSs that incorporates communication constraints, varying transmission intervals and varying delays is given in [47]. A first unifying framework on nonlinear NCSs and quantized control systems considering transmission interval bounds and controller design has been offered in [87].

Other results on nonlinear NCS analysis have been reviewed throughout this chapter. For instance, MPC was used in the context of nonlinear NCS in Section 2.6.2 and the MATI parameter is generically considered for nonlinear systems (Section 2.7.1.5).

2.9 Summary

In an NCS, the presence of a network in the feedback loops inevitably introduces new challenges for the control engineer. This chapter has gathered together recent results of linear NCS theory and discussed them in a comparative manner. After reviewing some fundamental notions and state of the art technology for real-time networks, several techniques for NCS modeling were shown. The analysis for control was partitioned into three classes:

- *control methods*, aiming at designing robust controllers to minimize the network effects
- *scheduling methods*, aiming at designing optimal communication scheduling protocols for a given controller
- *scheduling and controller codesign methods*, aiming at simultaneous design of a controller and a scheduler.

Structural and stability properties were discussed. Among the various techniques used for control of NCSs, the attention was focused to those that are of practical relevance and readily applicable to existing systems. Many of the results presented relied on Lyapunov-based techniques and only provided sufficient conditions for stability of the NCSs. For linear systems, linear matrix inequalities were often used. The conclusion is that the majority of work only addresses the stability issues of NCSs and performance has been neglected. Also, the use of Lyapunov techniques leads to conservative results.

3

A general framework for NCS modeling

CONTENTS

In the past few years, a considerable amount of work has been devoted in modeling different types of NCSs for analysis (stability, controllability, *etc.*) and design (communication scheduling, controllers, *etc.*) purposes. Some of the work, produced by different researchers even in different areas of control, presents so many similarities to other work that it is surprising how, so far, all this theory has not been organized in a general unified framework.

3.1 Introduction

The theoretical general framework introduced here is for modeling, analysis and design of a class of linear NCSs with time-triggered communication (contention-free protocols). Such communication protocols are deterministic in the sense that at each time instant the nodes that are accessing the communication medium are known exactly. Time-triggered communication is the preferred choice for safety-critical systems. In this framework, some of the already existing work (also presented in the previous chapter) is redefined under a graph theoretical viewpoint, reviewed, generalized and advanced.

Recently, considerable effort has been made in trying to unify various results and formulate a complete theory on NCSs. For example, [87] unifies some of the work on NCSs and quantized control systems regarding transmission interval bounds and controller design. The analysis is made for nonlinear systems. In [23], for the first time, results on time-varying sampling intervals, packet dropout and delay are considered under the same framework. The work includes LMI conditions for controller synthesis and Lyapunov-Krasovskii functions for stability analysis. In [47], communication constraints, varying transmission intervals and varying delays are modeled and analyzed under a unified framework. A more information-theoretical unifying approach has been taken by [113].

An NCS, as defined here, is a control system where the communication between a multivariable plant and the feedback controller is constrained. This is represented by a limited shared communication medium with nodes, intended as communication endpoints, attached to it. For the purpose of the analysis, a node, as intended so far, will be called a 'physical node' to distinguish it from the 'node' that will be defined later as the vertex of a graph and treated as an abstract object that can be any arbitrary collection of signals. A *node* can be a source of information if it provides sensor and/or demand signals or a means of action if it is connected to an actuator. A special node is the *controller node* which is where the control algorithm is computed. The policy that regulates the communication between nodes and the controller node is a communication sequence. The set of nodes with the communication sequence characterizes the communication network that will be modeled as a *scheduler*.

The NCS is modeled as a state space representation of the plant and the controller together with the state space model for the scheduling mechanism imposed by the communication network. Since the scheduling process is time-dependent, the result will be an LTV overall representation. If the communication sequence follows a periodic pattern, the LTV system can be 'lifted' to obtain an equivalent LTI representation. It will be shown that this LTI model is a generalization of multirate systems and also describes a class of switched and delayed systems. NCSs with subnetworks and task scheduling can be dealt with by this framework in a natural way.

The modeling procedures presented in this chapter define the general framework for NCSs that will be adopted in the rest of the book.

3.2 Limited communication and schedulers

The limited sensor-to-controller and controller-to-actuator communication is modeled as a scheduler. A scheduler can be visualized as a selector switch with as many positions as the number of signals to be scheduled. This concept is formalized using some definitions from graph theory and showing that the communication network is a special type of graph. From this graph, time-varying state space models that emulate the communication constraints can be derived.

3.2.1 A time-varying star graph

Here are some basic definitions from graph theory [31]. A *graph* is defined as a pair $G = (V, E)$ that consists of a set of *vertices* V and a set of *edges* $E \subseteq V \times V$ where the vertices are $v_m \in V$ and the edges are $e_{m,n} = (v_m, v_n) \in E$. A graph is called *undirected* if $e_{n,m} \in E \Rightarrow e_{m,n} \in E$ and *directed* otherwise. A *path* is a subgraph $\pi = (V, E_\pi) \subset G$ with distinct vertices $V = \{v_1, v_2, \ldots, v_q\}$ and $E_\pi \stackrel{\text{def}}{=} \{(v_1, v_2), (v_2, v_3), \ldots, (v_{q-1}, v_q)\}$. A *cycle* $C = (V, E_C)$ is a path (of length q) with an extra edge $(v_q, v_1) \in E$. A graph with no cycles is called *acyclic*. An undirected graph G is called *connected* if there exists a path π between any two distinct nodes of G (otherwise it is called *disconnected*). A *tree* is a connected acyclic graph. A *bigraph* is a graph whose vertices can be divided into two disjoint sets U_1 and U_2 such that every edge connects a vertex in U_1 to one in U_2, *i.e.* U_1 and U_2 are independent sets. A bigraph is complete if every vertex of U_1 is connected to every vertex of U_2. The complete bigraph with partitions of size $|U_1| = a$ and $|U_2| = b$, is denoted as $K_{a,b}$. A *star* S is the complete bigraph $K_{1,d}$, which is a tree with one *internal vertex* and d surrounding vertices called *leaves*.

Let us now define a special star $S^* = (V_d, \Sigma)$ where V_d is the set of leaves and $\Sigma : j \mapsto E(j)$. Σ is a policy used to determine the elements of a time-varying set of edges, $E(j)$, at each time step j where $j \in \mathbb{N}_0$ is a time index. S^* is a time-varying star as the set of edges $E(j)$ is not fixed (it is a function of j). More precisely, at time j, S^* is a star if all the possible edges are members of $E(j)$ (*i.e.* $|E(j)| = d$), a disconnected graph with a star as a subgraph if $0 < |E(j)| < d$, and a completely disconnected graph if $E(j) = \varnothing$ where $\varnothing$ is the empty set. An example of a time-varying star is shown in Figure 3.1. Vertex v_0 is the central vertex and the set of leaves is $V_5 = \{v_1, v_2, v_3, v_4, v_5\}$. At time step $j = 1$ the set of edges is $E(1) = \{(v_0, v_1), (v_0, v_2), (v_0, v_3), (v_0, v_4), (v_0, v_5)\}$, at $j = 2$ the set of edges

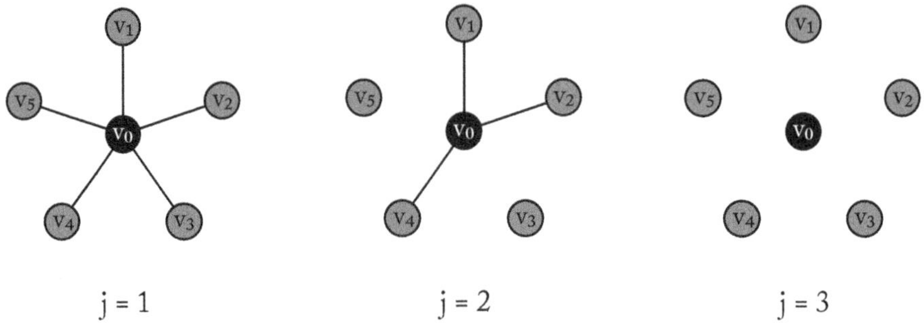

FIGURE 3.1: A time-varying star: at time $j = 1$ the set of edges is $E(1) = \{(v_0, v_1), (v_0, v_2), (v_0, v_3), (v_0, v_4), (v_0, v_5)\}$, at $j = 2$ the set of edges is $E(2) = \{(v_0, v_1), (v_0, v_2), (v_0, v_4)\}$ and at $j = 3$ the set of edges is $E(3) = \{\}$.

is $E(2) = \{(v_0, v_1), (v_0, v_2), (v_0, v_4)\}$ and at $j = 3$ the set of edges is $E(3) = \{\}$. For this example, a policy Σ can be defined by the set $\Sigma = \{\Sigma(1), \Sigma(2), \Sigma(3)\}$ where the subsets $\Sigma(1) = \{v_1, v_2, v_3, v_4, v_5\}$, $\Sigma(2) = \{v_1, v_2, v_4\}$ and $\Sigma(3) = \{\}$ indicate to which leaves v_0 is connected at each time step.

3.2.2 The scheduler

The communication between plant and controller with dedicated connections can be described by a star S where the internal vertex called v_K is the controller and the leaves v_i, $i = 1, 2, \ldots, d$, represent source of information (sensors and demands) or means of action (actuators). The communication would be better described by a directed star S because the direction of the information flows from v_K to v_i if v_i is a means of action, or from v_i to v_K if v_i is a source of information. However, it simplifies exposition to assume that S is an undirected star. This is not a limitation and the implications of this simplification will be discussed later (Remark 3.1).

On the other hand, the communication of an NCS can be described by the time-varying star $S^* = (V_d, \Sigma)$, where V_d is the set of all vertices and Σ is the policy to determine which element of V_d is connected to v_K at each time step. This means that the set of edges E, at each time step, can only admit some (normally only one) edge(s) from the set of all possible edges of S.

The next step is to obtain a dynamical (state space) model for S^*. The concept of 'node', 'communication sequence', and 'communication network' need to be formalized first.

Definition 3.1 *A node $\Delta(i)$, where i is the node identifier (its name), is a vertex of a graph. $\Delta(i)$ is also a collection of signals (a set). If an edge $(\Delta(i), v_K)$ exists, then all the signal information in $\Delta(i)$ is available to v_K (the controller node) and* vice versa. *The set of all nodes is $\Delta = \{\Delta(i) : \forall i\}$*

where $|\Delta| = d$ *(the cardinality) is the number of nodes and* $|\Delta(i)|$ *is the number of signals that belong to node* i. ▲

Note that the word 'node' has appeared previously to describe a physical object corresponding to a connection point of a network (like a microprocessor unit or an electronic control unit for automotive systems). From now on, the word 'node' will be used to describe the abstract concept introduced in Definition 3.1 while the words 'physical node' will be used to describe a node intended as a connection point in a network.

Let us use the following terminology: a node $\Delta(i)$ is said to be *scheduled* at time j if $(\Delta(i), v_K) \in E(j)$. Hence, the existence of an edge between the controller node and node i is equivalent to saying that node i is scheduled. Definition 3.1 implies the existence of two special nodes (they will be useful for later analysis): these are $\Delta(empty) \stackrel{\text{def}}{=} \varnothing$, the node with no signals associated to it and $\Delta(all) \stackrel{\text{def}}{=} \{all\ signals\}$, $|\Delta| = d$, the node with all the signals associated to it. These nodes are used when it is necessary not to schedule any signal or schedule all of them respectively.

Definition 3.2 *A communication sequence*

$$\sigma \stackrel{def}{=} \{\sigma(0), \sigma(1), \ldots\} \stackrel{def}{=} \{\sigma(j)\}_{j=0}^{\infty}, \tag{3.1}$$

where $\sigma(j) \in \Delta$ *is a node and* $j \in \mathbb{N}_0$, *is defined as an ordered list of nodes (a tuple).* ▲

It will be shown in the next definition that the communication sequence is an ordered list of nodes to determine, at each time step, which node is connected to the controller node (or in other words, which leaf is connected to the internal vertex). This definition of sequence, in essence, is equivalent to the one originally proposed by [18] and subsequently adopted in other work (*e.g.* [53, 105, 14]).

Definition 3.3 *A communication network is defined as the time-varying star*

$$N = (\Delta, \sigma), \tag{3.2}$$

where Δ *is the set of nodes of Definition 3.1 (which is the set of leaves of the star) and* $\sigma : j \mapsto E(j)$ *is the policy to determine the elements of the time-varying set of edges* $E(j) = (v_K, \sigma(j))$ *for all* j. v_K *is the controller node.*

▲

Since N is a star, every element of $E(j)$ is an edge between one vertex from the set of nodes Δ and the internal vertex v_K.

At this point, the limited time-triggered communication of the NCS is completely characterized. The set of nodes Δ determines all the possible connections between the controller and the sensors/actuators. The connection available at time step j is given by $\sigma(j)$. The assumption here is that only one node can be scheduled (or connected) at a time. This is not a limitation because a node, by definition, is any arbitrary set of signals.

Remark 3.1 *It was assumed that the NCS communication is described by an undirected graph; thus, for network N, $(v_K, \Delta(i)) \Rightarrow (\Delta(i), v_K)$, which practically means that when $\Delta(i)$ is scheduled then information flows both ways, from $\Delta(i)$ to v_K and from v_K to $\Delta(i)$. This assumption is only necessary if $\Delta(i)$ is both a source of information and a means of action. In many time-triggered communication protocols like the Time Triggered CAN with Shared Clock (TTC-SC) [9] the communication to the physical nodes is always initiated by a frame sent by the controller node. Hence, a two-way message exchange happens within each time tick. Under this and similar protocols, the undirected graph assumption is perfectly valid. If, however, a protocol does not support a two-way communication within a single time tick, nodes should be defined as a source of information or means of actuation only (and not both)* i.e. *at each time tick only one-directional communication is possible.* •

Definition 3.4 *For a given communication sequence σ, the sequence of scheduling matrices is a sequence of diagonal binary matrices $S_\sigma \overset{def}{=} \{\boldsymbol{S}(0), \boldsymbol{S}(1), \ldots\} \overset{def}{=} \{\boldsymbol{S}(j)\}_{j=0}^{\infty}$ where*

$$\boldsymbol{S}(j) \overset{def}{=} \operatorname{diag}(s_0(j), s_1(j), \ldots, s_{n_\nu}(j)), \tag{3.3}$$

and n_ν is the dimension of the signal vector subject to scheduling. The elements $s_i(j)$ satisfy

$$s_i(j) = \begin{cases} 1 & \textit{if signal } i \in \sigma(j) \\ 0 & \textit{otherwise} \end{cases} \quad \forall i = 1, 2, \ldots n_\nu. \tag{3.4}$$

Also, define $\bar{S}_\sigma \overset{def}{=} \{\bar{\boldsymbol{S}}(0), \bar{\boldsymbol{S}}(1), \ldots\} \overset{def}{=} \{\bar{\boldsymbol{S}}(j)\}_{j=0}^{\infty}$ where

$$\bar{\boldsymbol{S}}(j) = \boldsymbol{I} - \boldsymbol{S}(j), \tag{3.5}$$

and $\boldsymbol{I}$ is the identity matrix. ▲

Note that $\boldsymbol{S}(j) = \boldsymbol{S}(j)^T$ and $\bar{\boldsymbol{S}}(j) = \bar{\boldsymbol{S}}(j)^T$ have a 1 or a 0 in the diagonal and zeros elsewhere. Any permutation of the elements (signals) of $\Delta(i)$ will produce the same $\boldsymbol{S}(j)$ and $\bar{\boldsymbol{S}}(j)$. A scheduling matrix $\boldsymbol{S}(j)$ is a matrix that, when multiplied to a vector $\boldsymbol{S}(j)\boldsymbol{\nu}(j)$, selects particular elements of vector $\boldsymbol{\nu}(j)$ (the ones corresponding to the 1s in the matrix diagonal) and reset the rest to zero. Such matrices are used in [105, 14].

At this point, the general signal scheduler $\mathcal{S}$, characterized by N, can be defined. $\mathcal{S}$ is an LTV state space model that takes as an input a vector signal (with up-to-date information) and outputs the signal subject to scheduling. The scheduler model is

$$\mathcal{S}: \quad \begin{cases} \boldsymbol{\xi}(j+1) = \bar{\boldsymbol{S}}(j)\boldsymbol{\xi}(j) + \boldsymbol{S}(j)\boldsymbol{\nu}(j) \\ \hat{\boldsymbol{\nu}}(j) = \bar{\boldsymbol{S}}(j)\boldsymbol{\xi}(j) + \boldsymbol{S}(j)\boldsymbol{\nu}(j) \end{cases}, \tag{3.6}$$

where $\boldsymbol{\xi}(j) = \hat{\boldsymbol{\nu}}(j-1)$ is the state, $\boldsymbol{\nu}(j)$ is the input (the signal with up-to-date information), $\hat{\boldsymbol{\nu}}(j)$ is the output (the scheduled signal) and $\boldsymbol{\xi}(j), \boldsymbol{\nu}(j), \hat{\boldsymbol{\nu}}(j) \in \mathbb{R}^{n_\xi}$. The scheduler dynamics in (3.6) can be understood as follows: the signal vector $\hat{\boldsymbol{\nu}}(j)$ has all its entries equal to $\boldsymbol{\nu}(j-1)$ apart from some entries (given by the 1s in the diagonal of $\boldsymbol{S}(j)$) which will be equal to the elements of $\boldsymbol{\nu}(j)$ (the up-to-date signal vector). The assumption is that every signal latches so that if at time j there is no edge between the controller and a node (the node is not scheduled) the signals hold their previous value. This is the commonly used ZOH strategy. This scheduling mechanism can be seen as signals being delayed by multiples of the sampling time. The delays vary according to N and the whole vector is not to be delayed by the same amount of time but rather its individual components.

Remark 3.2 *This model is used in [105] (but only for single channel and only for actuator scheduling) and in [14]. The difference here is the reformulation of the problem under a graph theoretical framework and the novel definition of 'nodes' as an abstract collection of signals (see Definition 3.1). This allows for a more systematic modeling approach that naturally extends to other limited communication scenarios (task scheduling and subnetworks). Moreover, under this framework, it will be easy to show that the proposed models are special cases of delayed and switched systems and a generalization of multirate sampled-data systems.* •

3.3 NCS modeling

Consider the NCS shown in Figure 1.5 (p. 6) where the 'nodes' are physical nodes. The plant actuators and sensors are spatially distributed and connected to a network via physical nodes. The network is used to relay signals to each component, in contrast to traditional systems where communication takes place via dedicated connections. The communication medium is represented by a shared bus. Only a limited number of physical nodes (often only one) can access the communication medium at any time tick. The assumption is that a contention-free protocol is used where the sequence of physical node communication is solely dependent on the progression in time and therefore allows pre-planning of a particular periodic communication sequence at the design stage.

Under the above assumptions, the NCS of Figure 1.5 can be represented by the diagram in Figure 3.2 where the blocks and signals will be discussed next.

The plant, denoted by $\mathcal{P}$, is spatially distributed and can be described by

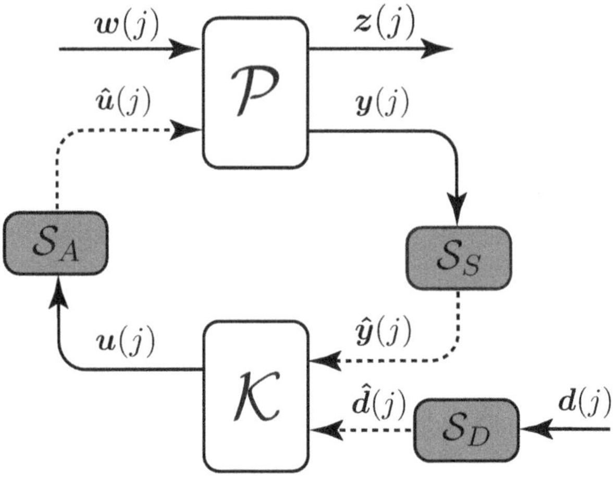

FIGURE 3.2: Generalized closed-loop system with schedulers in the loop: a model of an NCS with limited communication.

the discrete LTI state space model

$$\mathcal{P}: \quad \begin{cases} \boldsymbol{x}(j+1) = \boldsymbol{A}\boldsymbol{x}(j) + \boldsymbol{B}_1\boldsymbol{w}(j) + \boldsymbol{B}_2\hat{\boldsymbol{u}}(j) \\ \boldsymbol{z}(j) = \boldsymbol{C}_1\boldsymbol{x}(j) + \boldsymbol{D}_{11}\boldsymbol{w}(j) + \boldsymbol{D}_{12}\hat{\boldsymbol{u}}(j) \\ \boldsymbol{y}(j) = \boldsymbol{C}_2\boldsymbol{x}(j) + \boldsymbol{D}_{21}\boldsymbol{w}(j) + \boldsymbol{D}_{22}\hat{\boldsymbol{u}}(j) \end{cases}, \tag{3.7}$$

where $\boldsymbol{x}(j) \in \mathbb{R}^{n_x}$, $\boldsymbol{w}(j) \in \mathbb{R}^{n_w}$, $\hat{\boldsymbol{u}}(j) \in \mathbb{R}^{n_u}$, $\boldsymbol{z}(j) \in \mathbb{R}^{n_z}$, $\boldsymbol{y}(j) \in \mathbb{R}^{n_y}$ and j is the sampling instant. $\boldsymbol{A}, \boldsymbol{B}_1, \boldsymbol{B}_2, \boldsymbol{C}_1, \boldsymbol{C}_2, \boldsymbol{D}_{11}, \boldsymbol{D}_{12}, \boldsymbol{D}_{21}$ and $\boldsymbol{D}_{22}$ are constant matrices of appropriate dimensions. These matrices can be derived from the discretization of a continuous-time system (*e.g.* see [1]) or from the equivalent discretization for norm problems of [11].

$\mathcal{P}$ is connected via a network to the feedback controller $\mathcal{K}$ described by the discrete LTI state space model

$$\mathcal{K}: \quad \begin{cases} \boldsymbol{x}_K(j+1) = \boldsymbol{A}_K\boldsymbol{x}_K(j) + \boldsymbol{B}_{K1}\hat{\boldsymbol{y}}(j) + \boldsymbol{B}_{K2}\hat{\boldsymbol{d}}(j) \\ \boldsymbol{u}(j) = \boldsymbol{C}_K\boldsymbol{x}_K(j) + \boldsymbol{D}_{K1}\hat{\boldsymbol{y}}(j) + \boldsymbol{D}_{K2}\hat{\boldsymbol{d}}(j) \end{cases}, \tag{3.8}$$

where $\boldsymbol{x}_K(j) \in \mathbb{R}^{n_k}$, $\hat{\boldsymbol{y}}(j) \in \mathbb{R}^{n_y}$, $\hat{\boldsymbol{d}}(j) \in \mathbb{R}^{n_d}$, $\boldsymbol{u}(j) \in \mathbb{R}^{n_u}$ and $\boldsymbol{A}_K, \boldsymbol{B}_{K1}, \boldsymbol{B}_{K2}, \boldsymbol{C}_K, \boldsymbol{D}_{K1}$ and $\boldsymbol{D}_{K2}$ are constant matrices of appropriate dimensions. This controller could be given (scheduling design or system integration problem) or it needs to be designed (scheduling and controller codesign problem).

The generic signal scheduler was formulated in (3.6). Now, the more specific actuator, sensor and demand schedulers will be defined. They are all fully characterized by the communication network $N = (\Delta, \sigma)$.

Definition 3.5 *The sequence of* actuator scheduling matrices *is a sequence of diagonal binary matrices* $S_A \stackrel{def}{=} \{\boldsymbol{S}_A(0), \boldsymbol{S}_A(1), \ldots\} \stackrel{def}{=} \{\boldsymbol{S}_A(j)\}_{j=0}^{\infty}$ *where*

$$\boldsymbol{S}_A(j) \stackrel{def}{=} \mathrm{diag}(s_{A,1}(j), s_{A,2}(j), \ldots, s_{A,n_u}(j)). \tag{3.9}$$

The elements $s_{A,i}(j)$ *satisfy*

$$s_{A,i}(j) = \begin{cases} 1 & \textit{if actuator } i \in \sigma(j) \\ 0 & \textit{otherwise} \end{cases} \quad \forall i = 1, 2, \ldots n_u. \tag{3.10}$$

Also, $\bar{S}_A \stackrel{def}{=} \{\bar{\boldsymbol{S}}_A(0), \bar{\boldsymbol{S}}_A(1), \ldots\} \stackrel{def}{=} \{\bar{\boldsymbol{S}}_A(j)\}_{j=0}^{\infty}$ *where* $\bar{\boldsymbol{S}}_A(j) = \boldsymbol{I} - \boldsymbol{S}_A(j)$.

The sequence of sensor scheduling matrices *is defined as* $S_S \stackrel{def}{=} \{\boldsymbol{S}_S(0), \boldsymbol{S}_S(1), \ldots\} \stackrel{def}{=} \{\boldsymbol{S}_S(j)\}_{j=0}^{\infty}$ *where*

$$\boldsymbol{S}_S(j) \stackrel{def}{=} \mathrm{diag}(s_{S,1}(j), s_{S,2}(j), \ldots, s_{S,n_y}(j)). \tag{3.11}$$

The elements $s_{S,i}(j)$ *satisfy*

$$s_{S,i}(j) = \begin{cases} 1 & \textit{if sensor } i \in \sigma(j) \\ 0 & \textit{otherwise} \end{cases} \quad \forall i = 1, 2, \ldots n_y. \tag{3.12}$$

Also, $\bar{S}_S \stackrel{def}{=} \{\bar{\boldsymbol{S}}_S(0), \bar{\boldsymbol{S}}_S(1), \ldots\} \stackrel{def}{=} \{\bar{\boldsymbol{S}}_S(j)\}_{j=0}^{\infty}$ *where* $\bar{\boldsymbol{S}}_S(j) = \boldsymbol{I} - \boldsymbol{S}_S(j)$.

The sequence of demand scheduling matrices *is defined as* $S_D \stackrel{def}{=} \{\boldsymbol{S}_D(0), \boldsymbol{S}_D(1), \ldots\} \stackrel{def}{=} \{\boldsymbol{S}_D(j)\}_{j=0}^{\infty}$ *where*

$$\boldsymbol{S}_D(j) \stackrel{def}{=} \mathrm{diag}(s_{D,1}(j), s_{D,2}(j), \ldots, s_{D,n_d}(j)). \tag{3.13}$$

The elements $s_{D,i}(j)$ *satisfy*

$$s_{D,i}(j) = \begin{cases} 1 & \textit{if demand } i \in \sigma(j) \\ 0 & \textit{otherwise} \end{cases} \quad \forall i = 1, 2, \ldots n_d. \tag{3.14}$$

Also, $\bar{\boldsymbol{S}}_D \stackrel{def}{=} \{\bar{\boldsymbol{S}}_D(0), \bar{\boldsymbol{S}}_D(1), \ldots\} \stackrel{def}{=} \{\bar{\boldsymbol{S}}_D(j)\}_{j=0}^{\infty}$ *where* $\bar{\boldsymbol{S}}_D(j) = \boldsymbol{I} - \boldsymbol{S}_D(j)$. ▲

Thus, the actuator scheduler, $\mathcal{S}_A$, is represented by

$$\mathcal{S}_A : \quad \begin{cases} \boldsymbol{x}_A(j+1) = \bar{\boldsymbol{S}}_A(j)\boldsymbol{x}_A(j) + \boldsymbol{S}_A(j)\boldsymbol{u}(j) \\ \hat{\boldsymbol{u}}(j) = \bar{\boldsymbol{S}}_A(j)\boldsymbol{x}_A(j) + \boldsymbol{S}_A(j)\boldsymbol{u}(j) \end{cases}, \tag{3.15}$$

the sensor scheduler, $\mathcal{S}_S$, is represented by

$$\mathcal{S}_S : \quad \begin{cases} \boldsymbol{x}_S(j+1) = \bar{\boldsymbol{S}}_S(j)\boldsymbol{x}_S(j) + \boldsymbol{S}_S(j)\boldsymbol{y}(j) \\ \hat{\boldsymbol{y}}(j) = \bar{\boldsymbol{S}}_S(j)\boldsymbol{x}_S(j) + \boldsymbol{S}_S(j)\boldsymbol{y}(j) \end{cases}, \tag{3.16}$$

and the demand scheduler, $\mathcal{S}_D$, is represented by

$$\mathcal{S}_D : \quad \begin{cases} \boldsymbol{x}_D(j+1) = \bar{\boldsymbol{S}}_D(j)\boldsymbol{x}_D(j) + \boldsymbol{S}_D(j)\boldsymbol{d}(j) \\ \hat{\boldsymbol{d}}(j) = \bar{\boldsymbol{S}}_D(j)\boldsymbol{x}_D(j) + \boldsymbol{S}_D(j)\boldsymbol{d}(j) \end{cases}, \tag{3.17}$$

where

$$\boldsymbol{x}_A(j) = \hat{\boldsymbol{u}}(j-1), \quad \boldsymbol{x}_S(j) = \hat{\boldsymbol{y}}(j-1), \quad \boldsymbol{x}_D(j) = \hat{\boldsymbol{d}}(j-1). \tag{3.18}$$

Also, $\boldsymbol{x}_A(j) \in \mathbb{R}^{n_u}$, $\boldsymbol{x}_S(j) \in \mathbb{R}^{n_y}$ and $\boldsymbol{x}_D(j) \in \mathbb{R}^{n_d}$. The vectors $\boldsymbol{u}(j), \boldsymbol{y}(j)$ and $\boldsymbol{d}(j)$ are the ones containing all the up-to-date information while $\hat{\boldsymbol{u}}(j), \hat{\boldsymbol{y}}(j)$ and $\hat{\boldsymbol{d}}(j)$ are the ones containing some old and some new information depending on the communication network N.

The following example should further clarify the concept of nodes, sequence and scheduling matrices.

Example 3.1 *Let us model the system in Figure 1.4, p. 5. First, let the communication network be a time-varying star graph defined by* $N = (\Delta, \sigma)$*. A possible set of nodes can be defined as* $\Delta = \{\Delta(1), \Delta(2), \Delta(3), \Delta(4)\}$*, where*

$$\begin{aligned} &\Delta(1) = \{Actuator\ 1\}, \quad \Delta(2) = \{Actuator\ 2\}, \\ &\Delta(3) = \{Sensor\ 1,\ Sensor\ 3\}, \quad \Delta(4) = \{Sensor\ 2\}. \end{aligned} \tag{3.19}$$

Δ *captures all the possible connections allowed for communication according to the diagram in Figure 1.4. (Notice that the choice of the identifier* i *for node* $\Delta(i)$ *is arbitrary.) The communication sequence (from Figure 1.4) is*

$$\sigma = \{\Delta(3), \Delta(1), \Delta(2), \Delta(4), \Delta(3), \Delta(1), \Delta(2), \Delta(4), \Delta(3), \ldots\}, \tag{3.20}$$

which means that node $\Delta(3)$ *is scheduled at time step* $j = 1$*, node* $\Delta(2)$ *at* $j = 2$ *and so on. For conciseness of notation, the following convention might be used: instead of writing each node* $\Delta(i)$*, only the node identifiers* i *will be written in the sequence, hence, the sequence in* (3.20) *becomes*

$$\sigma = \{3, 1, 2, 4, 3, 1, 2, 4, 3, \ldots\}. \tag{3.21}$$

This is a special type of sequence because it is periodic with periodicity equal to 4. The time-varying graph N *is shown in Figure 3.3. Once the sequence is defined, the scheduling matrices follow naturally from Definition 3.4. They*

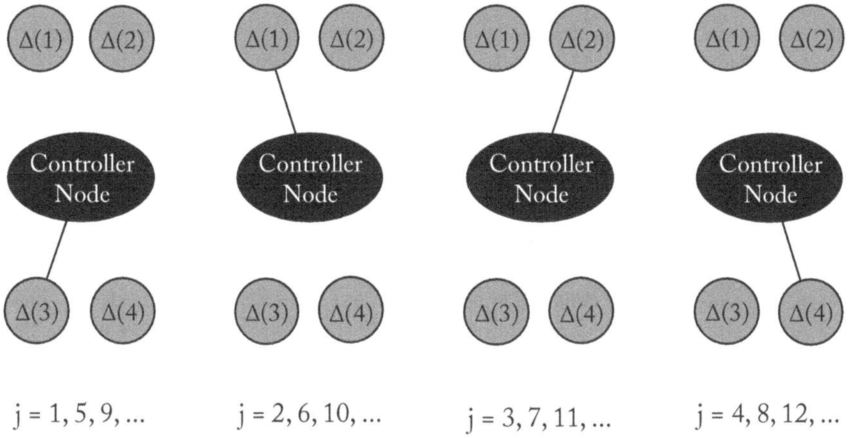

FIGURE 3.3: The communication network as a time-varying (periodic) graph: at each time step only one edge is allowed to connect the controller node to any other node.

are:

$$\begin{aligned}
S_A &= \{\boldsymbol{S}_A(1), \boldsymbol{S}_A(2), \boldsymbol{S}_A(3), \boldsymbol{S}_A(4), \boldsymbol{S}_A(5) = \boldsymbol{S}_A(1), \dots\} \\
&= \left\{\boldsymbol{0}_{2\times 2}, \begin{bmatrix} 1 & 0 \\ 0 & 0 \end{bmatrix}, \begin{bmatrix} 0 & 0 \\ 0 & 1 \end{bmatrix}, \boldsymbol{0}_{2\times 2}, \boldsymbol{0}_{2\times 2}, \dots \right\}, \\
\bar{S}_A &= \{\bar{\boldsymbol{S}}_A(1), \bar{\boldsymbol{S}}_A(2), \bar{\boldsymbol{S}}_A(3), \bar{\boldsymbol{S}}_A(4), \bar{\boldsymbol{S}}_A(5) = \bar{\boldsymbol{S}}_A(1), \dots\} \\
&= \left\{\boldsymbol{I}_2, \begin{bmatrix} 0 & 0 \\ 0 & 1 \end{bmatrix}, \begin{bmatrix} 1 & 0 \\ 0 & 0 \end{bmatrix}, \boldsymbol{I}_2, \boldsymbol{I}_2, \dots \right\}, \\
S_S &= \{\boldsymbol{S}_S(1), \boldsymbol{S}_S(2), \boldsymbol{S}_S(3), \boldsymbol{S}_S(4), \boldsymbol{S}_S(5) = \boldsymbol{S}_S(1), \dots\} \\
&= \left\{\begin{bmatrix} 1 & 0 & 0 \\ 0 & 0 & 0 \\ 0 & 0 & 1 \end{bmatrix}, \boldsymbol{0}_{3\times 3}, \boldsymbol{0}_{3\times 3}, \begin{bmatrix} 0 & 0 & 0 \\ 0 & 1 & 0 \\ 0 & 0 & 0 \end{bmatrix}, \begin{bmatrix} 1 & 0 & 0 \\ 0 & 0 & 0 \\ 0 & 0 & 1 \end{bmatrix}, \dots \right\}, \\
\bar{S}_S &= \{\bar{\boldsymbol{S}}_S(1), \bar{\boldsymbol{S}}_S(2), \bar{\boldsymbol{S}}_S(3), \bar{\boldsymbol{S}}_S(4), \bar{\boldsymbol{S}}_S(5) = \bar{\boldsymbol{S}}_S(1), \dots\} \\
&= \left\{\begin{bmatrix} 0 & 0 & 0 \\ 0 & 1 & 0 \\ 0 & 0 & 0 \end{bmatrix}, \boldsymbol{I}_3, \boldsymbol{I}_3, \begin{bmatrix} 1 & 0 & 0 \\ 0 & 0 & 0 \\ 0 & 0 & 1 \end{bmatrix}, \begin{bmatrix} 0 & 0 & 0 \\ 0 & 1 & 0 \\ 0 & 0 & 0 \end{bmatrix}, \dots \right\}.
\end{aligned} \tag{3.22}$$

(There is no S_D and $\bar{S}_D$ because there is no demand signal.) These scheduling matrices are those of the state space models $\mathcal{S}_A$ and $\mathcal{S}_S$ that emulate the scheduled controller-to-actuator and sensor-to-controller time-triggered communication assuming a ZOH strategy is used. ♦

Remark 3.3 *The proposed models can be readily used to include packet dropout. If, at time j, a packet is lost (information is corrupted or the packet* *does not reach destination) then this situation is equivalent to that signal being non-scheduled at time j. For the stochastic case, let $\theta_A(j), \theta_S(j), \theta_D(j) \in \{1, 0\}$ be an independent and identically distributed (iid) processes. Then, if*

$$\begin{aligned} \boldsymbol{S}_A(j) &\xrightarrow{sub.\ with} \theta_A(j)\boldsymbol{S}_A(j), \\ \boldsymbol{S}_S(j) &\xrightarrow{sub.\ with} \theta_S(j)\boldsymbol{S}_S(j), \\ \boldsymbol{S}_D(j) &\xrightarrow{sub.\ with} \theta_D(j)\boldsymbol{S}_D(j), \end{aligned} \tag{3.23}$$

where $\xrightarrow{sub.\ with}$ means substitute value on the LHS with the value of the RHS, stochastic packet dropout is accounted for [58, 107] (see Section 2.3.2.1, p. 18). •

Now that the communication protocol has been modeled, three different ways to proceed are shown:

(i) merge the schedulers into the plant model to obtain an *augmented plant* model

(ii) merge the schedulers into the controller model to obtain an *augmented controller* model

(iii) merge plant controller and schedulers together to obtain an *augmented closed-loop* model.

3.3.1 Augmented plant

Here, the schedulers are incorporated into the plant model to obtain the augmented plant $\hat{\mathcal{P}}$. A diagram of this configuration is shown in Figure 3.4. The plant $\mathcal{P}$ can only be augmented with the scheduler $\mathcal{S}_A$ and $\mathcal{S}_S$, therefore $\mathcal{S}_D$ will be omitted in this formulation. The equations for $\hat{\mathcal{P}}$ are

$$\hat{\mathcal{P}}: \quad \begin{cases} \hat{\boldsymbol{x}}(j+1) = \hat{\boldsymbol{A}}(j)\hat{\boldsymbol{x}}(j) + \hat{\boldsymbol{B}}_1(j)\boldsymbol{w}(j) + \hat{\boldsymbol{B}}_2(j)\boldsymbol{u}(j) \\ \boldsymbol{z}(j) = \hat{\boldsymbol{C}}_1(j)\hat{\boldsymbol{x}}(j) + \hat{\boldsymbol{D}}_{11}(j)\boldsymbol{w}(j) + \hat{\boldsymbol{D}}_{12}(j)\boldsymbol{u}(j) \\ \hat{\boldsymbol{y}}(j) = \hat{\boldsymbol{C}}_2(j)\hat{\boldsymbol{x}}(j) + \hat{\boldsymbol{D}}_{21}(j)\boldsymbol{w}(j) + \hat{\boldsymbol{D}}_{22}(j)\boldsymbol{u}(j) \end{cases}, \tag{3.24}$$

where

$$\hat{\boldsymbol{x}} = \begin{bmatrix} \boldsymbol{x}_A(j) \\ \boldsymbol{x}(j) \\ \boldsymbol{x}_S(j) \end{bmatrix} = \begin{bmatrix} \hat{\boldsymbol{u}}(j-1) \\ \boldsymbol{x}(j) \\ \hat{\boldsymbol{y}}(j-1) \end{bmatrix}, \tag{3.25}$$

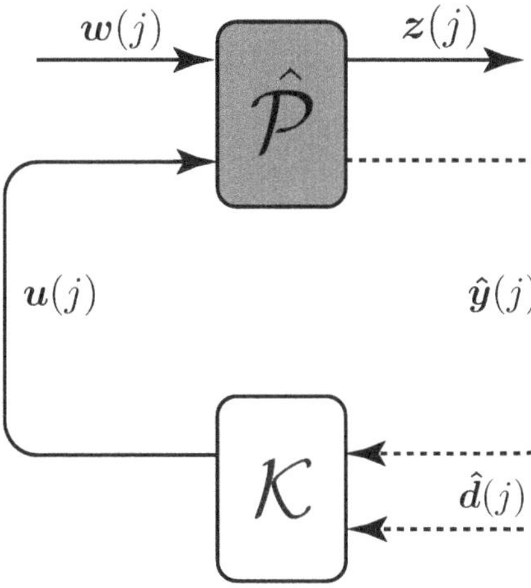

FIGURE 3.4: Augmented plant: the model of the plant and the network are merged together.

and

$$\hat{\boldsymbol{A}}(j) = \begin{bmatrix} \bar{\boldsymbol{S}}_A(j) & \boldsymbol{0} & \boldsymbol{0} \\ \boldsymbol{B}_2\bar{\boldsymbol{S}}_A(j) & \boldsymbol{A} & \boldsymbol{0} \\ \boldsymbol{S}_S(j)\boldsymbol{D}_{22}\bar{\boldsymbol{S}}_A(j) & \boldsymbol{S}_S(j)\boldsymbol{C}_2 & \bar{\boldsymbol{S}}_S(j) \end{bmatrix},$$

$$\hat{\boldsymbol{B}}_1(j) = \begin{bmatrix} \boldsymbol{0} \\ \boldsymbol{B}_1 \\ \boldsymbol{S}_S(j)\boldsymbol{D}_{21} \end{bmatrix}, \quad \hat{\boldsymbol{B}}_2(j) = \begin{bmatrix} \boldsymbol{S}_A(j) \\ \boldsymbol{B}_2\boldsymbol{S}_A(j) \\ \boldsymbol{S}_S(j)\boldsymbol{D}_{22}\boldsymbol{S}_A(j) \end{bmatrix},$$

$$\hat{\boldsymbol{C}}_1(j) = \begin{bmatrix} \boldsymbol{D}_{12}\bar{\boldsymbol{S}}_A(j) & \boldsymbol{C}_1 & \boldsymbol{0} \end{bmatrix},$$

$$\hat{\boldsymbol{C}}_2(j) = \begin{bmatrix} \boldsymbol{S}_S(j)\boldsymbol{D}_{22}\bar{\boldsymbol{S}}_A & \boldsymbol{S}_S(j)\boldsymbol{C}_2 & \bar{\boldsymbol{S}}_S(j) \end{bmatrix},$$

$$\hat{\boldsymbol{D}}_{11} = \boldsymbol{D}_{11}, \quad \hat{\boldsymbol{D}}_{12}(j) = \boldsymbol{D}_{12}\boldsymbol{S}_A(j),$$

$$\hat{\boldsymbol{D}}_{21}(j) = \boldsymbol{S}_S(j)\boldsymbol{D}_{21}, \quad \hat{\boldsymbol{D}}_{22}(j) = \boldsymbol{S}_S(j)\boldsymbol{D}_{22}\boldsymbol{S}_A(j), \tag{3.26}$$

which are easily derived. $\hat{\mathcal{P}}$ is a model of the plant to be controlled via the network. It can be used to analyze particular structural properties (controllability, observability, *etc.*) deriving from particular communication networks N or to design optimal/robust controllers which take into account the dynamics of the plant and the network simultaneously. However, it should be noticed that this is an LTV model. Also, recall that to be able to derive a model for the augmented plant, a communication network $N = (\Delta, \sigma)$ must be given. Hence, the controller will only be optimal/robust for that given N. This model is a generalization of the models used in [75, 80, 105, 14, 150, 56, 51, 52].

3.3.2 Augmented controller

Here, the schedulers are incorporated into the controller model to obtain the augmented controller $\hat{\mathcal{K}}$. A diagram of this configuration is shown in Figure 3.5. The controller $\mathcal{K}$ is augmented with the scheduler $\mathcal{S}_A$, $\mathcal{S}_S$ and $\mathcal{S}_D$.

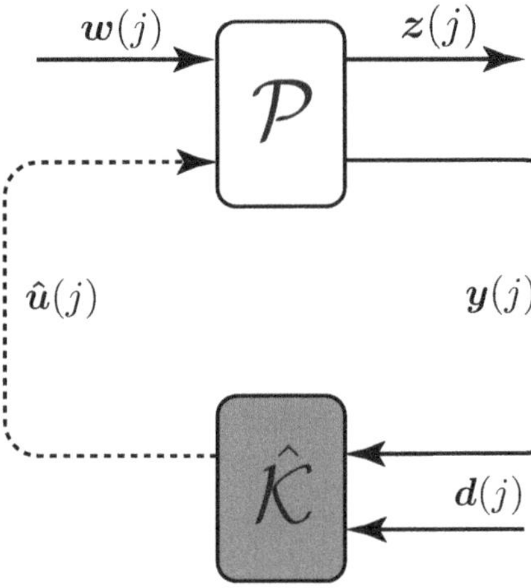

FIGURE 3.5: Augmented controller: the model of the controller and the network are merged together.

The equations for $\hat{\mathcal{K}}$ are

$$\hat{\mathcal{K}}: \quad \begin{cases} \hat{\boldsymbol{x}}_K(j+1) = \hat{\boldsymbol{A}}_K(j)\hat{\boldsymbol{x}}_K(j) + \hat{\boldsymbol{B}}_{K1}(j)\boldsymbol{y}(j) + \hat{\boldsymbol{B}}_{K2}(j)\boldsymbol{d}(j) \\ \hat{\boldsymbol{u}}(j) = \hat{\boldsymbol{C}}_K(j)\hat{\boldsymbol{x}}_K(j) + \hat{\boldsymbol{D}}_{K1}(j)\boldsymbol{y}(j) + \hat{\boldsymbol{D}}_{K2}(j)\boldsymbol{d}(j) \end{cases}, \tag{3.27}$$

where

$$\hat{\boldsymbol{x}}_K(j) = \begin{bmatrix} \boldsymbol{x}_S(j) \\ \boldsymbol{x}_D(j) \\ \boldsymbol{x}_K(j) \\ \boldsymbol{x}_A(j) \end{bmatrix} = \begin{bmatrix} \hat{\boldsymbol{y}}(j-1) \\ \hat{\boldsymbol{d}}(j-1) \\ \boldsymbol{x}_K(j) \\ \hat{\boldsymbol{u}}(j-1) \end{bmatrix}, \tag{3.28}$$

and

$$
\begin{aligned}
\hat{\boldsymbol{A}}_K(j) &= \begin{bmatrix} \bar{\boldsymbol{S}}_S(j) & \boldsymbol{0} & \boldsymbol{0} & \boldsymbol{0} \\ \boldsymbol{0} & \bar{\boldsymbol{S}}_D(j) & \boldsymbol{0} & \boldsymbol{0} \\ \boldsymbol{B}_{K1}\bar{\boldsymbol{S}}_S(j) & \boldsymbol{B}_{K2}\bar{\boldsymbol{S}}_D(j) & \boldsymbol{A}_K & \boldsymbol{0} \\ \boldsymbol{S}_A(j)\boldsymbol{D}_{K1}\bar{\boldsymbol{S}}_S(j) & \boldsymbol{S}_A(j)\boldsymbol{D}_{K2}\bar{\boldsymbol{S}}_D(j) & \boldsymbol{S}_A(j)\boldsymbol{C}_K & \bar{\boldsymbol{S}}_A(j) \end{bmatrix}, \\
\hat{\boldsymbol{B}}_{K1}(j) &= \begin{bmatrix} \boldsymbol{S}_S(j) \\ \boldsymbol{0} \\ \boldsymbol{B}_{K1}\boldsymbol{S}_S(j) \\ \boldsymbol{S}_A(j)\boldsymbol{D}_{K1}\boldsymbol{S}_S(j) \end{bmatrix}, \quad \hat{\boldsymbol{B}}_{K2}(j) = \begin{bmatrix} \boldsymbol{0} \\ \boldsymbol{S}_D(j) \\ \boldsymbol{B}_{K2}\boldsymbol{S}_D(j) \\ \boldsymbol{S}_A(j)\boldsymbol{D}_{K2}\boldsymbol{S}_D(j) \end{bmatrix}, \\
\hat{\boldsymbol{C}}_K(j) &= \begin{bmatrix} \boldsymbol{S}_A(j)\boldsymbol{D}_{K1}\bar{\boldsymbol{S}}_S(j) & \boldsymbol{S}_A(j)\boldsymbol{D}_{K2}\bar{\boldsymbol{S}}_D(j) & \boldsymbol{S}_A(j)\boldsymbol{C}_K & \bar{\boldsymbol{S}}_A(j) \end{bmatrix}, \\
\hat{\boldsymbol{D}}_{K1}(j) &= \boldsymbol{S}_A(j)\boldsymbol{D}_{K1}\boldsymbol{S}_S(j), \quad \hat{\boldsymbol{D}}_{K2}(j) = \boldsymbol{S}_A(j)\boldsymbol{D}_{K2}\boldsymbol{S}_D(j),
\end{aligned} \tag{3.29}
$$

which are easily derived. $\hat{\mathcal{K}}$ is a model of the controller and the network and can be used for analyzing the performance of a scheduling sequence for a given controller and plant. For example, in Chapter 8, the performance of communication sequences for hybrid systems (systems with a continuous plant dynamics and discrete controller-network dynamics) will be considered. This is also an LTV model.

3.3.3 Augmented closed-loop system

Finally, all the blocks are merged together. A diagram of this configuration is shown in Figure 3.6. The equations of the closed-loop system $\hat{\mathcal{P}}_{cl}$ from $\boldsymbol{w}(j)$

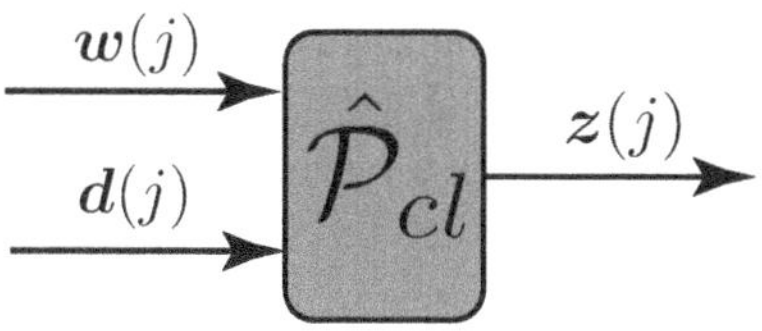

FIGURE 3.6: Augmented closed-loop system: the model of the plant, controller and the network are merged together.

and $\boldsymbol{d}(j)$ to $\boldsymbol{z}(j)$ is

$$
\hat{\mathcal{P}}_{cl}: \quad \begin{cases} \hat{\boldsymbol{x}}_{cl}(j+1) = \hat{\boldsymbol{A}}_{cl}(j)\hat{\boldsymbol{x}}_{cl}(j) + \hat{\boldsymbol{B}}_{cl,1}(j)\boldsymbol{w}(j) + \hat{\boldsymbol{B}}_{cl,2}(j)\boldsymbol{d}(j) \\ \boldsymbol{z}(j) = \hat{\boldsymbol{C}}_{cl}(j)\hat{\boldsymbol{x}}_{cl}(j) + \hat{\boldsymbol{D}}_{cl,1}(j)\boldsymbol{w}(j) + \hat{\boldsymbol{D}}_{cl,2}(j)\boldsymbol{d}(j) \end{cases}, \tag{3.30}
$$

where

$$
\hat{\boldsymbol{x}}_{cl}(j) = \begin{bmatrix} \boldsymbol{x}(j) \\ \hat{\boldsymbol{x}}_K(j) \end{bmatrix}, \tag{3.31}
$$

and

$$
\begin{aligned}
\hat{\boldsymbol{A}}_{cl}(j) &= \begin{bmatrix} \hat{\boldsymbol{A}}_{cl,11}(j) & \hat{\boldsymbol{A}}_{cl,12}(j) \\ \hat{\boldsymbol{A}}_{cl,21}(j) & \hat{\boldsymbol{A}}_{cl,22}(j) \end{bmatrix}, \\
\hat{\boldsymbol{B}}_{cl,1}(j) &= \begin{bmatrix} \boldsymbol{B}_1 + \boldsymbol{B}_2\hat{\boldsymbol{D}}_{K1}(j)\boldsymbol{L}_1(j)^{-1}\boldsymbol{D}_{21} \\ \hat{\boldsymbol{B}}_{K1}(j)\boldsymbol{D}_{21} + \hat{\boldsymbol{B}}_{K1}(j)\boldsymbol{D}_{22}\boldsymbol{L}_2(j)^{-1}\hat{\boldsymbol{D}}_{K1}(j)\boldsymbol{D}_{21} \end{bmatrix}, \\
\hat{\boldsymbol{B}}_{cl,2}(j) &= \begin{bmatrix} \boldsymbol{B}_2\hat{\boldsymbol{D}}_{K2}(j)\boldsymbol{B}_2\hat{\boldsymbol{D}}_{K1}(j)\boldsymbol{L}_1(j)^{-1}\boldsymbol{D}_{22}\hat{\boldsymbol{D}}_{K2}(j) \\ \hat{\boldsymbol{B}}_{K2}(j) + \hat{\boldsymbol{B}}_{K1}(j)\boldsymbol{D}_{22}\boldsymbol{L}_2(j)^{-1}\hat{\boldsymbol{D}}_{K2}(j) \end{bmatrix}, \\
\hat{\boldsymbol{C}}_{cl}(j) &= \begin{bmatrix} \boldsymbol{C}_1 + \boldsymbol{D}_{12}\hat{\boldsymbol{D}}_{K1}(j)\boldsymbol{L}_1(j)^{-1}\boldsymbol{C}_2 \\ \boldsymbol{D}_{12}\hat{\boldsymbol{C}}_K(j) + \boldsymbol{D}_{12}\hat{\boldsymbol{D}}_{K1}(j)\boldsymbol{L}_1(j)^{-1}\boldsymbol{D}_{22}\hat{\boldsymbol{C}}_K(j) \end{bmatrix}, \\
\hat{\boldsymbol{D}}_{cl,1}(j) &= \boldsymbol{D}_{11} + \boldsymbol{D}_{12}\hat{\boldsymbol{D}}_{K1}(j)\boldsymbol{L}_1(j)^{-1}\boldsymbol{D}_{21}, \\
\hat{\boldsymbol{D}}_{cl,2}(j) &= \boldsymbol{D}_{12}\hat{\boldsymbol{D}}_{K1}(j)\boldsymbol{L}_1(j)^{-1}\boldsymbol{D}_{22}\hat{\boldsymbol{D}}_{K2}(j) + \boldsymbol{D}_{12}\hat{\boldsymbol{D}}_{K2}(j),
\end{aligned}
\tag{3.32}
$$

$$
\begin{aligned}
\hat{\boldsymbol{A}}_{cl,11}(j) &= \boldsymbol{A} + \boldsymbol{B}_2\hat{\boldsymbol{D}}_{K1}(j)\boldsymbol{L}_1(j)^{-1}\boldsymbol{C}_2, \\
\hat{\boldsymbol{A}}_{cl,12}(j) &= \boldsymbol{B}_2\hat{\boldsymbol{C}}_K(j) + \boldsymbol{B}_2\hat{\boldsymbol{D}}_{K1}(j)\boldsymbol{L}_1(j)^{-1}\boldsymbol{D}_{22}\hat{\boldsymbol{C}}_K(j), \\
\hat{\boldsymbol{A}}_{cl,21}(j) &= \hat{\boldsymbol{B}}_{K1}(j)\boldsymbol{C}_2 + \hat{\boldsymbol{B}}_{K1}(j)\boldsymbol{D}_{22}\boldsymbol{L}_2(j)^{-1}\hat{\boldsymbol{D}}_{K1}(j)\boldsymbol{C}_2, \\
\hat{\boldsymbol{A}}_{cl,22}(j) &= \hat{\boldsymbol{A}}_K(j) + \hat{\boldsymbol{B}}_{K1}(j)\boldsymbol{D}_{22}\boldsymbol{L}_2(j)^{-1}\boldsymbol{C}_K(j), \\
\boldsymbol{L}_1(j) &= (\boldsymbol{I} - \boldsymbol{D}_{22}\hat{\boldsymbol{D}}_{K1}(j)), \quad \boldsymbol{L}_2(j) = (\boldsymbol{I} - \hat{\boldsymbol{D}}_{K1}\boldsymbol{D}_{22}).
\end{aligned}
\tag{3.33}
$$

Note that $\boldsymbol{L}_1(j)$ and $\boldsymbol{L}_2(j)$ must be nonsingular. This is not an issue as it will be assumed, for simplicity of exposition and without loss of generality, that $\boldsymbol{D}_{22} = \boldsymbol{0}$. It should also be noticed that the scheduling matrices are sparse[1] and therefore many other simplifications can be made. For instance, $\boldsymbol{S}_A(j)\bar{\boldsymbol{S}}_A(j) = \boldsymbol{0}$, $\boldsymbol{S}_S(j)\bar{\boldsymbol{S}}_S(j) = \boldsymbol{0}$, $\boldsymbol{S}_D(j)\bar{\boldsymbol{S}}_D(j) = \boldsymbol{0}$, for all j. This, and other considerations, will turn out to be extremely useful when analyzing the structural properties of the augmented system in the next chapter.

3.4 NCS without ZOH

Although the majority of practically implemented NCSs use a ZOH strategy to reconstruct continuous-time signals from discrete signals, some authors (see [149, 150, 56, 51]) claim that choosing to reset signals that are not scheduled to zero (instead of holding their old value) may benefit the system performance. Also, the resulting model is simpler. Such models are considered for completeness and for comparison of results in the next chapter.

The notion of nodes, communication sequence, communication network

[1] A matrix populated primarily with zeros [115, p. 619].

and scheduling matrices remains unchanged. The difference is only in the construction of the schedulers $\mathcal{S}_A$, $\mathcal{S}_S$ and $\mathcal{S}_D$. They become

$$\begin{aligned} \mathcal{S}_A: &\quad \hat{\boldsymbol{u}}(j) = \boldsymbol{S}_A(j)\boldsymbol{u}(j), \\ \mathcal{S}_S: &\quad \hat{\boldsymbol{y}}(j) = \boldsymbol{S}_S(j)\boldsymbol{y}(j), \\ \mathcal{S}_D: &\quad \hat{\boldsymbol{d}}(j) = \boldsymbol{S}_D(j)\boldsymbol{d}(j). \end{aligned} \tag{3.34}$$

The scheduler dynamics in (3.34) can be understood as follows: the signal vectors $\hat{\boldsymbol{u}}(j)$, $\hat{\boldsymbol{y}}(j)$ and $\hat{\boldsymbol{d}}(j)$ have all their entries equal to zero apart from some entries (given by the 1s in the diagonal of the matrices $\boldsymbol{S}_A(j)$, $\boldsymbol{S}_S(j)$ and $\boldsymbol{S}_D(j)$ respectively) which will be equal to the up-to-date signals $\boldsymbol{u}(j)$, $\boldsymbol{y}(j)$, and $\boldsymbol{d}(j)$. The assumption is that when the communication stops the node resets its signal values to zero until those signals are scheduled again.

For this non-ZOH model, only the case where the network dynamics are included into the plant dynamics is considered simply because the other configurations are obvious and are not used in the sequel. For the non-ZOH case, the schedulers are memoryless systems, then, state augmentation of the plant is not necessary. As before, $\mathcal{S}_D$ will be omitted in this formulation. The equations for the augmented non-ZOH plant $\check{\mathcal{P}}$ are

$$\check{\mathcal{P}}: \quad \begin{cases} \boldsymbol{x}(j+1) = \boldsymbol{A}\boldsymbol{x}(j) + \boldsymbol{B}_1\boldsymbol{w}(j) + \check{\boldsymbol{B}}_2(j)\boldsymbol{u}(j) \\ \boldsymbol{z}(j) = \boldsymbol{C}_1\boldsymbol{x}(j) + \boldsymbol{D}_{11}\boldsymbol{w}(j) + \check{\boldsymbol{D}}_{12}(j)\boldsymbol{u}(j) \\ \hat{\boldsymbol{y}}(j) = \check{\boldsymbol{C}}_2(j)\boldsymbol{x}(j) + \check{\boldsymbol{D}}_{21}(j)\boldsymbol{w}(j) + \check{\boldsymbol{D}}_{22}(j)\boldsymbol{u}(j) \end{cases}, \tag{3.35}$$

where

$$\begin{aligned} &\check{\boldsymbol{B}}_2(j) = \boldsymbol{B}_2\boldsymbol{S}_A(j), \quad \check{\boldsymbol{D}}_{12}(j) = \boldsymbol{D}_{12}\boldsymbol{S}_A(j), \\ &\check{\boldsymbol{C}}_2(j) = \boldsymbol{S}_S(j)\boldsymbol{C}_2, \quad \check{\boldsymbol{D}}_{21}(j) = \boldsymbol{S}_S(j)\boldsymbol{D}_{21}, \quad \check{\boldsymbol{D}}_{22}(j) = \boldsymbol{S}_S(j)\boldsymbol{D}_{22}\boldsymbol{S}_A(j), \end{aligned} \tag{3.36}$$

which are derived by merging the scheduler in (3.34) with the plant in (3.7). The model in (3.35) is a generalization of the model used in [149, 150, 56, 51, 54]. In the same way as for the ZOH case, (3.35) can be readily used to model stochastic packet dropout by applying the substitution in (3.23). By doing so, (3.35) becomes the model used in [58].

3.5 Periodicity and discrete-time lifting

The models presented so far, for the plant and controller with the loop closed via a network, are LTV models because the scheduling mechanism is a time-varying process. The main aim of this work is to design optimal fixed scheduling policies offline. Hence, from the practical point of view, it is reasonable to consider finite length communication sequences.

Definition 3.6 *A finite communication sequence*

$$\sigma \overset{def}{=} \{\sigma(0), \sigma(1), \ldots, \sigma(p-1)\} \overset{def}{=} \{\sigma(k)\}_{k=0}^{p-1}, \tag{3.37}$$

where $\sigma(k)$ is a node and $k = 0, 1, \ldots, p-1$, is defined as an ordered list of elements of size p. ▲

A finite communication sequence will define the finite sequences of scheduling matrices (in a similar way as for (3.15), (3.16), (3.17))

$$\begin{aligned}
\mathcal{S}_A &\overset{\text{def}}{=} \{\boldsymbol{S}_A(k)\}_{k=0}^{p-1}, & \bar{\mathcal{S}}_A &\overset{\text{def}}{=} \{\bar{\boldsymbol{S}}_A(k)\}_{k=0}^{p-1},\\
\mathcal{S}_S &\overset{\text{def}}{=} \{\boldsymbol{S}_S(k)\}_{k=0}^{p-1}, & \bar{\mathcal{S}}_S &\overset{\text{def}}{=} \{\bar{\boldsymbol{S}}_S(k)\}_{k=0}^{p-1},\\
\mathcal{S}_D &\overset{\text{def}}{=} \{\boldsymbol{S}_D(k)\}_{k=0}^{p-1}, & \bar{\mathcal{S}}_D &\overset{\text{def}}{=} \{\bar{\boldsymbol{S}}_D(k)\}_{k=0}^{p-1}.
\end{aligned} \tag{3.38}$$

The finite communication sequence and the finite sequence of scheduling matrices have length p but it is assumed that they are applied repeatedly in a circular fashion such that

$$\begin{aligned}
&\sigma = \{\sigma(0), \sigma(1), \ldots, \sigma(p-1), \sigma(p) = \sigma(0), \sigma(p+1) = \sigma(1), \ldots\},\\
&\mathcal{S}_A = \{\boldsymbol{S}_A(0), \boldsymbol{S}_A(1), \ldots, \boldsymbol{S}_A(p-1), \boldsymbol{S}_A(p) = \boldsymbol{S}_A(0), \boldsymbol{S}_A(p+1) = \boldsymbol{S}_A(1), \ldots\},
\end{aligned} \tag{3.39}$$

etc.. The results are p-periodic communication and scheduling sequences. For this reason let

$$k \overset{\text{def}}{=} \text{mod}(j, p), \tag{3.40}$$

where $\text{mod}(\cdot, \cdot)$ is the modulo operator and

$$j \overset{\text{def}}{=} pl + k, \quad l \in \mathbb{N}_0. \tag{3.41}$$

Now, the schedulers equations (3.15)-(3.17) can be rewritten, for p-periodic schedules, as

$$\mathcal{S}_A : \quad \begin{cases} \boldsymbol{x}_A(pl+k+1) = \bar{\boldsymbol{S}}_A(k)\boldsymbol{x}_A(pl+k) + \boldsymbol{S}_A(k)\boldsymbol{u}(pl+k) \\ \hat{\boldsymbol{u}}(pl+k) = \bar{\boldsymbol{S}}_A(k)\boldsymbol{x}_A(pl+k) + \boldsymbol{S}_A(k)\boldsymbol{u}(pl+k) \end{cases}, \tag{3.42}$$

$$\mathcal{S}_S : \quad \begin{cases} \boldsymbol{x}_S(pl+k+1) = \bar{\boldsymbol{S}}_S(k)\boldsymbol{x}_S(pl+k) + \boldsymbol{S}_S(k)\boldsymbol{y}(pl+k) \\ \hat{\boldsymbol{y}}(pl+k) = \bar{\boldsymbol{S}}_S(k)\boldsymbol{x}_S(pl+k) + \boldsymbol{S}_S(k)\boldsymbol{y}(pl+k) \end{cases}, \tag{3.43}$$

$$\mathcal{S}_D : \quad \begin{cases} \boldsymbol{x}_D(pl+k+1) = \bar{\boldsymbol{S}}_D(k)\boldsymbol{x}_D(pl+k) + \boldsymbol{S}_D(k)\boldsymbol{d}(pl+k) \\ \hat{\boldsymbol{d}}(pl+k) = \bar{\boldsymbol{S}}_D(k)\boldsymbol{x}_D(pl+k) + \boldsymbol{S}_D(k)\boldsymbol{d}(pl+k) \end{cases}. \tag{3.44}$$

3.5.1 Elimination of periodicity via lifting

Let us consider the augmented plant $\hat{\mathcal{P}}$. The equation in (3.24) for periodic schedules becomes

$$\hat{\mathcal{P}}: \begin{cases} \hat{\boldsymbol{x}}(pl+k+1) = \hat{\boldsymbol{A}}(k)\hat{\boldsymbol{x}}(pl+k) + \hat{\boldsymbol{B}}_1(k)\boldsymbol{w}(pl+k) + \hat{\boldsymbol{B}}_2(k)\boldsymbol{u}(pl+k) \\ \boldsymbol{z}(pl+k) = \hat{\boldsymbol{C}}_1(k)\hat{\boldsymbol{x}}(pl+k) + \hat{\boldsymbol{D}}_{11}(k)\boldsymbol{w}(pl+k) + \hat{\boldsymbol{D}}_{12}(k)\boldsymbol{u}(pl+k) \\ \hat{\boldsymbol{y}}(pl+k) = \hat{\boldsymbol{C}}_2(k)\hat{\boldsymbol{x}}(pl+k) + \hat{\boldsymbol{D}}_{21}(k)\boldsymbol{w}(pl+k) + \hat{\boldsymbol{D}}_{22}(k)\boldsymbol{u}(pl+k) \end{cases}. \tag{3.45}$$

This is now a periodic system with periodicity p. Note that $\hat{\boldsymbol{D}}_{11}(0) = \hat{\boldsymbol{D}}_{11}(1) = \ldots = \hat{\boldsymbol{D}}_{11}(p-1)$ but they are represented as k-dependent matrices to extend the analysis to any general periodic system. An equivalent system with matrices independent from k, (*i.e.* an LTI model) will now be presented. This is achieved by using the discrete-time lifting technique (called 'discrete-time' to differentiate it from the 'continuous-time' lifting that will be discussed in Section 8.4.1, p. 165). The discrete-time lifting is described in [20] and used in [105, 14] for similar purposes. The aim is to create a higher dimensional system by down-sampling the periodic system. Thus, the dependance on k can be eliminated. This can be obtained by considering that the state equation evolution, from (3.45), is

$$\begin{aligned}
\text{for } k=0: \quad \hat{\boldsymbol{x}}(pl+1) &= \hat{\boldsymbol{A}}(0)\hat{\boldsymbol{x}}(pl) + \hat{\boldsymbol{B}}_1(0)\boldsymbol{w}(pl) + \hat{\boldsymbol{B}}_2(0)\boldsymbol{u}(pl), \\
\text{for } k=1: \quad \hat{\boldsymbol{x}}(pl+2) &= \hat{\boldsymbol{A}}(1)\hat{\boldsymbol{A}}(0)\hat{\boldsymbol{x}}(pl) + \hat{\boldsymbol{A}}(1)\hat{\boldsymbol{B}}_1(0)\boldsymbol{w}(pl) \\
&\quad + \hat{\boldsymbol{B}}_1(1)\boldsymbol{w}(pl+1) + \hat{\boldsymbol{A}}(1)\hat{\boldsymbol{B}}_2(0)\boldsymbol{u}(pl) \\
&\quad + \hat{\boldsymbol{B}}_2(1)\boldsymbol{u}(pl+1), \\
&\ \vdots \\
\text{for } k=p-1: \quad \hat{\boldsymbol{x}}(pl+p) &= \hat{\boldsymbol{A}}(p-1)\ldots\hat{\boldsymbol{A}}(0)\hat{\boldsymbol{x}}(pl) \\
&\quad + \hat{\boldsymbol{A}}(p-1)\ldots\hat{\boldsymbol{A}}(1)\hat{\boldsymbol{B}}_1(0)\boldsymbol{w}(pl) + \ldots \\
&\quad + \hat{\boldsymbol{B}}_1(p-1)\boldsymbol{w}(pl+p-1) \\
&\quad + \hat{\boldsymbol{A}}(p-1)\ldots\hat{\boldsymbol{A}}(1)\hat{\boldsymbol{B}}_2(0)\boldsymbol{u}(pl) + \ldots \\
&\quad + \hat{\boldsymbol{B}}_2(p-1)\boldsymbol{u}(pl+p-1),
\end{aligned} \tag{3.46}$$

which can be written more compactly as

$$\bar{\boldsymbol{x}}(pl+p) = \bar{\boldsymbol{A}}\hat{\boldsymbol{x}}(pl) + \bar{\boldsymbol{B}}_1\bar{\boldsymbol{w}}(pl) + \bar{\boldsymbol{B}}_2\bar{\boldsymbol{u}}(pl), \tag{3.47}$$

where

$$\bar{\boldsymbol{x}}(pl) = \begin{bmatrix} \hat{\boldsymbol{x}}(pl) \\ \hat{\boldsymbol{x}}(pl+1) \\ \vdots \\ \hat{\boldsymbol{x}}(pl+p-1) \end{bmatrix}, \quad \bar{\boldsymbol{u}}(pl) = \begin{bmatrix} \boldsymbol{u}(pl) \\ \boldsymbol{u}(pl+1) \\ \vdots \\ \boldsymbol{u}(pl+p-1) \end{bmatrix},$$
$$\bar{\boldsymbol{w}}(pl) = \begin{bmatrix} \boldsymbol{w}(pl) \\ \boldsymbol{w}(pl+1) \\ \vdots \\ \boldsymbol{w}(pl+p-1) \end{bmatrix}, \tag{3.48}$$

and

$$\bar{\boldsymbol{A}} = \begin{bmatrix} \prod_{i=1}^{1} \hat{\boldsymbol{A}}(1-i) \\ \prod_{i=1}^{2} \hat{\boldsymbol{A}}(2-i) \\ \vdots \\ \prod_{i=1}^{p} \hat{\boldsymbol{A}}(p-i) \end{bmatrix},$$
$$\bar{\boldsymbol{B}}_1 = \begin{bmatrix} \hat{\boldsymbol{B}}_1(0) & \boldsymbol{0} & \cdots & \boldsymbol{0} \\ \left(\prod_{i=1}^{1} \hat{\boldsymbol{A}}(2-i)\right)\hat{\boldsymbol{B}}_1(0) & \hat{\boldsymbol{B}}_1(1) & \cdots & \boldsymbol{0} \\ \left(\prod_{i=1}^{2} \hat{\boldsymbol{A}}(3-i)\right)\hat{\boldsymbol{B}}_1(0) & \left(\prod_{i=1}^{1} \hat{\boldsymbol{A}}(3-i)\right)\hat{\boldsymbol{B}}_1(1) & \cdots & \boldsymbol{0} \\ \vdots & \vdots & \ddots & \vdots \\ \left(\prod_{i=1}^{p-1} \hat{\boldsymbol{A}}(p-i)\right)\hat{\boldsymbol{B}}_1(0) & \left(\prod_{i=1}^{p-2} \hat{\boldsymbol{A}}(p-i)\right)\hat{\boldsymbol{B}}_1(1) & \cdots & \hat{\boldsymbol{B}}_1(p-1) \end{bmatrix},$$
$$\bar{\boldsymbol{B}}_2 = \begin{bmatrix} \hat{\boldsymbol{B}}_2(0) & \boldsymbol{0} & \cdots & \boldsymbol{0} \\ \left(\prod_{i=1}^{1} \hat{\boldsymbol{A}}(2-i)\right)\hat{\boldsymbol{B}}_2(0) & \hat{\boldsymbol{B}}_2(1) & \cdots & \boldsymbol{0} \\ \left(\prod_{i=1}^{2} \hat{\boldsymbol{A}}(3-i)\right)\hat{\boldsymbol{B}}_2(0) & \left(\prod_{i=1}^{1} \hat{\boldsymbol{A}}(3-i)\right)\hat{\boldsymbol{B}}_2(1) & \cdots & \boldsymbol{0} \\ \vdots & \vdots & \ddots & \vdots \\ \left(\prod_{i=1}^{p-1} \hat{\boldsymbol{A}}(p-i)\right)\hat{\boldsymbol{B}}_2(0) & \left(\prod_{i=1}^{p-2} \hat{\boldsymbol{A}}(p-i)\right)\hat{\boldsymbol{B}}_2(1) & \cdots & \hat{\boldsymbol{B}}_2(p-1) \end{bmatrix}. \tag{3.49}$$

Equation 3.47 is a system of lifted systems. A single lifted system can be written as

$$\hat{\boldsymbol{x}}(pl+k) = \tilde{\boldsymbol{A}}_k \hat{\boldsymbol{x}}(pl) + \tilde{\boldsymbol{B}}_{1,k} \bar{\boldsymbol{w}}(pl) + \tilde{\boldsymbol{B}}_{2,k} \bar{\boldsymbol{u}}(pl), \tag{3.50}$$

where

$$\tilde{\boldsymbol{A}}_k = (\boldsymbol{e}_k^T \otimes \boldsymbol{I})\bar{\boldsymbol{A}}, \quad \tilde{\boldsymbol{B}}_{1,k} = (\boldsymbol{e}_k^T \otimes \boldsymbol{I})\bar{\boldsymbol{B}}_1, \quad \tilde{\boldsymbol{B}}_{2,k} = (\boldsymbol{e}_k^T \otimes \boldsymbol{I})\bar{\boldsymbol{B}}_2. \tag{3.51}$$

$\boldsymbol{I}$ has the appropriate dimensions; $\boldsymbol{e}_k$, $1 \le k \le p$, is the column p-dimensional standard basis vector with a 1 in the kth coordinate and 0 elsewhere; and $\otimes$ is the Kronecker product[2]. The matrix $(\boldsymbol{e}_k^T \otimes \boldsymbol{I})$ is used to select the kth matrix row from the larger block matrices, hence the subscript k in $\tilde{\boldsymbol{A}}_k$, $\tilde{\boldsymbol{B}}_{1,k}$ and $\tilde{\boldsymbol{B}}_{2,k}$ indicates the k-lifted system.

Once the state equation has been lifted, the output equations can be treated in a similar way. The controlled variable equation (*i.e.* the equation for $\boldsymbol{z}(j) = \boldsymbol{z}(pl+k)$), from (3.45), becomes

$$
\begin{aligned}
\boldsymbol{z}(pl) &= \hat{\boldsymbol{C}}_1(0)\hat{\boldsymbol{x}}(pl) + \hat{\boldsymbol{D}}_{11}(0)\boldsymbol{w}(pl) + \hat{\boldsymbol{D}}_{12}(0)\boldsymbol{u}(pl), \\
\boldsymbol{z}(pl+1) &= \hat{\boldsymbol{C}}_1(1)\hat{\boldsymbol{x}}(pl+1) + \hat{\boldsymbol{D}}_{11}(1)\boldsymbol{w}(pl+1) + \hat{\boldsymbol{D}}_{12}(1)\boldsymbol{u}(pl+1) \\
&= \hat{\boldsymbol{C}}_1(1)\tilde{\boldsymbol{A}}_1\hat{\boldsymbol{x}}(pl) + \hat{\boldsymbol{C}}_1(1)\tilde{\boldsymbol{B}}_{1,1}\bar{\boldsymbol{w}}(pl) + \hat{\boldsymbol{C}}_1(1)\tilde{\boldsymbol{B}}_{2,1}\bar{\boldsymbol{u}}(pl), \\
&\quad + \hat{\boldsymbol{D}}_{11}(1)\boldsymbol{w}(pl+1) + \hat{\boldsymbol{D}}_{12}(1)\boldsymbol{u}(pl+1), \\
&\vdots \\
\boldsymbol{z}(pl+p-1) &= \hat{\boldsymbol{C}}_1(p-1)\tilde{\boldsymbol{A}}_{p-1}\hat{\boldsymbol{x}}(pl) + \hat{\boldsymbol{C}}_1(p-1)\tilde{\boldsymbol{B}}_{1,p-1}\bar{\boldsymbol{w}}(pl) \\
&\quad + \hat{\boldsymbol{C}}_1(p-1)\tilde{\boldsymbol{B}}_{2,p-1}\bar{\boldsymbol{u}}(pl) + \hat{\boldsymbol{D}}_{11}(p-1)\boldsymbol{w}(pl+p-1) \\
&\quad + \hat{\boldsymbol{D}}_{12}(p-1)\boldsymbol{u}(pl+p-1),
\end{aligned}
\tag{3.52}
$$

which can be written more compactly as

$$
\bar{\boldsymbol{z}}(pl) = \bar{\boldsymbol{C}}_1\hat{\boldsymbol{x}}(pl) + \bar{\boldsymbol{D}}_{11}\bar{\boldsymbol{w}}(pl) + \bar{\boldsymbol{D}}_{12}\bar{\boldsymbol{u}}(pl), \tag{3.53}
$$

where

$$
\bar{\boldsymbol{z}}(pl) = \begin{bmatrix} \boldsymbol{z}(pl) \\ \boldsymbol{z}(pl+1) \\ \vdots \\ \boldsymbol{z}(pl+p-1) \end{bmatrix}, \tag{3.54}
$$

---

[2] $\boldsymbol{A} \otimes \boldsymbol{B} = \begin{bmatrix} a_{11}\boldsymbol{B} & \cdots & a_{1n}\boldsymbol{B} \\ & \ddots & \\ a_{m1}\boldsymbol{B} & \cdots & a_{mn}\boldsymbol{B} \end{bmatrix}.$

and

$$\bar{\boldsymbol{C}}_1 = \begin{bmatrix} \hat{\boldsymbol{C}}_1(0) \\ \hat{\boldsymbol{C}}_1(1)\tilde{\boldsymbol{A}}_1 \\ \vdots \\ \hat{\boldsymbol{C}}_1(p-1)\tilde{\boldsymbol{A}}_{p-1} \end{bmatrix},$$

$$\bar{\boldsymbol{D}}_{11} = \begin{bmatrix} \boldsymbol{e}_1^T \otimes \hat{\boldsymbol{D}}_{11}(0) \\ \hat{\boldsymbol{C}}_1(1)\tilde{\boldsymbol{B}}_{1,1} + \boldsymbol{e}_2^T \otimes \hat{\boldsymbol{D}}_{11}(1) \\ \vdots \\ \hat{\boldsymbol{C}}_1(p-1)\tilde{\boldsymbol{B}}_{1,p-1} + \boldsymbol{e}_{p-1}^T \otimes \hat{\boldsymbol{D}}_{11}(p-1) \end{bmatrix},$$

$$\bar{\boldsymbol{D}}_{12} = \begin{bmatrix} \boldsymbol{e}_1^T \otimes \hat{\boldsymbol{D}}_{12}(0) \\ \hat{\boldsymbol{C}}_1(1)\tilde{\boldsymbol{B}}_{2,1} + \boldsymbol{e}_2^T \otimes \hat{\boldsymbol{D}}_{12}(1) \\ \vdots \\ \hat{\boldsymbol{C}}_1(p-1)\tilde{\boldsymbol{B}}_{2,p-1} + \boldsymbol{e}_{p-1}^T \otimes \hat{\boldsymbol{D}}_{12}(p-1) \end{bmatrix}. \tag{3.55}$$

The same lifting procedure can be applied to the measurement equation (*i.e.* the equation for $\hat{\boldsymbol{y}}(j) = \hat{\boldsymbol{y}}(pl+k)$) by defining the vector

$$\bar{\boldsymbol{y}}(pl) = \begin{bmatrix} \boldsymbol{y}(pl) \\ \boldsymbol{y}(pl+1) \\ \vdots \\ \boldsymbol{y}(pl+p-1) \end{bmatrix}, \tag{3.56}$$

and deriving the matrices

$$\bar{\boldsymbol{C}}_2 = \begin{bmatrix} \hat{\boldsymbol{C}}_2(0) \\ \hat{\boldsymbol{C}}_2(1)\tilde{\boldsymbol{A}}_1 \\ \vdots \\ \hat{\boldsymbol{C}}_2(p-1)\tilde{\boldsymbol{A}}_{p-1} \end{bmatrix},$$

$$\bar{\boldsymbol{D}}_{21} = \begin{bmatrix} \boldsymbol{e}_1^T \otimes \hat{\boldsymbol{D}}_{21}(0) \\ \hat{\boldsymbol{C}}_2(1)\tilde{\boldsymbol{B}}_{1,1} + \boldsymbol{e}_2^T \otimes \hat{\boldsymbol{D}}_{21}(1) \\ \vdots \\ \hat{\boldsymbol{C}}_2(p-1)\tilde{\boldsymbol{B}}_{1,p-1} + \boldsymbol{e}_{p-1}^T \otimes \hat{\boldsymbol{D}}_{21}(p-1) \end{bmatrix},$$

$$\bar{\boldsymbol{D}}_{22} = \begin{bmatrix} \boldsymbol{e}_1^T \otimes \hat{\boldsymbol{D}}_{22}(0) \\ \hat{\boldsymbol{C}}_2(1)\tilde{\boldsymbol{B}}_{2,1} + \boldsymbol{e}_2^T \otimes \hat{\boldsymbol{D}}_{22}(1) \\ \vdots \\ \hat{\boldsymbol{C}}_2(p-1)\tilde{\boldsymbol{B}}_{2,p-1} + \boldsymbol{e}_{p-1}^T \otimes \hat{\boldsymbol{D}}_{22}(p-1) \end{bmatrix}. \tag{3.57}$$

Finally, the equations for the p-lifted plant $\bar{\mathcal{P}}$ are

$$\bar{\mathcal{P}}: \quad \begin{cases} \hat{\boldsymbol{x}}(pl+p) = \tilde{\boldsymbol{A}}_p\hat{\boldsymbol{x}}(pl) + \tilde{\boldsymbol{B}}_{1,p}\bar{\boldsymbol{w}}(pl) + \tilde{\boldsymbol{B}}_{2,p}\bar{\mathbf{u}}(j) \\ \bar{\boldsymbol{z}}(pl) = \bar{\boldsymbol{C}}_1\hat{\boldsymbol{x}}(pl) + \bar{\boldsymbol{D}}_{11}\bar{\boldsymbol{w}}(pl) + \bar{\boldsymbol{D}}_{12}\bar{\mathbf{u}}(j) \\ \bar{\boldsymbol{y}}(pl) = \bar{\boldsymbol{C}}_2\hat{\boldsymbol{x}}(pl) + \bar{\boldsymbol{D}}_{21}\bar{\boldsymbol{w}}(j) + \bar{\boldsymbol{D}}_{22}\bar{\mathbf{u}}(j) \end{cases} . \tag{3.58}$$

System $\bar{\mathcal{P}}$ in (3.58) is equivalent to system $\hat{\mathcal{P}}$ in (3.24) in the sense that $\bar{\mathcal{P}}$ is a down-sampled version of $\hat{\mathcal{P}}$ with a sampling time of ph. Since $\hat{\mathcal{P}}$ was periodic in p, then, its p-lifted version $\bar{\mathcal{P}}$ has dropped this periodicity and it has reduced to an LTI model. The discrete-time lifting technique just described will be often applied, thus, it may be useful to define a lifting operator.

Definition 3.7 *The discrete-time lifting operator* $\mathbb{L}_D\{\cdot,\cdot\}$ *is defined as*

$$\bar{\mathcal{G}} = \mathbb{L}_D\{\hat{\mathcal{G}}, g\}, \tag{3.59}$$

where $\hat{\mathcal{G}}$ *is the discrete system to be lifted with sampling-time* h, *and* $\bar{\mathcal{G}}$ *is the higher-dimensional* g*-lifted version of* $\hat{\mathcal{P}}$, *where* $g \in \mathbb{N}$. $\bar{\mathcal{G}}$ *is the down-sampled version of* $\hat{\mathcal{G}}$ *with sampling time* ph *and input/output dimension increased by a factor of* g. ▲

Remark 3.4 *The discrete-time lifting technique is not restricted to periodic systems. For example, a* p*-lifted version of the LTI system* $\mathcal{P}$ *in* (3.7), *is given by*

$$\bar{\mathcal{P}} = \mathbb{L}_D\{\mathcal{P}, p\}, \tag{3.60}$$

where

$$\bar{\mathcal{P}}: \quad \begin{cases} \boldsymbol{x}(pl+p) = \tilde{\boldsymbol{A}}_p\boldsymbol{x}(pl) + \tilde{\boldsymbol{B}}_{1,p}\bar{\boldsymbol{w}}(pl) + \tilde{\boldsymbol{B}}_{2,p}\bar{\mathbf{u}}(j) \\ \bar{\boldsymbol{z}}(pl) = \bar{\boldsymbol{C}}_1\boldsymbol{x}(pl) + \bar{\boldsymbol{D}}_{11}\bar{\boldsymbol{w}}(pl) + \bar{\boldsymbol{D}}_{12}\bar{\mathbf{u}}(j) \\ \bar{\boldsymbol{y}}(pl) = \bar{\boldsymbol{C}}_2\boldsymbol{x}(pl) + \bar{\boldsymbol{D}}_{21}\bar{\boldsymbol{w}}(pl) + \bar{\boldsymbol{D}}_{22}\bar{\mathbf{u}}(j) \end{cases} . \tag{3.61}$$

The signal vectors are given in (3.48), (3.54) *and* (3.56), *and the system ma-*

trices are

$$
\begin{aligned}
&\tilde{\boldsymbol{A}}_p = \boldsymbol{A}^p,\\
&\tilde{\boldsymbol{B}}_{1,p} = \begin{bmatrix}\boldsymbol{A}^{p-1}\boldsymbol{B}_1 & \boldsymbol{A}^{p-2}\boldsymbol{B}_1 & \cdots & \boldsymbol{A}^{0}\boldsymbol{B}_1\end{bmatrix},\\
&\tilde{\boldsymbol{B}}_{2,p} = \begin{bmatrix}\boldsymbol{A}^{p-1}\boldsymbol{B}_2 & \boldsymbol{A}^{p-2}\boldsymbol{B}_2 & \cdots & \boldsymbol{A}^{0}\boldsymbol{B}_2\end{bmatrix},\\
&\bar{\boldsymbol{C}}_1 = \begin{bmatrix}\boldsymbol{C}_1\\ \boldsymbol{C}_1\boldsymbol{A}\\ \cdots\\ \boldsymbol{C}_1\boldsymbol{A}^{p-1}\end{bmatrix}, \quad \bar{\boldsymbol{C}}_2 = \begin{bmatrix}\boldsymbol{C}_2\\ \boldsymbol{C}_2\boldsymbol{A}\\ \cdots\\ \boldsymbol{C}_2\boldsymbol{A}^{p-1}\end{bmatrix},\\
&\bar{\boldsymbol{D}}_{11} = \begin{bmatrix}\boldsymbol{e}_1^T\otimes\boldsymbol{D}_{11}\\ \boldsymbol{C}_1\boldsymbol{A}^{p-1}\boldsymbol{B}_1 + \boldsymbol{e}_2^T\otimes\boldsymbol{D}_{11}\\ \vdots\\ \boldsymbol{C}_1\boldsymbol{A}^{0}\boldsymbol{B}_1 + \boldsymbol{e}_{p-1}^T\otimes\boldsymbol{D}_{11}\end{bmatrix}, \quad \bar{\boldsymbol{D}}_{12} = \begin{bmatrix}\boldsymbol{e}_1^T\otimes\boldsymbol{D}_{12}\\ \boldsymbol{C}_1\boldsymbol{A}^{p-1}\boldsymbol{B}_2 + \boldsymbol{e}_2^T\otimes\boldsymbol{D}_{12}\\ \vdots\\ \boldsymbol{C}_1\boldsymbol{A}^{0}\boldsymbol{B}_2 + \boldsymbol{e}_{p-1}^T\otimes\boldsymbol{D}_{12}\end{bmatrix},\\
&\bar{\boldsymbol{D}}_{21} = \begin{bmatrix}\boldsymbol{e}_1^T\otimes\boldsymbol{D}_{21}\\ \boldsymbol{C}_2\boldsymbol{A}^{p-1}\boldsymbol{B}_1 + \boldsymbol{e}_2^T\otimes\boldsymbol{D}_{21}\\ \vdots\\ \boldsymbol{C}_2\boldsymbol{A}^{0}\boldsymbol{B}_1 + \boldsymbol{e}_{p-1}^T\otimes\boldsymbol{D}_{21}\end{bmatrix}, \quad \bar{\boldsymbol{D}}_{22} = \begin{bmatrix}\boldsymbol{e}_1^T\otimes\boldsymbol{D}_{22}\\ \boldsymbol{C}_2\boldsymbol{A}^{p-1}\boldsymbol{B}_2 + \boldsymbol{e}_2^T\otimes\boldsymbol{D}_{22}\\ \vdots\\ \boldsymbol{C}_2\boldsymbol{A}^{0}\boldsymbol{B}_2 + \boldsymbol{e}_{p-1}^T\otimes\boldsymbol{D}_{22}\end{bmatrix}.
\end{aligned}
\tag{3.62}
$$

($\boldsymbol{A}^0 = \boldsymbol{I}$ has been added only to show continuity of the power series throughout the dots.) •

3.6 Extension to multi-networks, subnetworks and task scheduling

The modeling approach presented so far assumes that only one network is used as a communication medium to connect the physical nodes. Also, it is implicitly assumed that each physical node has a dedicated connection to its sensors, actuators and/or demand signals and can perform all the operations within the sampling time. Although this is a reasonable assumption, it may not always be the case. Nonetheless, the NCS models proposed are still applicable if some reconsideration for the schedulers are made.

3.6.1 Multi-networks

Instead of a single communication network N, let us assume now that there are n 'parallel' networks which may or may not share some or all the nodes. For instance, a wireless NCS where two distinct frequency bands are used for communication is an example of an NCS with two parallel networks. This can be represented as a *multigraph.* A multigraph is a graph where multiple edges (*i.e.* edges having the same end vertices) are permitted. Figure 3.7 shows an

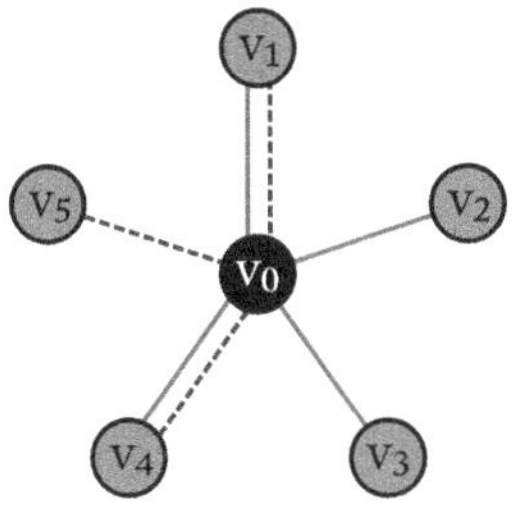

FIGURE 3.7: An example of multigraph: two sets of edges, one dashed line and one solid line, define two parallel communication networks.

example of a multigraph in star configuration where two sets of edges, one dashed line and one solid line, define two parallel communication networks. Nodes v_1 and v_4 are shared between the two networks.

The communication network (Definition 3.3) characterizes the signal scheduling according to a set of nodes and a communication sequence that is assumed to be periodic in this case. Let us say that, for a network made of n parallel networks, $N^i = (\Delta^i, \sigma^i)$ characterizes network i. Δ^i and σ^i are the set of nodes and the communication sequence of the ith network and $i = 1, 2, \ldots, n$. If the communication in an NCS is via a network composed by n parallel networks N^i, then this NCS will be called a *multi-network.*

It was shown in Section 3.2.2 that a model for the signal scheduler can be derived for a given network N. Hence, to model the multi-network as a signal scheduler, the n networks $N^i = (\Delta^i, \sigma^i)$, $i = 1, 2, \ldots, n$ must be reduced to a single network $N = (\Delta, \sigma)$. This is achieved by defining a unique set of nodes Δ and a unique sequence σ. Δ is the set of all the combination of nodes obtained by selecting one node from each of the n sets of nodes. The total number of combinations is the size of Δ which is $|\Delta| = \prod_{i=1}^{n} |\Delta^i| = d$. For example, if $\Delta^1 = \{1, 3\}$ and $\Delta^2 = \{2, 4\}$, then $\Delta = \{\{1, 2\}, \{1, 4\}, \{3, 2\}, \{3, 4\}\}$. The unique sequence is $\sigma = \{\sigma(j)\}_{j=0}^{\infty}$ where $\sigma(j)$ is the node $\Delta(a)$ defined as

$$\Delta(a) \stackrel{\text{def}}{=} \bigcup_{i=1}^{n} \sigma^i(j). \tag{3.63}$$

$\Delta(a) \in \Delta$ for all j because Δ, by definition, includes all the possible combinations of nodes among the n networks. In σ, the element at location j is the node which is formed by the union of the n nodes at location j of all the n sequences.

Hence, multi-networks can be reformulated in terms of the generic communication network $N = (\Delta, \sigma)$ by a special definition of Δ and σ.

3.6.2 Subnetworks and task scheduling

An NCS with *subnetworks* and/or *task scheduling* in the microprocessor unit is also a type of multi-network but, in this case, the networks are not in parallel but rather in 'serial' with each other. For instance, an NCS where a physical node is a gateway to another network is an example of a network with a subnetwork.

Subnetworks are widely used in practical applications as it will be shown in Example 3.2. Task scheduling arises because operations on signals (A/D and D/A conversion), control algorithms *etc.* are often performed in embedded microprocessor units with limited computational power and/or resources. The result is that the numerous tasks to be performed in the microprocessor cannot run concurrently and terminate before the next sampling time. Thus, tasks in the microprocessor need to be scheduled. Subnetworks and task scheduling, under this framework, can be treated in exactly the same way.

In the case of subnetworks (or task scheduling), the NCS is described by a time-varying tree rather than a time-varying star. More precisely, a network with subnetworks is a *rooted tree* where the *root* is the controller node. Recall that, in a graph $G = (V, E)$, $v_a \in V$ is a vertex and E the set of edges. The *distance* $d_{a,b}$ between vertices v_a and v_b is the number of edges, in the shortest path, connecting them. For a rooted tree, which is a special type of graph, the distance from the root to any vertex is simply the number of edges connecting them. A rooted tree can be divided into *subtrees*. Let v_0 be the root of the tree and v_c the internal vertex of a subtree. Then, the subtree of interest here is the star with internal vertex v_c and leaves from the set $\{v_l : d_{c,l} = 1, d_{c,0} < d_{l,0}\}$. These stars can be classified according to the distance from their internal vertices to the root of the tree. Hence, the 0th *level* star would be the one having the root as its internal vertex. The 1st *level* star(s) would be the one(s) with a distance from their internal vertex to the root equal to 1 and so on. See Figure 3.8 for an illustration of the network

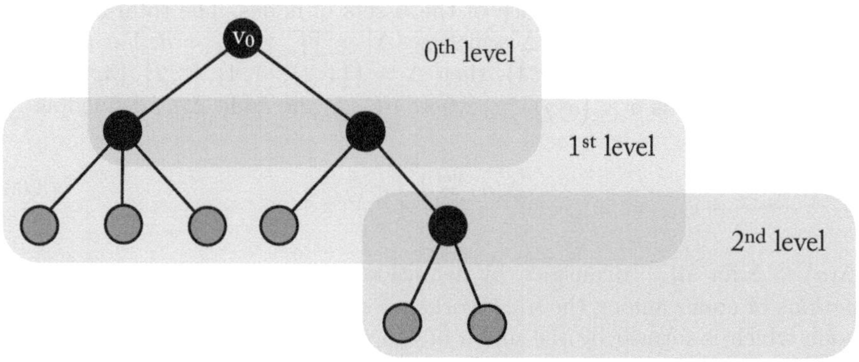

FIGURE 3.8: A network with subnetworks represented by a rooted tree: the distance from the root determines the network level.

as a tree partitioned into star subtrees with level classification. Note that the internal vertex of the ath level star is a leaf of the $(a-1)^{th}$ level star. Also, a root of the ath level star, where $a > 0$, is a node that acts like a gateway to relay information from the root to a leaf if that leaf is associated to a means of action, and from a leaf to the root if that leaf is associated to a source of information. Hence, the root of the ath level star is a node of which elements are the union of all the signals of the nodes that are leaves generating from that root (plus other signals, if any, directly associated to it).

For an NCS with subnetworks, the communication networks can be defined as $N^i = (\Delta^i, \sigma^i)$ where $i = 0, 1, \dots n-1$ this time is the level of the network (*i.e.* $i = 0$ for the 0th level network, $i = 1$ for the 1st level subnetwork(s), *etc.*) and n is the maximum distance between a vertex (or a node) to the root vertex (or the controller node). The n networks form a hierarchical structure. For $i > 0$, the ith level network can have as many subtrees as the number of leaves of the subtrees of the $(i-1)^{th}$ level network. These subtrees are parallel networks (in the sense defined for multi-networks). Hence, they can be defined by a single network as shown in Section 3.6.1.

Each N^i characterizes a generic signal scheduler, $\mathcal{S}^i$, which is a state space model (see Section 3.2.2). Let $\mathcal{S}^0, \mathcal{S}^1, \dots, \mathcal{S}^{n-1}$ be generic signal schedulers deriving from the communication networks $N^0, N^1, \dots, N^{n-1}$. Because of the hierarchical structure of the subnetworks, if $\mathcal{S}^0, \mathcal{S}^1, \dots, \mathcal{S}^{n-1}$ are actuator schedulers, then, the ith scheduler, $\mathcal{S}^i$, takes as an input the output signal of scheduler $\mathcal{S}^{i-1}$ which takes as an input the output of scheduler $\mathcal{S}^{i-2}$ and so on until scheduler $\mathcal{S}^0$ is reached. $\mathcal{S}^0$ is the scheduler that takes as an input the up-to-date signal coming from the controller node. For sensor/demand schedulers, the process is reversed *i.e.* for the ith sensor/demand scheduler, $\mathcal{S}^i$ takes as an input the output signal of scheduler $\mathcal{S}^{i+1}$ and so on until scheduler $\mathcal{S}^{n-1}$ is reached. $\mathcal{S}^{n-1}$ is the scheduler that takes as an input the up-to-date signal coming from the sensor/demand.

Remark 3.5 *If it happens that, at a given time instant, a signal is scheduled through more than one scheduler in cascade (*e.g.* the time-varying tree has a path from a leaf to the root where their distance is bigger than 2), then the signal is received at destination as if there was a dedicated connection between the two points. This is unrealistic, as a microprocessor will take some finite time to process that signal (need of decoding, A/D conversion, *etc.*). It would be more realistic to assume that information updates are available only some finite time after the signal has been scheduled and therefore it can only be passed onto the next vertex in the tree the next time an edge exists between the two vertices. This complication will not be considered here.* •

Definition 3.8 *It is convenient now to use the following shorthand notation. Let $\mathcal{G}^0, \mathcal{G}^1, \dots, \mathcal{G}^{n-1}$ be systems where the output $\hat{\boldsymbol{\nu}}_{i+1}(j)$ of $\mathcal{G}^{i+1}$ is equal to the input $\boldsymbol{\nu}_i(j)$ of $\mathcal{G}^i$, $i = 0, 2, \dots, n-2$,* i.e.

$$\mathcal{G}^i : \quad \boldsymbol{\nu}_i(j) \mapsto \hat{\boldsymbol{\nu}}_i(j), \quad i = 0, 1, \dots, n-1, \tag{3.64}$$

and

$$\hat{\boldsymbol{\nu}}_i(j) = \boldsymbol{\nu}_{i-1}(j), \quad i = 1, 2, \ldots, n-1. \tag{3.65}$$

Then,

$$\mathcal{G} = \mathcal{G}^0\mathcal{G}^1 \cdots \mathcal{G}^{n-1}: \quad \boldsymbol{\nu}_{n-1}(j) \mapsto \hat{\boldsymbol{\nu}}_0(j), \tag{3.66}$$

is the cascaded connection of subsystem $\mathcal{G}^0$ with subsystem $\mathcal{G}^1$ and so on until $\mathcal{G}^{n-1}$. In other words, $\mathcal{G}$ is the system with the output of $\mathcal{G}^0$ and the inputs of $\mathcal{G}^{n-1}$. ▲

Let us construct, from N^i, the actuator, sensor and demand schedulers $\mathcal{S}_A^i$, $\mathcal{S}_S^i$ and $\mathcal{S}_D^i$ respectively ($i = 0, 1, \ldots, n-1$). Then, using the convention of cascaded systems in Definition 3.8, the signal scheduler of the network with subnetworks can be written as

$$\begin{aligned} \mathcal{S}_A &= \mathcal{S}_A^0\mathcal{S}_A^1 \cdots \mathcal{S}_A^{n-1}, \\ \mathcal{S}_S &= \mathcal{S}_S^{n-1}\mathcal{S}_S^{n-2} \cdots \mathcal{S}_S^0, \\ \mathcal{S}_D &= \mathcal{S}_D^{n-1}\mathcal{S}_D^{n-2} \cdots \mathcal{S}_D^0. \end{aligned} \tag{3.67}$$

The model of Figure 3.2 can be built by using the schedulers in (3.67).

In conclusion, it has been shown how the model depicted in Figure 3.2 can account for multi-networks, subnetworks and task scheduling by carefully identifying and modeling the various networks. A final remark should be made on the period of the resulting closed-loop system when periodic communication sequences are used for the networks/tasks. If the sequences are periodic (so that the discrete-time lifting operator can be applied), the period p of the sequence σ will be

$$p = \text{LCM}(p_0, p_1, \ldots, p_{n-1}), \tag{3.68}$$

where $\text{LCM}(\cdot, \cdot, \ldots, \cdot)$ is the least common multiple operator and $p_0, p_1, \ldots, p_{n-1}$ are the periods of sequences $\sigma^0, \sigma^1, \ldots, \sigma^{n-1}$ respectively. This is true for multi-networks, subnetworks and task scheduling.

The following, is an example to show how to model subnetworks and task scheduling under the proposed framework.

Example 3.2 *Consider the door control of a typical vehicle [86] where a Controller Area Network (CAN) bus is used for communication between Electronic Control Units (ECUs) as shown in Figure 3.9. There are four ECUs associated to each door and one associated to the dashboard. The functions associated to the door ECUs are the window, mirror and seat position control (three actuators) according to the driver/passenger request. Since the window, mirror and seat actuators are physically distant from each other, they are supported by three other ECUs connected on a Local Interconnect Network (LIN) bus. The door ECU information is transmitted (through the CAN bus) to the dashboard*

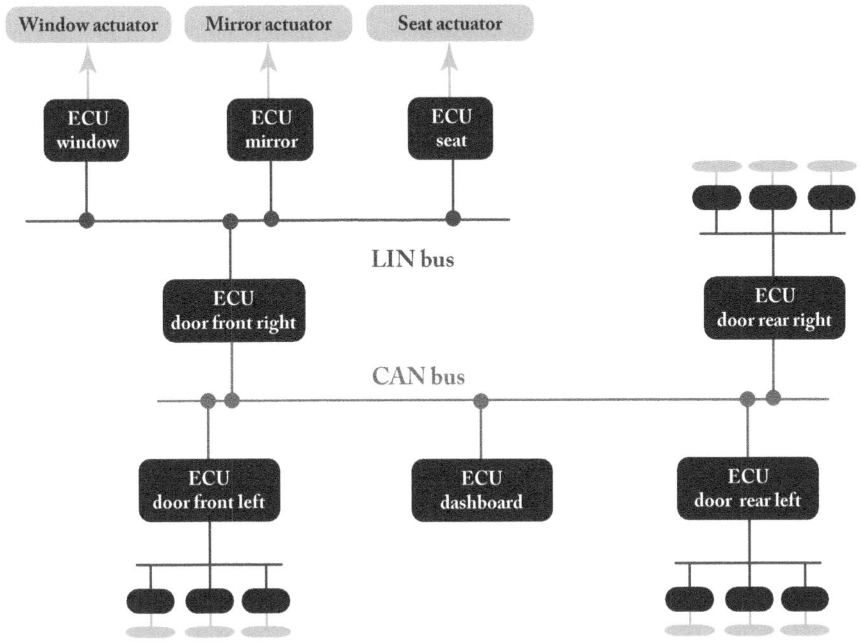

FIGURE 3.9: Example of door control in a vehicle: a CAN bus (the main network) is used for the doors ECUs communication while a LIN bus connects local ECUs.

ECU to present to the driver the status of the door ECUs. This network with subnetworks scenario can be modeled using the techniques presented above. The dashboard ECU is the controller node (the root of the tree) that also acts as a scheduler. The four door ECUs are physical nodes (leaves of the 0th *level network and internal vertices of the* 1st *level network) and the twelve window/mirrow/seat ECUs (three for each door) are other physical nodes (leaves of the* 1st *level network). Hence, there are two levels, given by the communication networks* $N^0 = (\Delta^0, \sigma^0)$ *and* $N^1 = (\Delta^1, \sigma^1)$*, associated to the CAN and LIN buses respectively. Note that* N^1 *is a multi-network (as defined in Section 3.6.1) made of four parallel networks that, in this example, do not share any node.*

Remark 3.6 *Consider the vehicle door control again, but this time, assume that window, mirror and seat actuators are not supported by individual ECUs but have a dedicated wire connection to the respective door ECUs. Also, assume that the microprocessor of the door ECU is not powerful (or fast) enough and can only deal with one actuator at every time tick. Hence, there are three separate tasks within the door ECU for window, mirror and seat actuator control. These tasks need to be scheduled. To model this scenario with the discussed method, two levels with communication networks,* $N^0 = (\Delta^0, \sigma^0)$ *and* $N^1 = (\Delta^1, \sigma^1)$ *need to be considered.* N^0 *describes the CAN bus communication between the doors and dashboard ECUs and* N^1 *describes the task scheduling inside each door ECU (as before,* N^1 *comprises the four parallel networks).* •

The communication network N^1 *derives from the four networks* $N^1_\kappa = (\Delta^1_\kappa, \sigma^1_\kappa)$*, where* $\kappa \in \Theta$ *and*

$$\begin{aligned}\Theta =\{&\text{ECU door front right, ECU door front left,}\\ &\text{ECU door rear right, ECU door rear left}\}.\end{aligned} \tag{3.69}$$

Δ^1_κ *and* σ^1_κ *are the four sets of nodes and communication sequences respectively for the four door ECUs. (Note that in this case, the abstract concept of node (Definition 3.1) actually corresponds to a physical node or ECU.) It was shown in Section 3.6.1 that a unique set of nodes* Δ^1 *and a unique sequence* σ^1 *can be defined for the four* N^1_κ*. Hence,* $N^1 = (\Delta^1, \sigma^1)$ *where each element of* σ^1 *is a node in* Δ^1*, and* Δ^1 *is the set of the union of the four nodes scheduled in each* σ^1_κ*.*

Let us assume that the chosen communication sequence for the CAN bus and the four LIN buses is the simple round robin schedule (all the nodes in the periodic sequence appear only once). Then, the period of σ^0 *is 4 (because of the five nodes connected by the CAN bus, only four need to be scheduled since one, the dashboard ECU, is the actual scheduler) and the period of* σ^1 *is* $\mathrm{LCM}(3,3,3,3) = 3$*. From* N^0 *and* N^1 *obtain* $\mathcal{S}^0$ *and* $\mathcal{S}^1$ *respectively. Finally, since there are only actuators,* $\mathcal{S}_A = \mathcal{S}^1\mathcal{S}^2$ *(there are no schedulers* $\mathcal{S}_S$ *and* $\mathcal{S}_D$ *in this case). The period of* $\mathcal{S}_A$ *is* $p = \mathrm{LCM}(4,3) = 12$*.* ♦

3.7 Multirate systems, a special case of NCSs

In multirate systems, signals are sampled with different sampling rates. This has huge practical benefits as technological or economical constraints may not allow the controller to sample all its inputs or calculate all its output values at the same sampling rate. It is also reasonable, in large multivariable systems, from an optimization point of view, to use different A/D and D/A conversion rates for signals with different bandwidths. Also, multirate controllers can achieve simultaneous stabilization, gain margin improvements and distributed control, that are impossible for single rate controllers. Detailed analysis of systems and the design of controllers with multirate characteristics are given in [26, 21, 131, 2, 154] and in the references therein.

Following the framework for NCSs so far discussed, it can be easily seen how multirate systems are a subset of NCSs. In multirate systems, the sampling period of each signal may differ from the one of other signals but it is constant with time *i.e.* if $\nu_1(\sum_{\iota=0}^{j} h_1(\iota)) \stackrel{\text{def}}{=} \nu_1(jh_1(j))$ and $\nu_2(\sum_{\iota=0}^{j} h_2(\iota)) \stackrel{\text{def}}{=} \nu_2(jh_2(j))$ are two signals with time-varying sampling periods $h_1(j)$ and $h_2(j)$ (j is the sampling instant) then, $h_1(j)$ might not be equal to $h_2(j)$ for all $j \in \mathbb{N}_0$ but $h_1(j) = h_1(j+i)$ and $h_2(j) = h_2(j+i)$ for all $j, i \in \mathbb{N}_0$. The sampling periods in multirate systems are not time-dependent. In an NCS, the sampling period of each signal can be time-varying *i.e.* $\exists i, j$ such that $h_1(j) \neq h_1(j+i)$ and $h_2(j) \neq h_2(j+i)$. Since the interest here is in periodic systems, the special case of periodically time-varying sampling periods will be considered *i.e.* $h_1(k_1)$ and $h_2(k_2)$ where $k_1 \stackrel{\text{def}}{=} \text{mod}(j, p_1)$ and $k_2 \stackrel{\text{def}}{=} \text{mod}(j, p_2)$, and p_1 and p_2 are the periods of signal 1 and 2 respectively.

It was previously shown how the network $N = (\Delta, \sigma)$ completely characterizes the communication of an NCS. Thus, in order to model a multirate system under this framework, a set of nodes Δ and a sequence σ need to be defined. One way to proceed is to construct a sequence with period $p = \text{LCM}(h_1, h_2, \ldots, h_n)$ where $h_i, i = 1, 2, \ldots, n$, are the sampling periods of signals $1, 2, \ldots, n$ respectively of a multirate system. Then, associate a node to each element of a sequence of length p *i.e.*

$$\sigma \stackrel{\text{def}}{=} \{\Delta(1), \Delta(2), \ldots, \Delta(p)\} \stackrel{\text{def}}{=} \{\Delta(\iota)\}_{\iota=1}^{p}. \tag{3.70}$$

The signals associated to each node $\Delta(\iota)$ satisfy the condition

$$\textit{signal } i \in \Delta(\iota) \quad \text{if} \quad \text{mod}(\iota, h_i) = 0. \tag{3.71}$$

(Notice that some nodes might be empty. Also, since every h_i is a divisor of p, the last node, $\Delta(p)$, will always schedule all the signals.) Hence, this demonstrates that the proposed models for NCSs with scheduled communication generalize those of multirate systems.

The following example shows how the pair (Δ, σ) can be derived for the formulation of a multirate system in terms of NCSs.

Example 3.3 *Let us consider a multirate system where a plant with 2 sensors and 2 actuators is controlled by a periodically time-varying feedback controller implemented in an embedded microprocessor with a tick frequency of $\frac{1}{h}$ Hz. Sensor 1 is measuring a fast process and therefore is sampled every $h_{s1} = h$ seconds while sensor 2 is associated to a slower process where a sampling time of $h_{s2} = 2h$ seconds is sufficient. The controller cannot run its algorithms fast enough and therefore actuator 1 signal is updated every $h_{a1} = 3h$ and actuator 2 signal every $h_{a2} = 6h$ seconds.*

To model this scenario as an NCS, a set of nodes Δ and a sequence σ for the communication network $N = (\Delta, \sigma)$ needs to be derived. The period of σ is

$$p = \text{LCM}(h_{s1}, h_{s2}, h_{a1}, h_{a2}) = \text{LCM}(h, 2h, 3h, 6h) = 6h, \tag{3.72}$$

and therefore $\sigma = \{\Delta(1), \Delta(2), \ldots, \Delta(6)\}$. Finally, using (3.71), the following set of nodes is constructed:

$$\begin{aligned}&\Delta(1) = \{sensor\ 1\}, \quad \Delta(2) = \{sensor\ 1, sensor\ 2\},\\ &\Delta(3) = \{sensor\ 1, actuator\ 1\}, \quad \Delta(4) = \{sensor\ 1, sensor\ 2\},\\ &\Delta(5) = \{sensor\ 1\}, \quad \Delta(6) = \{sensor\ 1, sensor\ 2, actuator\ 1, actuator\ 2\}.\end{aligned} \tag{3.73}$$

(Notice that $\Delta(1) = \Delta(5)$ and $\Delta(2) = \Delta(4)$ hence only four nodes would have sufficed in this case.) The communication network N just defined can be used to find a model for the schedulers $\mathcal{S}_A$ and $\mathcal{S}_S$ (Section 3.2.2) that, together with the plant and the controller model, form an NCS which is in fact a multirate system. ♦

3.8 NCSs, a special case of switched and delayed systems

3.8.1 Switched systems

Switched systems are commonly defined as hybrid dynamical systems composed of a family of subsystems together with a switching rule coordinating the switching between subsystems [118, 117]. Consider the special class of discrete switched linear control systems. They can be described by

$$\begin{aligned}\boldsymbol{x}(j+1) &= \hat{\boldsymbol{A}}_\varsigma \boldsymbol{x}(j) + \hat{\boldsymbol{B}}_\varsigma \boldsymbol{u}(j)\\ \boldsymbol{y}(j) &= \hat{\boldsymbol{C}}_\varsigma \boldsymbol{x}(j),\end{aligned} \tag{3.74}$$

where the signal vectors are the usual and ς is the piecewise constant switching signal taking value from the finite index set $\mathfrak{D} = \{1, 2, \ldots, d_\varsigma\}$. The behavior of the system depends on the switching signal that may depend on its past values, the time and the state/output [117]

$$\varsigma(j+) = \Psi(j, \boldsymbol{x}(j)/\boldsymbol{y}(j), \varsigma(j-)), \quad \forall j, \tag{3.75}$$

where $\varsigma(j+)$ denotes future switchings, $\varsigma(j-)$ are past switchings and $\boldsymbol{x}(j)/\boldsymbol{y}(j)$ is the state/output.

Let us consider the augmented systems of Section 3.3. By inspecting the augmented plant model $\hat{\mathcal{P}}$ in (3.24), for instance, it can be noticed that the matrix $\boldsymbol{A}(j)$ can only take values from the set

$$\{\boldsymbol{A}_{\Delta(1)}, \boldsymbol{A}_{\Delta(2)}, \ldots, \boldsymbol{A}_{\Delta(d)}\}, \tag{3.76}$$

where $\boldsymbol{A}_{\Delta(i)}, i = 1, 2, \ldots, d$ is the matrix resulting from scheduling node $\Delta(i)$. Therefore, $\boldsymbol{A}(j) = \boldsymbol{A}_{\sigma(j)}$ because, from Definition 3.2, $\sigma(j)$ is the node scheduled at time j. A similar argument applies for the other matrices in (3.24) ($\boldsymbol{B}_1 = \{\boldsymbol{B}_{1,\Delta(i)}\}_{i=1}^{d}$, $\boldsymbol{C}_1 = \{\boldsymbol{C}_{1,\Delta(i)}\}_{i=1}^{d}$, *etc.*) and for all the other augmented models presented in Section 3.3.

Consider the subclass of switching functions $\Psi(j)$ where the switching signals only depend on time and a switching occurs every sampling time, *i.e.* $\varsigma(j+1) = \Psi(j)$, and let the switching function be $\Psi(j) = \sigma(j+1)$. Then, this special class of switched system is in fact an NCS. Hence, an NCS is a discrete switched linear control system with switchings occurring at every sampling where the switching rule is the communication sequence σ and the switching signal takes values from the set of nodes $\Delta = \{\Delta(1), \Delta(2), \ldots, \Delta(d)\}$. In other words, NCSs are special cases of switched systems.

3.8.2 Delayed systems

The effect the communication scheduling has on the closed-loop system is equivalent to the one of a time-varying delay equal to multiples of the sampling time that may vary for each signal component (no delays in the states are present). The following definition is needed to show how to construct the delayed system equations for a given communication network $N = (\Delta, \sigma)$.

Definition 3.9 *For a communication sequence* $\sigma = \{\sigma(j)\}_{j=0}^{\infty}$, *the* actuator delay vector *is defined as*

$$\boldsymbol{\tau}^A(j) \overset{def}{=} \begin{bmatrix}\tau_1^A(j) & \tau_2^A(j) & \cdots & \tau_{n_u}^A(j)\end{bmatrix}, \quad j = 1, 2, \ldots, \tag{3.77}$$

where

$$\tau_i^A(j) = \min\Big\{j - j^* : \quad \text{actuator } i \in \sigma(j^*), \quad j^* = 0, 1, \ldots, j\Big\},$$
$$i = 1, 2, \ldots, n_u. \tag{3.78}$$

Similarly, the sensor delay vector is defined as

$$\boldsymbol{\tau}^S(j) \overset{def}{=} \begin{bmatrix}\tau_1^S(j) & \tau_2^S(j) & \cdots & \tau_{n_y}^S(j)\end{bmatrix}, \quad j = 1, 2, \ldots, \tag{3.79}$$

where

$$\tau_i^S(j) = \min\Big\{j - j^* : \quad \text{sensor } i \in \sigma(j^*), \quad j^* = 0, 1, \ldots, j\Big\},$$
$$i = 1, 2, \ldots, n_y. \tag{3.80}$$

▲

For the actuators, the scalar delay $\tau_i^A(j)$ represents the distance, in terms of j and going backward, from the sequence element $\sigma(j)$ to the sequence element where *actuator* i was last scheduled. Therefore, $\tau_i^A(j) = 0$ if *actuator* i is scheduled at time j (*i.e. actuator* $i \in \sigma(j)$), $\tau_i^A(j) = 1$ if *actuator* i is scheduled at $j-1$ and not at time j (*i.e. actuator* $i \in \sigma(j-1)$, *actuator* $i \notin \sigma(j)$), and so on. A similar argument applies for $\tau_i^S(j)$.

The model of [57] can now be used to represent the NCS as a delayed system. First, consider the non-delayed system

$$\begin{aligned} \boldsymbol{x}(j+1) &= \boldsymbol{A}\boldsymbol{x}(j) + \boldsymbol{B}\boldsymbol{u}(j) \\ \boldsymbol{y}(j) &= \boldsymbol{C}\boldsymbol{x}(j), \end{aligned} \tag{3.81}$$

where

$$\boldsymbol{u}(j) = \begin{bmatrix} u_1(j) & u_2(j) & \ldots & u_{n_u}(j)\end{bmatrix}^T, \quad \boldsymbol{y}(j) = \begin{bmatrix} y_1(j) & y_2(j) & \ldots & y_{n_y}(j)\end{bmatrix}^T. \tag{3.82}$$

The delayed version is

$$\boldsymbol{x}(j+1) = \boldsymbol{A}\boldsymbol{x}(j) + \boldsymbol{B}\boldsymbol{u}(j - \boldsymbol{\tau}^A(j)), \tag{3.83}$$

where, if the ZOH strategy is used,

$$\boldsymbol{u}(j - \boldsymbol{\tau}^A(j)) = \begin{bmatrix} u_1(j - \tau_1^A(j)) & u_2(j - \tau_2^A(j)) & \ldots & u_{n_u}(j - \tau_{n_u}^A(j))\end{bmatrix}^T. \tag{3.84}$$

The delayed output equation is

$$\boldsymbol{y}(j - \boldsymbol{\tau}^S(j)) = \begin{bmatrix} y_1(j - \tau_1^S(j)) & y_2(j - \tau_2^S(j)) & \ldots & y_{n_y}(j - \tau_{n_y}^S(j))\end{bmatrix}^T. \tag{3.85}$$

If the non-scheduled signals are reset to zero (non-ZOH case), then (3.84)-(3.85) still apply but with

$$\begin{aligned} u_i(j - \tau_i^A(j)) = 0 \quad &\text{if} \quad \tau_i^A(j) \neq 0, \\ y_i(j - \tau_i^S(j)) = 0 \quad &\text{if} \quad \tau_i^S(j) \neq 0. \end{aligned} \tag{3.86}$$

Remark 3.7 *In [57] the assumption is made that the delay may only remain constant or increase with unitary increments. This assumption does not need to be taken into consideration here because, since the delay vector is derived from the scheduling sequence, the delay will automatically increase unitarily. Also, [57] assumes that each channel delay is bounded by a finite value. Here, if periodic sequences are used, the delay is always bounded by* p *(the sequence period).* •

For periodic sequences, the number of delay vectors (Definition 3.9) is p and they are defined in a similar way. The 'backward counting' still applies assuming that, when the first element is reached, the counting continues from the last element in a circular fashion. It is necessary that each sensor and actuator is scheduled at least once to bound the delay by p, however, this will be mandatory in order to guarantee 'feasibility' (this issue will be discussed in the next chapter).

It should be noticed that the delayed plant model of (3.83) and (3.85) used by [57] is equivalent to the model of the plant augmented with the actuator and sensor schedulers. The difference is only in the formulation. More specifically, in the delayed model of [57], the delay vector is time-varying and it is applied explicitly in the input/output vectors; in the augmented model presented here the input/output vectors are left unchanged but the plant is augmented with extra states corresponding to the delayed versions of the input/output vectors. The advantage of the augmented model discussed here, which is derived from an NCS formulation, is that it is in the standard state space form and therefore better suited for analysis and design purposes when compared to the model of [57].

3.9 Application to a vehicle brake-by-wire control system

It will be shown how the framework for NCS modeling discussed earlier can be applied to a realistic automotive system. The system considered is the Hardware-In-the-Loop (HIL) Adaptive Cruise Controller System (ACCS) developed by TTE Systems Ltd[3]. This system consists of ten physical microprocessor nodes that use the time-triggered communication protocol of [9] over a CAN bus. Each physical node is an Electronic Control Unit (ECU). For the sake of clarity, only the brake-by-wire system, which comprises seven ECUs, will be modeled. The function of each ECU, shown in Figure 3.10, is the following.

- Four wheel ECUs connected to the local wheel velocity sensor and a brake actuator. Each wheel ECU (Front Right (FR), Front Left (FL), Rear Right (RR) and Rear Left (RL)) runs three tasks: (i) measuring the wheel angular velocity, (ii) performing an Anti-lock Braking System (ABS) algorithm to control the brake actuator and (iii) checking that everything is running correctly (a status update task).
- One ECU (LV), connected to a radar, measures the vehicle Linear Velocity relative to the ground.

[3] `www.tte-systems.com`.

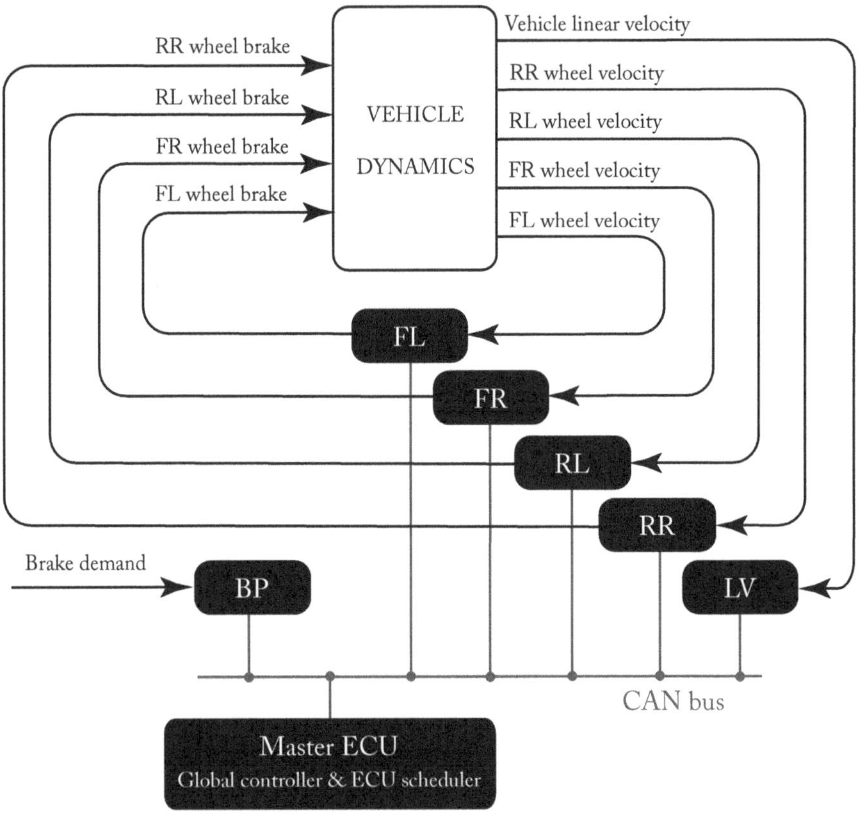

FIGURE 3.10: Diagram of brake-by-wire system: each ECU closes its local loop but the global controller can only communicate via a CAN bus. In 'XY wheel velocity/Brake' X is Rear/Front and Y is Right/Left.

- One ECU (BP), connected to the Brake Pedal position transducer, detects the amount of braking power required by the driver.
- One Master ECU runs a global control algorithm and is also responsible for the communication scheduling between the other ECUs.

The tasks in each wheel ECU cannot run at the same time, hence, they need to be scheduled locally. The 'status update task', in the wheel ECUs does not affect the system dynamics, however, it needs to run (and therefore be scheduled) for safety reasons. Communication between ECUs is via a single CAN bus and therefore only one ECU can access the communication medium at any time tick.

The control system works as follows: the global controller (implemented in the Master ECU) runs an algorithm which takes as an input the individual wheel velocities, the linear velocity of the vehicle with respect to the ground, and the braking power demanded by the driver, and outputs a global control signal. The wheel ECUs run a local ABS algorithm which takes as an input the global control signal, the linear velocity, and the local wheel velocity, and output a local control signal to the brake actuator.

Two types of scheduling are needed to capture this scenario. ECU scheduling (via the bus) and task scheduling inside each wheel node. Synchronization is provided by the Master ECU.

3.9.1 Modeling the bus and task scheduling

To model the ECU scheduling, a set of nodes, that map to the set of signals that can be scheduled simultaneously, needs to be defined. It is easy to see that these nodes correspond to the six ECUs (excluding the Master ECU) connected to the bus. The Master ECU does not need scheduling because it is analogous to the 'controller node' in Figure 3.3. The set of nodes for ECU scheduling is

$$\Delta^E = \{\Delta^E(\kappa), \Delta^E(V), \Delta^E(D)\}, \tag{3.87}$$

where

$$\begin{aligned} \Delta^E(\kappa) &= \{\kappa \textit{ wheel velocity},\ \kappa \textit{ brake control}\}, \quad \text{for } \forall \kappa \in \Theta, \\ &\qquad\qquad\qquad\qquad\qquad\qquad \Theta = \{FL,FR,RL,RR\}, \\ \Delta^E(V) &= \{\textit{linear velocity}\}, \\ \Delta^E(D) &= \{\textit{brake demand}\}. \end{aligned} \tag{3.88}$$

The superscript 'E' is to indicate that these nodes are used for *ECU* scheduling, and κ, V and D are the nodes identifiers. The task scheduling inside each wheel ECU, is modeled by defining some other nodes. Since only one task can run at each time tick, each node will correspond to a task as follows:

$$\begin{aligned}
&\Delta^t(\kappa, v) = \{\kappa \textit{ wheel velocity}\}, \\
&\Delta^t(\kappa, b) = \{\ \kappa \textit{ brake control}\}, \\
&\Delta^t(\kappa, u) = \{\kappa \textit{ status update}\}, \\
&\forall \kappa \in \Theta, \quad \Theta = \{FL,FR,RL,RR\},
\end{aligned} \tag{3.89}$$

where the superscript 't' is to indicate that these nodes are used for *task* scheduling. Now, let us define five periodic scheduling sequences (one for the ECU scheduling and four for each wheel task scheduling) as follows:

$$\begin{aligned}
&\sigma^E = \{\sigma^E(0), \sigma^E(1), \ldots, \sigma^E(p_E - 1)\}, \\
&\sigma^t_\kappa = \{\sigma^t_\kappa(0), \sigma^t_\kappa(1), \ldots, \sigma^t_\kappa(p_\kappa - 1)\}, \quad \forall \kappa \in \Theta, \quad \Theta = \{FL,FR,RL,RR\},
\end{aligned} \tag{3.90}$$

where p_E and p_κ, for all $\kappa \in \Theta$, are the periods of the bus scheduling and task scheduling sequences respectively.

Similar to Example 3.2, this is an NCS with multi-networks (tasks scheduling) that are subnetworks of the ECU network. Since the task scheduling forms a multi-network, the four sequences and the four nodes sets can be reduced into one sequence, σ^t, and one set of nodes, Δ^t, respectively as shown in Section 3.6. The period of σ^t is $p_t = \text{LCM}(p_{FL}, p_{FR}, p_{RL}, p_{RR})$. The communication networks $N^E = (\Delta^E, \sigma^E)$ and $N^t = (\Delta^t, \sigma^t)$ fully characterize the ECU scheduling and the task scheduling (in each ECU) respectively. N^E is the 0th level network and N^t the 1st level network according to the definition of subnetwork levels in Section 3.6.2.

Given N^E, actuator, sensor and demand schedulers can be constructed as shown in Definition 3.5. They will be called $\mathcal{S}^E_A$, $\mathcal{S}^E_S$ and $\mathcal{S}^E_D$ respectively. These are the schedulers that model the bus access dynamics. Similarly, actuator and sensor schedulers can be constructed for N^t. They are $\mathcal{S}^t_A$ and $\mathcal{S}^t_S$ respectively (there is no demand scheduler in this case as there is no task in the ECUs which deals with demand signals). These are the schedulers that model the task execution dynamics inside each wheel node. The closed-loop block of the brake-by-wire systems with bus and task scheduling is shown in Figure 3.11. $\mathcal{P}$ is the vehicle dynamics, $\mathcal{K}_G$ is the global controller (in the Master ECU) and $\mathcal{K}_L$ is the local ABS controllers in each wheel node. The four *Wheel velocity* signals are scheduled by the local task scheduler $\mathcal{S}^t_S$. Then, together with the *vehicle linear velocity* signal, they are scheduled by the bus scheduler $\mathcal{S}^E_S$. This and the bus scheduled *Brake demand* signal (via $\mathcal{S}^E_D$) are the input of the global controller $\mathcal{K}_G$. The *Global control* signal, before reaching the local controllers $\mathcal{K}_L$ (which is in fact a diagonally structured controller), is scheduled by the bus scheduler $\mathcal{S}^E_A$. The local controller also takes as an input the *Wheel velocity* that, since it is measured in the same ECU, is only subject to the task scheduler $\mathcal{S}^t_S$. Finally, the four *Wheel brake* signals are subject to the task scheduler $\mathcal{S}^t_A$ before reaching the vehicle, $\mathcal{P}$.

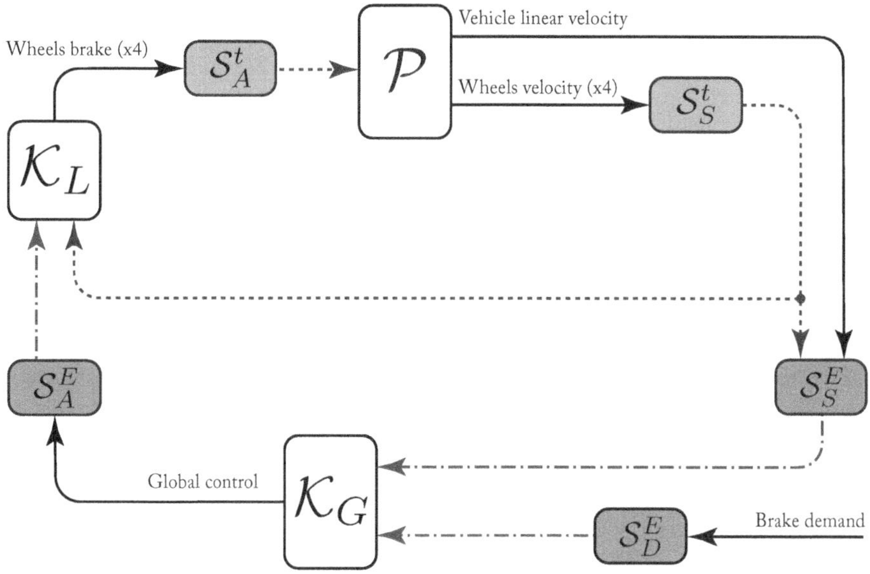

FIGURE 3.11: Block diagram of the brake-by-wire system with scheduling: the dashed lines are the signals subject to task scheduling and the dash-dot lines are the signals subject to bus scheduling.

Figure 3.11 shows how the distributed networked system with task scheduling in Figure 3.10 can be represented by standard block diagrams. The different dashed lines in Figure 3.11 are only to highlight signals subject to different scheduling sequences, nonetheless they are standard signals. Since all the blocks in Figure 3.11 can be represented as state space models, it is only a manipulation exercise to transform the diagram in Figure 3.11 into a more standard closed-loop system. For example, the local feedback signal (dashed line) can be fed to $\mathcal{K}_L$ via $\mathcal{S}_S^E$, $\mathcal{K}_G$, and $\mathcal{S}_A^E$, by appropriately augmenting $\mathcal{S}_S^E$, $\mathcal{K}_G$ and $\mathcal{S}_A^E$ system matrices. The task scheduler $\mathcal{S}_S^t$ can also be augmented to incorporate the *vehicle linear velocity* and the four *Wheel velocity* signals.

It should be noticed that modeling the closed-loop networked system in Figure 3.11 is only possible if the communication sequences in (3.90) are given. Methods for optimal/robust design of such sequences is the topic of Chapters 6, 7 and 8. Moreover, in the next chapter, the structural properties of NCSs subject to communication sequencing will be investigated.

3.10 Summary

Recently, a considerable effort has been made in trying to unify various results and formulate a complete theory for NCSs. This chapter has presented a general unified framework for the modeling of the limited communication in NCSs. It was shown how the communication dynamics can be described by a special time-varying graph characterized by the pair (Δ, σ) where Δ is the set of nodes associated to the system's sensors and actuators and σ is the communication sequence. This framework allowed the analysis for standard networks to be extended to the case of multi-networks, subnetworks and task scheduling in a natural way. It turned out that the method proposed efficiently models special cases of systems with deterministic switchings and systems with delays. Examples were used to show how to apply the method to practical situations.

4

Controllability and observability

CONTENTS

Loosely speaking, the ability to arbitrarily move a system around a space is known as controllability. For example, can a juggler balance two independent sticks on his hand by moving the hand back-forward, left-right? For a sufficiently skilled juggler, the answer is yes... but only if the two sticks have different lengths! The problem is that the two-stick system becomes uncontrollable if the sticks have the same length.

If a particular combination of parameters (the stick length in the above example) can render a system uncontrollable, what happens when systems are connected to controllers via a network with limited communication? Are there particular communication policies that must be avoided to be able to retain some desirable structural properties of the control system?

4.1 Introduction

In the previous chapters, the limited communication arising in NCSs was discussed. In the type of NCSs considered, sensors and actuators are connected to a feedback controller through a shared communication medium with time-triggered (contention-free) communication protocol. Hence, a communication sequence needs to be assigned to the NCS, before run-time, to schedule the medium access of the system's components.

Recently, significant attention has been given to the stabilization problem of NCSs under medium access constraints. Because of the scheduled communication, not only does a stabilizing controller need to be designed but also a communication sequence that preserves stability when the plant is controlled through a network. Important results on control and communication scheduling in linear NCSs have been achieved in [150, 56, 51, 52]. It was proved that such communication sequences exist, they can be periodic, have an upper bound on their period and can be found by a simple iterative algorithm. The communication and control codesign problem is solved by first designing a stabilizing periodic communication sequence and afterwards a controller. Structural properties (reachability and observability) of linear NCSs are discussed in [150, 56] where no Zero-Order-Hold (ZOH) element is placed between the controller and the plant (non-communicating sensor and actuator signals are reset to zero). In [51, 52], the structural analysis is also extended to include a ZOH strategy and improved results in terms of a shorter period of the communication sequence are presented. The approach of [150, 56, 51, 52] is to provide iterative algorithms which allow successive construction of a communication sequence that guarantees controllability (observability) of the NCS for a given controllable (observable) plant. An upper bound for the length of the constructed sequence is given.

It was shown in the previous chapter that multirate systems are a special case of NCSs (Section 3.7, p. 77). The stabilizability and detectability properties of a multirate sampled-data system are discussed in [26].

In this chapter, the controllability and observability (and the weaker stabilizability and detectability) properties of a class of NCSs are reconsidered and advanced. The focus is mainly on the more realistic ZOH case where all the non-updated control signals remain the same until next updated. The main question to answer is: given a discrete-time stabilizable plant linear model, what are those communication sequences that preserve the system's controllability properties when the system is implemented as an NCS? Once these properties are defined, it is easy to find a scheduler with a fixed and periodic communication sequence that preserves controllability. It turns out that the approach is readily extendable to the case where no ZOH element is placed between controller and actuators and to the case where also the sensor signals are scheduled.

The problem of finding a sensor communication scheduling sequence that preserves observability is the dual problem and the equivalent result will be discussed. Thus, an improved solution compared to [150, 56, 51, 52] is presented. More specifically, some rules will be formulated for the appropriate selection of communication sequences. It will be shown that this approach can provide sequences that are 'minimal' in length (and still preserve the structural properties) and therefore shorter than previously established by [150, 56, 51, 52].

4.2 NCSs with ZOH

The controllability analysis of NCSs is first considered and then the results are extended to the weaker stabilizability problem. Observability (and the weaker detectability) is the dual problem that can be formulated in a similar way, therefore, the results naturally follow. The assumption is that only actuator signals are subject to bus communication constraints. Moreover, the actuators signals latch so that, when communication is broken, the signal value is retained (ZOH strategy).

Let the distributed plant $\mathcal{P}$ be the discrete LTI system described by

$$\mathcal{P}: \quad \boldsymbol{x}(j+1) = \boldsymbol{A}\boldsymbol{x}(j) + \boldsymbol{B}\hat{\boldsymbol{u}}(j), \tag{4.1}$$

where $\boldsymbol{x}(j) \in \mathbb{R}^{n_x}$, $\hat{\boldsymbol{u}}(j) \in \mathbb{R}^{n_u}$ and j is the sampling instant. $\boldsymbol{A}$ and $\boldsymbol{B}$ are matrices of appropriate dimensions. Notice that the exogenous input and the performance/measurement output equations are not considered because they are irrelevant for the controllability problem.

The next step is to obtain a model of the scheduler that emulates the limited communication channel. This is described in Section 3.3, p. 53. For practical reasons, the sequence is assumed to be periodic. Because the analysis is on controllability, only the actuator scheduler is considered. This can be viewed as a system where sensor and demand signals are directly wired to the controller and therefore are not subject to communication constraints. This networked configuration can be modeled as a set of nodes, $\Delta = \{\Delta(i)\}_{i=1}^{d}$, where each node, $\Delta(i)$, for all i, includes every demand and sensor signal but only a limited number of actuator signals. Hence, for a communication network $N = (\Delta, \sigma)$, the actuator (input) scheduler, $\mathcal{S}_A$, is represented by the periodic system

$$\mathcal{S}_A: \quad \begin{cases} \boldsymbol{x}_A(j+1) = \bar{\boldsymbol{S}}_A(k)\boldsymbol{x}_A(j) + \boldsymbol{S}_A(k)\boldsymbol{u}(j) \\ \hat{\boldsymbol{u}}(j) = \bar{\boldsymbol{S}}_A(k)\boldsymbol{x}_A(j) + \boldsymbol{S}_A(k)\boldsymbol{u}(j) \end{cases}, \tag{4.2}$$

where $\boldsymbol{x}_A(j) = \hat{\boldsymbol{u}}(j-1)$, $\boldsymbol{x}_A(j) \in \mathbb{R}^{n_u}$, $k \stackrel{\text{def}}{=} \text{mod}(j,p)$ and p is the sequence period. Recall that $\sigma = \{\sigma(k)\}_{k=0}^{p-1}$ is the p-periodic communication sequence (a sequence of nodes) and the pair N completely characterizes the network (scheduler) dynamics.

The plant $\mathcal{P}$ state evolution, with the limited communication, can be described by the *augmented plant*

$$\hat{\mathcal{P}}: \quad \hat{\boldsymbol{x}}(j+1) = \hat{\boldsymbol{A}}(k)\hat{\boldsymbol{x}}(j) + \hat{\boldsymbol{B}}(k)\boldsymbol{u}(j), \tag{4.3}$$

where

$$\hat{\boldsymbol{x}} = \begin{bmatrix} \boldsymbol{x}(j) \\ \boldsymbol{x}_A(j) \end{bmatrix} = \begin{bmatrix} \boldsymbol{x}(j) \\ \hat{\boldsymbol{u}}(j-1) \end{bmatrix},$$

$$\hat{\boldsymbol{A}}(k) = \begin{bmatrix} \boldsymbol{A} & \boldsymbol{B}\bar{\boldsymbol{S}}_A(k) \\ \boldsymbol{0} & \bar{\boldsymbol{S}}_A(k) \end{bmatrix}, \quad \hat{\boldsymbol{B}}(k) = \begin{bmatrix} \boldsymbol{B}\boldsymbol{S}_A(k) \\ \boldsymbol{S}_A(k) \end{bmatrix}, \tag{4.4}$$

which is periodic in p (see Section 3.3.1, p. 58, for more detail). Periodicity is removed by applying the discrete-time lifting operator

$$\bar{\mathcal{P}} = \mathbb{L}_D\{\hat{\mathcal{P}}, p\}. \tag{4.5}$$

$\bar{\mathcal{P}}$ is an LTI equivalent system that models the original plant plus the actuator's limited communication. It is given by

$$\bar{\mathcal{P}}: \quad \bar{\boldsymbol{x}}(pl+p) = \tilde{\boldsymbol{A}}_p \hat{\boldsymbol{x}}(pl) + \tilde{\boldsymbol{B}}_p \bar{\boldsymbol{u}}(pl), \tag{4.6}$$

where

$$\bar{\boldsymbol{u}}(pl) = \begin{bmatrix} \boldsymbol{u}(pl) \\ \boldsymbol{u}(pl+1) \\ \vdots \\ \boldsymbol{u}(pl+p-1) \end{bmatrix},$$

$$\tilde{\boldsymbol{A}}_p = \prod_{\iota=1}^{p} \hat{\boldsymbol{A}}(p-\iota), \quad \tilde{\boldsymbol{B}}_p = \begin{bmatrix} \boldsymbol{G}(0) & \boldsymbol{G}(1) & \cdots & \boldsymbol{G}(p-1) \end{bmatrix}, \tag{4.7}$$

and[1]

$$\boldsymbol{G}(k) = \left(\prod_{\iota=1}^{p-k-1} \hat{\boldsymbol{A}}(p-\iota) \right) \hat{\boldsymbol{B}}(k), \quad k = 0, 1, \ldots, p-1, \tag{4.8}$$

(see Section 3.5, p. 63, for the detail of the discrete-time lifting).

4.2.1 Controllability and stabilizability

At this point, the conditions for a communication sequence that guarantees the preservation of controllability of the controllable modes of the lifted augmented system in (4.6) can be stated. Stabilizability can be easily proved by ensuring that at least all the unstable modes retain controllability. The aim is to define those communication sequences, such that, for a given controllable pair $(\boldsymbol{A}, \boldsymbol{B})$ (the original plant)

$$(\boldsymbol{A}, \boldsymbol{B}) \textit{ controllable} \quad \Rightarrow \quad (\tilde{\boldsymbol{A}}_p, \tilde{\boldsymbol{B}}_p) \textit{ controllable} \tag{4.9}$$

where the pair $(\tilde{\boldsymbol{A}}_p, \tilde{\boldsymbol{B}}_p)$ is the equivalent system that includes the plant and the communication network dynamics.

Sufficient conditions will be presented for which the following definitions are needed first.

[1]Note that in this equation, when $k = p-1$, the product is undefined. For this reason, to allow for this short notation it must be defined, in the context of this book, $\prod_{j=1}^{i} f(j) = 1$ for $i < j$.

Definition 4.1 *The matrices*

$$\boldsymbol{S}_A(k) \overset{def}{=} \text{diag}(s_{A,1}(k), s_{A,1}(k), \ldots, s_{A,n_u}(k)) \tag{4.10}$$

and

$$\bar{\boldsymbol{S}}_A(k) \overset{def}{=} \text{diag}(\bar{s}_{A,1}(k), \bar{s}_{A,1}(k), \ldots, \bar{s}_{A,n_u}(k)) \tag{4.11}$$

can be decomposed into the matrices

$$\boldsymbol{Z}^{(i)}(k) \overset{def}{=} \text{diag}(z_1^{(i)}(k), z_1^{(i)}(k), \ldots, z_{n_u}^{(i)}(k)) \tag{4.12}$$

and

$$\bar{\boldsymbol{Z}}^{(i)}(k) \overset{def}{=} \text{diag}(\bar{z}_1^{(i)}(k), \bar{z}_1^{(i)}(k), \ldots, \bar{z}_{n_u}^{(i)}(k)) \tag{4.13}$$

respectively where

$$\begin{aligned} z_r^{(i)}(k) &= \begin{cases} s_{A,r}(k) & \textit{if} \quad r = i \\ 0 & \textit{otherwise} \end{cases}, \\ \bar{z}_r^{(i)}(k) &= \begin{cases} \bar{s}_{A,r}(k) & \textit{if} \quad r = i \\ 0 & \textit{otherwise} \end{cases}, \quad \forall r, i = 1, 2, \ldots, n_u. \end{aligned} \tag{4.14}$$

▲

Note that $\sum_{i=1}^{n_u} \boldsymbol{Z}^{(i)}(k) = \boldsymbol{S}_A(k)$ and $\sum_{i=1}^{n_u} \bar{\boldsymbol{Z}}^{(i)}(k) = \bar{\boldsymbol{S}}_A(k)$. Decomposing the scheduling matrix $\boldsymbol{S}_A(k)$ into n_u matrices $\boldsymbol{Z}^{(i)}(k)$ allows for a sequence of matrices for each actuator with only one 1 in the diagonal, corresponding to the scheduled actuator. This will be useful for the formulation of the controllability conditions.

Definition 4.2 *For a sequence σ, $k_f^{(i)}$ are the sequence indices where actuator i is scheduled for the last time in the sequence,* i.e. *when*

$$\left\{n : n = 1, 2, \ldots, p - k_f^{(i)} - 1; s_{A,i}(k_f^{(i)}) = s_{A,i}(k_f^{(i)} + n) = 1, \forall i\right\} = \varnothing. \tag{4.15}$$

▲

Remark 4.1 *From Definition 4.2, it follows that, if*

$$\text{rank}\left(\sum_{k=0}^{p-1} \boldsymbol{S}_A(k)\right) = n_u, \tag{4.16}$$

*there will be n_u (not necessarily distinct) $k_f^{(i)}$ associated with n_u distinct matrices $\boldsymbol{Z}^{(i)}(k_f^{(i)})$ (*i.e. rank $\left(\sum_{i=1}^{n_u} \boldsymbol{Z}^{(i)}(k_f^{(i)})\right) = n_u$*).* •

Definition 4.3 *If* $\operatorname{rank}\left(\sum_{k=0}^{p-1} \boldsymbol{S}_A(k)\right) = n_u$*, then*

$$\bar{q}^{(i)} = \min_{\{\iota : \prod_{l=0}^{\iota-1} \bar{\boldsymbol{Z}}^{(i)}(l)=\boldsymbol{0}\}} \iota, \qquad \hat{q}^{(i)}(k) = \min\left\{\bar{q}^{(i)} - 1, k\right\}. \tag{4.17}$$

▲

The following example should help clarify the matter.

Example 4.1 *Consider a communication network* $N = (\Delta, \sigma)$*. The set of nodes is* $\Delta = \{\Delta(1), \Delta(2), \Delta(3)\}$ *where* $\Delta(i)$ *is the node associated to actuator* i *and* $i = 1, 2, 3$ *(*n_u*=3). The communication sequence is* $\sigma = \{\Delta(1), \Delta(2), \Delta(3), \Delta(1), \Delta(2)\}$ *(*$p = 5$*). In this case, each node is associated to one signal only (*i.e. *only one actuator is controlled at any time tick), therefore*

$$\boldsymbol{S}(0) = \boldsymbol{Z}^{(1)}(0) = \begin{bmatrix} 1 & 0 & 0 \\ 0 & 0 & 0 \\ 0 & 0 & 0 \end{bmatrix}, \quad \boldsymbol{S}(1) = \boldsymbol{Z}^{(2)}(1) = \begin{bmatrix} 0 & 0 & 0 \\ 0 & 1 & 0 \\ 0 & 0 & 0 \end{bmatrix},$$

$$\ldots, \quad \boldsymbol{S}(4) = \boldsymbol{Z}^{(2)}(4) = \begin{bmatrix} 0 & 0 & 0 \\ 0 & 1 & 0 \\ 0 & 0 & 0 \end{bmatrix},$$

$$\boldsymbol{Z}^{(2)}(0) = \boldsymbol{Z}^{(3)}(0) = \boldsymbol{Z}^{(1)}(1) = \boldsymbol{Z}^{(3)}(1) = \ldots = \boldsymbol{Z}^{(3)}(4) = \boldsymbol{0},$$

$$\bar{\boldsymbol{S}}(0) = \begin{bmatrix} 0 & 0 & 0 \\ 0 & 1 & 0 \\ 0 & 0 & 1 \end{bmatrix}, \quad \bar{\boldsymbol{S}}(1) = \begin{bmatrix} 1 & 0 & 0 \\ 0 & 0 & 0 \\ 0 & 0 & 1 \end{bmatrix}, \quad \ldots, \quad \bar{\boldsymbol{S}}(4) = \begin{bmatrix} 1 & 0 & 0 \\ 0 & 0 & 0 \\ 0 & 0 & 1 \end{bmatrix},$$

$$\bar{\boldsymbol{Z}}^{(2)}(0) = \begin{bmatrix} 0 & 0 & 0 \\ 0 & 1 & 0 \\ 0 & 0 & 0 \end{bmatrix}, \quad \bar{\boldsymbol{Z}}^{(3)}(0) = \begin{bmatrix} 0 & 0 & 0 \\ 0 & 0 & 0 \\ 0 & 0 & 1 \end{bmatrix},$$

$$\ldots, \quad \bar{\boldsymbol{Z}}^{(3)}(4) = \begin{bmatrix} 0 & 0 & 0 \\ 0 & 0 & 0 \\ 0 & 0 & 1 \end{bmatrix},$$

$$\bar{\boldsymbol{Z}}^{(1)}(0) = \bar{\boldsymbol{Z}}^{(2)}(1) = \ldots = \bar{\boldsymbol{Z}}^{(2)}(4) = \boldsymbol{0}. \tag{4.18}$$

Also, $\bar{q}^{(i)} = 3$ *for* $i = 3$ *according to Definitions 4.1 and 4.3. For* $i = 3$*, according to Definition 4.2,* $k_f^{(i)} = 2$ *and consequently* $\hat{q}^{(i)}(k_f^{(i)}) = 2$*.* ♦

The following theorem gives conditions for the controllability of the system in (4.6).

Theorem 4.1 (Controllability with ZOH) *Suppose that the pair* $(\boldsymbol{A}, \boldsymbol{B})$ *in (4.1) is controllable,* $\boldsymbol{A}$ *is invertible and* $\operatorname{rank}(\boldsymbol{B}) = n_u$*. Moreover, all* n_u *actuators are required for controllability of the plant,* i.e. $(\boldsymbol{A}, \boldsymbol{B}_r)$ *is not controllable where* $\boldsymbol{B}_r$ *is an input distribution matrix covering only part of the input range of* $\boldsymbol{B}$*. Under these conditions the following can be stated:*

(i) the pair $(\tilde{\boldsymbol{A}}_p, \tilde{\boldsymbol{B}}_p)$ (the lifted augmented system) is controllable if

$$\operatorname{rank}\left(\sum_{k=0}^{p-1} \boldsymbol{S}_A(k)\right) = n_u, \tag{4.19}$$

$$\lambda \neq \exp\left(\frac{2\pi l\sqrt{-1}}{p - k_f^{(i)} + \hat{q}^{(i)}(k_f^{(i)})}\right),$$
$$\forall i = 1, 2, \ldots, n_u, \quad l = 1, 2, \ldots, p - k_f^{(i)} + \hat{q}^{(i)}(k_f^{(i)}) - 1, \tag{4.20}$$

for any arbitrary eigenvalue λ of $\boldsymbol{A}$ and[2] *assuming that, for any pair of eigenvalues, (λ_1, λ_2) of $\boldsymbol{A}$, the following holds*

$$\lambda_1 \neq \lambda_2 \exp\left(\frac{2\pi l\sqrt{-1}}{p}\right), \quad l = \ldots, -1, 0, 1, 2, \ldots \tag{4.21}$$

(ii) the pair $(\tilde{\boldsymbol{A}}_p, \tilde{\boldsymbol{B}}_p)$ is not controllable if

$$\operatorname{rank}\left(\sum_{k=0}^{p-1} \boldsymbol{S}_A(k)\right) < n_u. \tag{4.22}$$

Proof. Sufficient conditions to guarantee controllability of the lifted augmented system $\bar{\mathcal{P}}$ will be established. The following definitions are needed first.

Definition 4.4 *Given the sequence of matrices*

$$Z^{(i)} = \left\{\boldsymbol{Z}^{(i)}(0), \boldsymbol{Z}^{(i)}(1), \ldots, \boldsymbol{Z}^{(i)}(p-1)\right\}, \tag{4.23}$$

(Definition 4.1), let $Q_d^{(i)}(k)$ be the set of the sequence position distances from the matrix element $z_i^{(i)}(k)$ to the next repeated element $z_i^{(i)}(k+n)$ (if one exists), i.e.

$$Q_d^{(i)}(k) = \{n : n = 1, 2, \ldots, p-k-1; z_i^{(i)}(k) = z_i^{(i)}(k+n) = 1\}. \tag{4.24}$$

Then, define $q^{(i)}(k) = 1, 2, \ldots, p-k-1$ for all $k = 0, 1, \ldots, p-2$ as

$$q^{(i)}(k) = \begin{cases} \min_{q \in Q_d^{(i)}(k)} q & \text{if } Q_d^{(i)}(k) \neq \varnothing \\ p - k & \text{otherwise} \end{cases}. \tag{4.25}$$

▲

[2] Note that condition (4.20) is void for $p - k_f^{(i)} + \hat{q}^{(i)}(k_f^{(i)}) \leq 1$.

Note that $Q_d^{(i)}(p-1) = \varnothing \Rightarrow q^{(i)}(p-1) = p - k = 1$ for all i.

Definition 4.5 *With reference to Definitions 4.1 and 4.4, let*

$$\hat{\boldsymbol{Z}}^{(i)}(k) = \begin{cases} \boldsymbol{Z}^{(i)}(k) & \text{if } Q_d^{(i)}(k) = \varnothing \\ \boldsymbol{0}_{n_u \times n_u} & \text{otherwise} \end{cases}, \quad i = 1, 2, \ldots, n_u. \tag{4.26}$$

▲

Moreover, the following auxiliary Lemma is required.

Lemma 4.1 *If the pair $(\boldsymbol{A}, \boldsymbol{B})$ in (4.1) is controllable and any pair of eigenvalues, (λ_1, λ_2) of $\boldsymbol{A}$ satisfies (4.21) then:*

(i) The pair $(\boldsymbol{A}^p, \boldsymbol{B})$ is controllable[3]

(ii) The left hand nonzero eigenvectors of $\boldsymbol{A}^p$ are also the nonzero eigenvectors, $\boldsymbol{v}$, of $\boldsymbol{A}$ satisfying

$$\boldsymbol{v}\boldsymbol{A} = \lambda \boldsymbol{v}, \quad \boldsymbol{v}\boldsymbol{A}^p = \lambda^p \boldsymbol{v}, \quad \boldsymbol{v}\Gamma \neq 0. \tag{4.27}$$

The Lemma is a direct consequence of Remark 4.5 and the proof of [20, Theorem 3.2.1].

For the proof, the structure of $\tilde{\boldsymbol{A}}_p$ and $\tilde{\boldsymbol{B}}_p$ will be first defined. Then, the Popov-Belevitch-Hautus (PBH) eigenvector test is used to demonstrate the controllability of the system by defining some characteristic polynomials. If every actuator signal is updated during the sequence at least once, the result is that the system is proven to be controllable except when some particular eigenvalues of $\boldsymbol{A}$ appear.

Part (i). From (4.7) and (4.4) it follows that

$$\tilde{\boldsymbol{A}}_p = \prod_{\iota=1}^{p} \hat{\boldsymbol{A}}(p-\iota) = \begin{bmatrix} \boldsymbol{A}^p & \tilde{\boldsymbol{A}}_{p,12} \\ \boldsymbol{0} & \prod_{k=1}^{p} \bar{\boldsymbol{S}}_A(p-k) \end{bmatrix}, \tag{4.28}$$

where

$$\tilde{\boldsymbol{A}}_{p,12} = \sum_{\iota=1}^{p} \left(\boldsymbol{A}^{p-\iota} \boldsymbol{B} \prod_{l=0}^{\iota-1} \bar{\boldsymbol{S}}_A(l) \right). \tag{4.29}$$

Since $\bar{\boldsymbol{S}}_A(k)$ is diagonal, (4.29) can be decomposed in a further summation giving (see Definition 4.1)

$$\tilde{\boldsymbol{A}}_{p,12} = \sum_{i=1}^{n_u} \sum_{\iota=1}^{p} \left(\boldsymbol{A}^{p-\iota} \boldsymbol{B} \prod_{l=0}^{\iota-1} \bar{\boldsymbol{Z}}^{(i)}(l) \right), \tag{4.30}$$

[3] $\boldsymbol{A}^p = \underbrace{\boldsymbol{A}\boldsymbol{A}\cdots\boldsymbol{A}}_{p \text{ times}}$

but, because of Definition 4.3[4]

$$\tilde{\boldsymbol{A}}_{p,12} = \sum_{i=1}^{n_u} \sum_{\iota=1}^{\bar{q}^{(i)}-1} \left(\boldsymbol{A}^{p-\iota} \boldsymbol{B} \prod_{l=0}^{\iota-1} \bar{\boldsymbol{Z}}^{(i)}(l) \right). \tag{4.31}$$

Also, since rank $\left(\sum_{k=0}^{p-1} \boldsymbol{S}_A(k)\right) = n_u$, the bottom right of the matrix in (4.28) becomes

$$\prod_{k=1}^{p} \bar{\boldsymbol{S}}_A(p-k) = 0, \tag{4.32}$$

and therefore, from (4.28),

$$\tilde{\boldsymbol{A}}_p = \begin{bmatrix} \boldsymbol{A}^p & \tilde{\boldsymbol{A}}_{p,12} \\ \boldsymbol{0} & \boldsymbol{0} \end{bmatrix}. \tag{4.33}$$

Now, consider the input distribution matrix in (4.7) together with the augmented system matrices in (4.4) to obtain

$$\begin{aligned} \boldsymbol{G}(k) &= \left(\prod_{\iota=1}^{p-k-1} \hat{\boldsymbol{A}}(p-\iota) \right) \hat{\boldsymbol{B}}(k) \\ &= \begin{bmatrix} \sum_{\iota=1}^{p-k} \left(\boldsymbol{A}^{p-k-\iota} \boldsymbol{B} \prod_{l=1}^{\iota-1} \bar{\boldsymbol{S}}_A(k+\iota-l) \right) \boldsymbol{S}_A(k) \\ \left(\prod_{l=1}^{p-k-1} \bar{\boldsymbol{S}}_A(p-l) \right) \boldsymbol{S}_A(k) \end{bmatrix}. \end{aligned} \tag{4.34}$$

The matrix $\boldsymbol{G}(k)$ can be also decomposed and written as

$$\boldsymbol{G}(k) = \sum_{i=1}^{n_u} \begin{bmatrix} \sum_{\iota=1}^{p-k} \left(\boldsymbol{A}^{p-k-\iota} \boldsymbol{B} \prod_{l=1}^{\iota-1} \bar{\boldsymbol{Z}}^{(i)}(k+\iota-l) \right) \boldsymbol{Z}^{(i)}(k) \\ \left(\prod_{l=1}^{p-k-1} \bar{\boldsymbol{Z}}^{(i)}(p-l) \right) \boldsymbol{Z}^{(i)}(k) \end{bmatrix}. \tag{4.35}$$

Because of Definitions 4.1, 4.4 and 4.5, if $z_i^{(i)}(k) = z_i^{(i)}(k + q^{(i)}(k)) = 1$ then $\bar{\boldsymbol{Z}}^{(i)}(k + q^{(i)}(k))\boldsymbol{Z}^{(i)}(k) = \boldsymbol{0}$ and $\bar{\boldsymbol{Z}}^{(i)}(n)\bar{\boldsymbol{Z}}^{(i)}(n-1)\ldots\bar{\boldsymbol{Z}}^{(i)}(k + q^{(i)}(k))\ldots\bar{\boldsymbol{Z}}^{(i)}(k) = \boldsymbol{0}$, hence

$$\boldsymbol{G}(k) = \sum_{i=1}^{n_u} \begin{bmatrix} \sum_{\iota=1}^{q^{(i)}(k)} \boldsymbol{A}^{p-k-\iota} \boldsymbol{B} \boldsymbol{Z}^{(i)}(k) \\ \hat{\boldsymbol{Z}}^{(i)}(k) \end{bmatrix} \tag{4.36}$$

and

$$\tilde{\boldsymbol{B}}_p = \sum_{i=1}^{n_u} \begin{bmatrix} \sum_{\iota=1}^{q^{(i)}(0)} \boldsymbol{A}^{p-\iota} \boldsymbol{B} \boldsymbol{Z}^{(i)}(0) & \sum_{\iota=1}^{q^{(i)}(1)} \boldsymbol{A}^{p-1-\iota} \boldsymbol{B} \boldsymbol{Z}^{(i)}(1) & \cdots & \boldsymbol{B}\boldsymbol{Z}^{(i)}(p-1) \\ \hat{\boldsymbol{Z}}^{(i)}(0) & \hat{\boldsymbol{Z}}^{(i)}(1) & \cdots & \hat{\boldsymbol{Z}}^{(i)}(p-1) \end{bmatrix}. \tag{4.37}$$

[4]Note that in (4.31), (4.34) and (4.35) and later in (4.44) and (4.45) the summation $\sum_{\iota=1}^{i} f(\iota)$ will be undefined if $i < 1$. For this reason, to allow for this short notation, it must be defined in the context of this book, $\sum_{\iota=1}^{i} f(\iota) = 0$ for $i < 1$.

From the PBH eigenvector test [61, p. 135], it follows that the pair $(\tilde{\boldsymbol{A}}_p, \tilde{\boldsymbol{B}}_p)$ is not controllable if and only if there exists a row vector $\bar{\boldsymbol{v}} \neq \boldsymbol{0}$ such that

$$\bar{\boldsymbol{v}}\tilde{\boldsymbol{A}}_p = \bar{\lambda}\bar{\boldsymbol{v}} \quad \text{and} \quad \bar{\boldsymbol{v}}\tilde{\boldsymbol{B}}_p = \boldsymbol{0}. \tag{4.38}$$

What is needed to be proven is that there is no left eigenvector of $\tilde{\boldsymbol{A}}_p$ that is orthogonal to $\tilde{\boldsymbol{B}}_p$. Because the lower part of $\tilde{\boldsymbol{A}}_p$ is zero, the nonzero eigenvalues λ_p of $\tilde{\boldsymbol{A}}_p$ are given by the eigenvalues of $\boldsymbol{A}^p$. First consider the nonzero eigenvalues, λ_p. Recall that any eigenvector $\boldsymbol{v}$ of $\boldsymbol{A}$ is also an eigenvector of $\boldsymbol{A}^r$, $r \in \mathbb{Z}$ and $\boldsymbol{v}\boldsymbol{A}^r = \lambda^r\boldsymbol{v}$, where λ is the eigenvalue of $\boldsymbol{A}$. Hence, by Lemma 4.1, the eigenvectors of $\tilde{\boldsymbol{A}}_p$ are given by $\bar{\boldsymbol{v}}_1 = [\boldsymbol{v}_1 \ \boldsymbol{v}_2]$. This implies

$$\bar{\boldsymbol{v}}_1\tilde{\boldsymbol{A}}_p = \bar{\lambda}\bar{\boldsymbol{v}}_1. \tag{4.39}$$

Since

$$\bar{\boldsymbol{v}}_1\tilde{\boldsymbol{A}}_p = \begin{bmatrix} \boldsymbol{v}_1 & \boldsymbol{v}_2 \end{bmatrix} \begin{bmatrix} \boldsymbol{A}^p & \boldsymbol{A}_{p,12} \\ \boldsymbol{0} & \boldsymbol{0} \end{bmatrix} = \begin{bmatrix} \boldsymbol{v}_1\boldsymbol{A}^p & \boldsymbol{v}_1\boldsymbol{A}_{p,12} \end{bmatrix} = \lambda^p \begin{bmatrix} \boldsymbol{v}_1 & \frac{\boldsymbol{v}_1\boldsymbol{A}_{p,12}}{\lambda^p} \end{bmatrix}, \tag{4.40}$$

and

$$\bar{\lambda}\bar{\boldsymbol{v}}_1 = \bar{\lambda}\begin{bmatrix} \boldsymbol{v}_1 & \boldsymbol{v}_2 \end{bmatrix}, \tag{4.41}$$

it follows that $\lambda^p = \bar{\lambda}$ and

$$\boldsymbol{v}_2 = \frac{\boldsymbol{v}_1}{\lambda^p}\sum_{i=1}^{n_u}\sum_{\iota=1}^{\bar{q}^{(i)}-1}\left(\boldsymbol{A}^{p-\iota}\boldsymbol{B}\prod_{l=0}^{\iota-1}\bar{\boldsymbol{Z}}^{(i)}(l)\right) = \sum_{i=1}^{n_u}\sum_{\iota=1}^{\bar{q}^{(i)}-1}\left(\lambda^{-\iota}\boldsymbol{v}_1\boldsymbol{B}\prod_{l=0}^{\iota-1}\bar{\boldsymbol{Z}}^{(i)}(l)\right). \tag{4.42}$$

Now, consider

$$\bar{\boldsymbol{v}}_1\tilde{\boldsymbol{B}}_p = \begin{bmatrix}\boldsymbol{U}(0) & \boldsymbol{U}(1) & \cdots & \boldsymbol{U}(p-1)\end{bmatrix}, \tag{4.43}$$

and let $k = 0, 1, \ldots, p-1$. Then, every block $\boldsymbol{U}(k)$ has the structure

$$\begin{aligned} \boldsymbol{U}(k) &= \begin{bmatrix}\boldsymbol{v}_1 & \boldsymbol{v}_2\end{bmatrix}\boldsymbol{G}(k) \\ &= \boldsymbol{v}_1\sum_{i=1}^{n_u}\sum_{\iota=1}^{q^{(i)}(k)}\boldsymbol{A}^{p-k-\iota}\boldsymbol{B}\boldsymbol{Z}^{(i)}(k) \\ &\quad + \left(\sum_{i=1}^{n_u}\sum_{\iota=1}^{\bar{q}^{(i)}-1}\left(\lambda^{-\iota}\boldsymbol{v}_1\boldsymbol{B}\prod_{l=0}^{\iota-1}\bar{\boldsymbol{Z}}^{(i)}(l)\right)\right)\hat{\boldsymbol{Z}}^{(i)}(k). \end{aligned} \tag{4.44}$$

Because of Definitions 4.1, 4.3 and 4.5, $\boldsymbol{U}(k)$ can be written as

$$\begin{aligned} \boldsymbol{U}(k) &= \sum_{i=1}^{n_u}\left(\left(\sum_{\iota=1}^{q^{(i)}(k)}\lambda^{p-k-\iota}\right)\boldsymbol{v}_1\boldsymbol{B}\boldsymbol{Z}^{(i)}(k) + \left(\sum_{\iota=1}^{\hat{q}^{(i)}(k)}\lambda^{-\iota}\right)\boldsymbol{v}_1\boldsymbol{B}\hat{\boldsymbol{Z}}^{(i)}(k)\right) \\ &= \sum_{i=1}^{n_u}\mu_{k,i}(\lambda)\boldsymbol{v}_1\boldsymbol{B}\boldsymbol{Z}^{(i)}(k), \end{aligned} \tag{4.45}$$

where

$$\mu_{k,i}(\lambda) = \begin{cases} \sum_{\iota=1}^{q^{(i)}(k)} \lambda^{p-k-\iota} & \text{if} \quad \hat{\boldsymbol{Z}}^{(i)}(k) = 0 \\ \sum_{\iota=1}^{q^{(i)}(k)} \lambda^{p-k-\iota} + \sum_{\iota=1}^{\hat{q}^{(i)}(k)} \lambda^{-\iota} & \text{if} \quad \hat{\boldsymbol{Z}}^{(i)}(k) = \boldsymbol{Z}^{(i)}(k) \end{cases} . \tag{4.46}$$

The vector $\bar{\boldsymbol{v}}_1 \tilde{\boldsymbol{B}}_p$ can be rewritten as

$$\bar{\boldsymbol{v}}_1 \tilde{\boldsymbol{B}}_p = \sum_{i=1}^{n_u} \big[\; \mu_{0,i}(\lambda) \boldsymbol{v}_1 \boldsymbol{B} \boldsymbol{Z}^{(i)}(0) \quad \mu_{1,i}(\lambda) \boldsymbol{v}_1 \boldsymbol{B} \boldsymbol{Z}^{(i)}(1) \\ \cdots \quad \mu_{p-1,i}(\lambda) \boldsymbol{v}_1 \boldsymbol{B} \boldsymbol{Z}^{(i)}(p-1) \; \big] . \tag{4.47}$$

The matrices $\boldsymbol{Z}^{(i)}(k)$ act as selectors of the column vectors of the input distribution matrix $\boldsymbol{B}$. For $\bar{\boldsymbol{v}}_1 \tilde{\boldsymbol{B}}_p \neq 0$, it must be ensured that each column vector is selected at least once via $\boldsymbol{Z}^{(i)}(k)$ and that $\mu_{k,i} \neq 0$.

Consider some specific polynomials which always exist. These polynomials are the ones where $k = k_f^{(i)}$. According to Definitions 4.2, 4.4 and 4.5, $Q_d^{(i)}(k_f^{(i)}) = \varnothing$ and consequently $q^{(i)}(k_f^{(i)}) = p - k_f^{(i)}$ and $\hat{\boldsymbol{Z}}^{(i)}(k_f^{(i)}) = \boldsymbol{Z}^{(i)}(k_f^{(i)})$. Therefore, from (4.46),

$$\begin{aligned} \mu_{k_f^{(i)},i} &= \sum_{\iota=1}^{p-k_f^{(i)}} \lambda^{p-k_f^{(i)}-\iota} + \sum_{\iota=1}^{\hat{q}^{(i)}(k_f^{(i)})} \lambda^{-\iota} = \sum_{\iota=1}^{p-k_f^{(i)}+\hat{q}^{(i)}(k_f^{(i)})} \lambda^{p-k_f^{(i)}-\iota} \\ &= \lambda^{-\hat{q}^{(i)}(k_f^{(i)})} \sum_{\iota=1}^{p-k_f^{(i)}+\hat{q}^{(i)}(k_f^{(i)})} \lambda^{p-k_f^{(i)}+\hat{q}^{(i)}(k_f^{(i)})-\iota} \\ &= \frac{1-\lambda^{p-k_f^{(i)}+\hat{q}^{(i)}(k_f^{(i)})}}{1-\lambda} \lambda^{-\hat{q}^{(i)}(k_f^{(i)})} . \end{aligned} \tag{4.48}$$

Note that for $\lambda = 1$ the term $\mu_{k_f^{(i)},i}$ is not zero, *i.e.* $\mu_{k_f^{(i)},i} = p - k_f^{(i)} + \hat{q}^{(i)}(k_f^{(i)})$. Hence, the fractional expression as on the last line of (4.48) is permissible. Thus, with the assumption of (4.20), the requirement $\mu_{k_f^{(i)},i} \neq 0$ is always satisfied. Because of Remark 4.1, every column vector of $\boldsymbol{B}$ has been selected for $\tilde{\boldsymbol{B}}_p$. This implies $\bar{\boldsymbol{v}}_1 \tilde{\boldsymbol{B}}_p \neq \boldsymbol{0}$ since $\boldsymbol{v}_1 \boldsymbol{B} \neq \boldsymbol{0}$ for controllable $(\boldsymbol{A}, \boldsymbol{B})$, Lemma 4.1 and condition (4.21).

Now, consider the zero eigenvalues of $\tilde{\boldsymbol{A}}_p$ where the eigenvectors are given by $\bar{\boldsymbol{v}}_2 = [\boldsymbol{0} \quad \boldsymbol{v}_2]$ and $\boldsymbol{v}_2$ is any arbitrary nonzero vector. Let

$$\bar{\boldsymbol{v}}_2 \tilde{\boldsymbol{B}}_p = \big[\boldsymbol{V}(0) \quad \boldsymbol{V}(1) \quad \cdots \quad \boldsymbol{V}(p-1) \big] , \tag{4.49}$$

then every block $\boldsymbol{V}(k)$ has the structure

$$\boldsymbol{V}(k) = \big[\boldsymbol{0} \quad \boldsymbol{v}_2 \big] \boldsymbol{G}(k) = \sum_{i=1}^{n_u} \boldsymbol{v}_2 \hat{\boldsymbol{Z}}^{(i)}(k) . \tag{4.50}$$

There will be n_u occasions where $Q_d^{(i)}(k) = \varnothing$ and $\hat{\boldsymbol{Z}}^{(i)}(k) = \boldsymbol{Z}^{(i)}(k) \neq 0$ with n_u distinct $\boldsymbol{Z}^{(i)}(k)$ (this is when $k = k_f^{(i)}$) and therefore $\bar{\boldsymbol{v}}_2\tilde{\boldsymbol{B}}_p \neq 0$. Thus, it is impossible to find a row eigenvector of $\tilde{\boldsymbol{A}}_p$ that is orthogonal to $\tilde{\boldsymbol{B}}_p$ and the pair $(\tilde{\boldsymbol{A}}_p, \tilde{\boldsymbol{B}}_p)$, under the above assumptions, will always be controllable.

Part (ii). From the proof of Part (i), it easily follows that the system is rendered uncontrollable for rank $\left(\sum_{k=0}^{p-1} \boldsymbol{S}(k)\right) < n_u$, *i.e.* the sequence is not feasible according to Definition 4.6. ■

In Theorem 4.1, it is assumed that the plant matrix $\boldsymbol{A}$ has no zero eigenvalues. The following Lemma is for non-invertible plant matrices.

Lemma 4.2 (Controllability with ZOH for non-invertible A) *Suppose that the pair $(\boldsymbol{A}, \boldsymbol{B})$ in (4.1) is controllable. If there exist eigenvalues $\lambda = 0$ of $\boldsymbol{A}$ and a ZOH strategy is used, then the pair $(\tilde{\boldsymbol{A}}_p, \tilde{\boldsymbol{B}}_p)$ in (4.6) is not controllable for the eigenvalue $\lambda = 0$ under any communication sequence.*

Proof. This proof is a natural extension of the proof of Theorem 4.1. Because $\lambda = 0$ is now an eigenvalue of $\boldsymbol{A}$, $\boldsymbol{v}_1\boldsymbol{A} = \boldsymbol{v}_1\boldsymbol{A}^r = \lambda\boldsymbol{v}_1 = \boldsymbol{0}$ for $\lambda = 0$, where $\boldsymbol{v}_1$ is the associated eigenvector and $r \in \mathbb{Z}$. Also, $\boldsymbol{v}_1\boldsymbol{A}_{p,12} = \boldsymbol{0}$ because $\boldsymbol{v}_1\boldsymbol{A}^{p-\iota} = \boldsymbol{0}$ for $1 \leq \iota < p$ and $\prod_{l=0}^{\iota-1} \boldsymbol{S}_A(l) = 0$ for $\iota = p$ (refer to (4.29)). It follows from (4.40) that

$$\bar{\boldsymbol{v}}_1\tilde{\boldsymbol{A}}_p = \begin{bmatrix} \boldsymbol{v}_1 & \boldsymbol{v}_2 \end{bmatrix} \begin{bmatrix} \boldsymbol{A}^p & \tilde{\boldsymbol{A}}_{p,12} \\ \boldsymbol{0} & \boldsymbol{0} \end{bmatrix} = \begin{bmatrix} \boldsymbol{0} & \boldsymbol{0} \end{bmatrix}, \tag{4.51}$$

which implies that $\boldsymbol{v}_2$ can be any arbitrary vector for $[\boldsymbol{v}_1\ \boldsymbol{v}_2]$ to be an eigenvector of $\tilde{\boldsymbol{A}}_p$. Now consider

$$\bar{\boldsymbol{v}}_1\tilde{\boldsymbol{B}}_p = \begin{bmatrix} \boldsymbol{U}(0) & \boldsymbol{U}(1) & \cdots & \boldsymbol{U}(p-1) \end{bmatrix}, \tag{4.52}$$

and let $k = 0, 1, \ldots, p-1$. Then, every block $\boldsymbol{U}(k)$ has the structure

$$\boldsymbol{U}(k) = \begin{bmatrix} \boldsymbol{v}_1 & \boldsymbol{v}_2 \end{bmatrix} \boldsymbol{G}(k) = \boldsymbol{v}_1 \sum_{i=1}^{n_u} \sum_{\iota=1}^{q^{(i)}(k)} \boldsymbol{A}^{p-k-\iota}\boldsymbol{B}\boldsymbol{Z}^{(i)}(k) + \boldsymbol{v}_2\hat{\boldsymbol{Z}}^{(i)}(k). \tag{4.53}$$

It can be easily found that, for $\lambda = 0$,

$$\boldsymbol{U}(k) = \sum_{i=1}^{n_u} \boldsymbol{v}_1\boldsymbol{B}\hat{\boldsymbol{Z}}^{(i)}(k) + \boldsymbol{v}_2\hat{\boldsymbol{Z}}^{(i)}(k) = \sum_{i=1}^{n_u} (\boldsymbol{v}_1\boldsymbol{B} + \boldsymbol{v}_2)\,\hat{\boldsymbol{Z}}^{(i)}(k), \tag{4.54}$$

and it may be chosen $\boldsymbol{v}_2 = -\boldsymbol{v}_1\boldsymbol{B}$. In this case $\boldsymbol{U}(k) = \boldsymbol{0}$. Hence, the pair $(\tilde{\boldsymbol{A}}_p, \tilde{\boldsymbol{B}}_p)$, under the above assumptions, will lose controllability of the mode corresponding to the zero eigenvalue. ■

The following corollary gives conditions for the stabilizability of the system in (4.6).

Corollary 4.1 (Stabilizability with ZOH) *Suppose that the pair* $(\boldsymbol{A}, \boldsymbol{B})$ *in (4.1) is stabilizable and* $\text{rank}(\boldsymbol{B}) = n_u$. *Moreover,* n_u *actuators are required for stabilizability of the plant. Under these conditions the following can be stated:*

(i) the pair $(\tilde{\boldsymbol{A}}_p, \tilde{\boldsymbol{B}}_p)$ *is stabilizable if condition* (4.19) *is satisfied, condition* (4.20) *is satisfied for any arbitrary eigenvalue* $\mid \lambda \mid \geq 1$ *of* $\boldsymbol{A}$, *and condition* (4.21) *is satisfied for any pair of eigenvalues,* (λ_1, λ_2), $\mid \lambda_1 \mid \geq 1$, $\mid \lambda_2 \mid \geq 1$ *of*[5] $\boldsymbol{A}$

(ii) the pair $(\tilde{\boldsymbol{A}}_p, \tilde{\boldsymbol{B}}_p)$ *is not stabilizable if* (4.22) *holds.*

Based on Theorem 4.1 and Corollary 4.1, the idea of 'feasible control sequences' can be now introduced.

Definition 4.6 *A p-periodic sequence* $\sigma = \{\sigma(k)\}_{k=0}^{p-1}$ *will be called feasible if it generates a sequence of matrices* $S_A = \{\boldsymbol{S}_A(k)\}_{k=0}^{p-1}$ *such that*

$$\text{rank}\left(\sum_{k=0}^{p-1} \boldsymbol{S}_A(k)\right) = n_u. \tag{4.55}$$

▲

Condition (4.55) ensures that each actuator control signal is updated at least once during the control sequence.

Definition 4.7 *A p-periodic sequence* $\sigma = \{\sigma(k)\}_{k=0}^{p-1}$ *will be called minimum feasible if the actuators are controlled only once in the sequence,* i.e. *the sequence is feasible according to Definition 4.6 and* $\left|\bigcup_{k=0}^{p-1} \sigma(k)\right| = n_u$. ▲

From Definition 4.3 it follows that, for a minimum feasible sequence, $\bar{q}^{(i)} = k_f^{(i)} + 1 \Rightarrow \hat{q}^{(i)}(k_f^{(i)}) = k_f^{(i)}$. The condition on the eigenvalues, λ, in (4.20), for controllability/stabilizability with minimum feasible sequences, reduces to

$$\lambda \neq \exp\left(\frac{2\pi l\sqrt{-1}}{p}\right), \quad l = 1, 2, \ldots, p-1. \tag{4.56}$$

This applies to round robin schedules where the control signal delay is in fact constant for all actuators.

Remark 4.2 *The minimum feasible sequence provides a tight lower bound on the sequence length p to preserve controllability or stabilizability of the NCS as long as* (4.20) *and* (4.21) *are satisfied. By assuming* $|\sigma(k)| = c$, *for all k, where c is a constant (*i.e. *the number of actuators controlled at any time is constant over the sequence) this lower bound is given by*

$$\beta_l = \left\lceil \frac{n_u}{c} \right\rceil, \tag{4.57}$$

[5] $\mid \lambda \mid$ denotes the complex modulus of λ.

where $\lceil\cdot\rceil$ *is the ceiling function. This strict lower bound on* p*, together with the proof of existence of a sequence at this length, is an improvement to the results of [51, 52, 150, 56] where sequences of length* $\beta_u = \lceil \frac{n_x}{c} \rceil n_x$ *or* $\beta_u = n_x$*,* $(\beta_u \geq \beta_l)$ *are considered*[6] *as it will be shown with some examples in Section 4.5.* •

If only one control signal is updated at any time tick (single-channel case) then $\beta_l = n_u$.

Observation 4.1 *Let* $\boldsymbol{H}_r \in \mathbb{R}^{n_u \times n_u}$*, be a binary diagonal matrix able to select* $r < n_u$ *columns from the full input distribution matrix* $\boldsymbol{B}$ *and let* $\{\boldsymbol{H}_r : (\boldsymbol{A}, \boldsymbol{B}\boldsymbol{H}_r)$ *controllable*$\}$ *be the set of all* $\boldsymbol{H}_r$ *such that the pair* $(\boldsymbol{A}, \boldsymbol{B}\boldsymbol{H}_r)$ *is controllable. The sequence of scheduling matrices* $S_H = \{\boldsymbol{S}(0)\boldsymbol{H}_r, \boldsymbol{S}(1)\boldsymbol{H}_r, \ldots, \boldsymbol{S}(p-1)\boldsymbol{H}_r\}$ *will also preserve controllability if* $\operatorname{rank}\left(\sum_{k=0}^{p-1} \boldsymbol{S}_A(k)\boldsymbol{H}_r\right) = r$ *and conditions* (4.20) *and* (4.21) *are satisfied. This means that, to preserve controllability, the assumption that all* n_u *actuators are required for controllability can be relaxed if the system is redundant.* ▼

4.3 NCSs without ZOH

In this section, the previous results are extended to the problem formulation of [150]. Consider an NCS where the communication constraints apply to both input and output and no ZOH element is placed between the controller and the plant. This model was discussed for a generalized plant in Section 3.4, p. 62.

Let the spatially distributed plant $\mathcal{P}$ be the discrete LTI system described by

$$\mathcal{P}: \quad \begin{cases} \boldsymbol{x}(j+1) = \boldsymbol{A}\boldsymbol{x}(j) + \boldsymbol{B}\hat{\boldsymbol{u}}(j) \\ \boldsymbol{y}(j) = \boldsymbol{C}\boldsymbol{x}(j) \end{cases}, \tag{4.58}$$

where $\boldsymbol{x}(j) \in \mathbb{R}^{n_x}$, $\hat{\boldsymbol{u}}(j) \in \mathbb{R}^{n_u}$, $\boldsymbol{y}(j) \in \mathbb{R}^{n_y}$ and j is the sampling instant. $\boldsymbol{A}$, $\boldsymbol{B}$ and $\boldsymbol{C}$ are matrices of appropriate dimensions. The actuator scheduler $\mathcal{S}_A$ and the sensor scheduler $\mathcal{S}_S$ is defined in the usual way (see (3.15)–(3.16)).

The plant $\mathcal{P}$, with the limited communication, can be described by the *augmented plant*

$$\check{\mathcal{P}}: \quad \begin{cases} \boldsymbol{x}(j+1) = \boldsymbol{A}\boldsymbol{x}(j) + \check{\boldsymbol{B}}(k)\boldsymbol{u}(j) \\ \hat{\boldsymbol{y}}(j) = \check{\boldsymbol{C}}(k)\boldsymbol{x}(j) \end{cases}, \tag{4.59}$$

where

$$\check{\boldsymbol{B}}(k) = \boldsymbol{B}\boldsymbol{S}_A(k), \quad \check{\boldsymbol{C}}(k) = \boldsymbol{S}_S(k)\boldsymbol{C}. \tag{4.60}$$

[6] Recall that n_x is the number of states of the plant.

Contrary to the ZOH case, here the number of states of the augmented plant will not increase as the measurement and control signals are simply reset to zero when not updated.

As usual, periodicity is removed by applying the discrete-time lifting operator $\breve{\mathcal{P}} = \mathbb{L}_D\{\mathcal{P}, p\}$. $\breve{\mathcal{P}}$ is the LTI equivalent system that models the original plant plus the actuator's limited communication without ZOH. It is given by

$$\breve{\mathcal{P}}: \quad \begin{cases} \boldsymbol{x}(pl+p) = \breve{\boldsymbol{A}}\boldsymbol{x}(pl) + \breve{\boldsymbol{B}}\bar{\boldsymbol{u}}(pl) \\ \bar{\boldsymbol{y}}(pl) = \breve{\boldsymbol{C}}\boldsymbol{x}(pl), \end{cases} , \tag{4.61}$$

where

$$\bar{\boldsymbol{u}}(pl) = \begin{bmatrix} \boldsymbol{u}(pl) \\ \boldsymbol{u}(pl+1) \\ \cdots \\ \boldsymbol{u}(pl+p-1) \end{bmatrix}, \quad \bar{\boldsymbol{y}}(pl) = \begin{bmatrix} \boldsymbol{y}(pl) \\ \boldsymbol{y}(pl+1) \\ \cdots \\ \boldsymbol{y}(pl+p-1) \end{bmatrix}, \tag{4.62}$$

and

$$\breve{\boldsymbol{B}} = \begin{bmatrix} \boldsymbol{A}^{p-1}\boldsymbol{B}\boldsymbol{S}_A(0) & \boldsymbol{A}^{p-2}\boldsymbol{B}\boldsymbol{S}_A(1) & \cdots & \boldsymbol{A}^{0}\boldsymbol{B}\boldsymbol{S}_A(p-1) \end{bmatrix},$$
$$\breve{\boldsymbol{C}} = \begin{bmatrix} \boldsymbol{S}_S(0)\boldsymbol{C}\boldsymbol{A}^0 \\ \boldsymbol{S}_S(1)\boldsymbol{C}\boldsymbol{A}^1 \\ \vdots \\ \boldsymbol{S}_S(p-1)\boldsymbol{C}\boldsymbol{A}^{p-1} \end{bmatrix}, \quad \breve{\boldsymbol{A}} = \boldsymbol{A}^p, \tag{4.63}$$

(see Section 3.5, p. 63, for the detail of the discrete-time lifting).

4.3.1 Stabilizability and detectability

In a similar fashion as for the ZOH case, the conditions for a communication sequence that guarantees the preservation of controllability (observability) of the controllable (observable) modes in the lifted system $\breve{\mathcal{P}}$ in (4.61) can now be stated. The following theorem gives conditions for the controllability.

Theorem 4.2 (Controllability without ZOH) *Suppose that the pair* $(\boldsymbol{A}, \boldsymbol{B})$ *in (4.58) is controllable and* $\operatorname{rank}(\boldsymbol{B}) = n_u$. *The system matrix* $\boldsymbol{A}$ *is invertible. Moreover, all* n_u *actuators are required for controllability of the plant,* i.e. $(\boldsymbol{A}, \boldsymbol{B}_r)$ *is not controllable where* $\boldsymbol{B}_r$ *is an input distribution matrix covering only part of the input range of* $\boldsymbol{B}$. *Under these conditions the following can be stated:*

(i) the pair $(\breve{\boldsymbol{A}}, \breve{\boldsymbol{B}})$ *(the lifted system) is controllable if conditions* (4.19) *and* (4.21) *are satisfied*

(ii) the pair $(\breve{\boldsymbol{A}}, \breve{\boldsymbol{B}})$ *is not controllable if* (4.22) *holds.*

Proof. As for the previous proofs, the PBH eigenvector test will be used to prove controllability of the lifted augmented system without ZOH. In this case, the proof is straightforward.

Part (i). What is needed to be proven is that there is no left hand eigenvector of $\breve{\boldsymbol{A}}$ that is orthogonal to $\breve{\boldsymbol{B}}$ [61, p. 135]. By Lemma 4.1, the eigenvectors $\boldsymbol{v}$ of $\breve{\boldsymbol{A}} = \boldsymbol{A}^p$ are given by the eigenvectors of $\boldsymbol{A}$ and $\boldsymbol{vB} \neq 0$. Then, from (4.63),

$$\begin{aligned} \boldsymbol{v}\breve{\boldsymbol{B}} &= \begin{bmatrix} \boldsymbol{vA}^{p-1}\boldsymbol{BS}_A(0) & \boldsymbol{vA}^{p-2}\boldsymbol{BS}_A(1) & \cdots & \boldsymbol{vA}^{0}\boldsymbol{BS}_A(p-1) \end{bmatrix} \\ &= \begin{bmatrix} \lambda^{p-1}\boldsymbol{vBS}_A(0) & \lambda^{p-2}\boldsymbol{vBS}_A(1) & \cdots & \lambda^{0}\boldsymbol{vBS}_A(p-1) \end{bmatrix}. \end{aligned} \tag{4.64}$$

Note that $\boldsymbol{A}$ is assumed to be invertible and, for each eigenvalue $\lambda \neq 0$ of $\boldsymbol{A}$, the matrices $\boldsymbol{S}_A(k)$ will be able to select every column on $\boldsymbol{B}$ because $\operatorname{rank}\left(\sum_{k=0}^{p-1} \boldsymbol{S}_A(k)\right) = \operatorname{rank}(\boldsymbol{B}) = n_u$. By Lemma 4.1, $(\boldsymbol{A}^p, \boldsymbol{B})$ is controllable and $\boldsymbol{vB} \neq 0$, which implies $\boldsymbol{v}\breve{\boldsymbol{B}} \neq 0$.

Part (ii). If the sequence does not select every column of $\boldsymbol{B}$ then $\boldsymbol{vB} \neq 0$ is not satisfied and it will not be possible to guarantee $\boldsymbol{v}\breve{\boldsymbol{B}} \neq 0$. ■

In Theorem 4.2, it is assumed that the matrix $\boldsymbol{A}$ has no zero eigenvalues. The following Lemma is for non-invertible plant matrices.

Lemma 4.3 (Controllability without ZOH for non-invertible $\boldsymbol{A}$) *Suppose that the pair $(\boldsymbol{A}, \boldsymbol{B})$ in (4.58) is controllable. If there exist eigenvalues $\lambda = 0$ of $\boldsymbol{A}$ and a ZOH strategy is not used, then the pair $(\breve{\boldsymbol{A}}, \breve{\boldsymbol{B}})$ in (4.61) is controllable for $\lambda = 0$ if $\sigma(p-1)$ (the last element of the sequence) selects all the actuators corresponding to the mode of the zero eigenvalue.*

Proof. For an eigenvalue $\lambda = 0$ of $\boldsymbol{A}$, (4.64) reduces to

$$\boldsymbol{v}\breve{\boldsymbol{B}} = \begin{bmatrix} \boldsymbol{0} & \boldsymbol{0} & \cdots & \boldsymbol{vBS}_A(p-1) \end{bmatrix}. \tag{4.65}$$

If the scheduling matrix $\boldsymbol{S}_A(p-1)$ selects the actuators associated with the mode of the zero eigenvalue, then $\boldsymbol{vB} \neq 0$ because the pair $(\boldsymbol{A}, \boldsymbol{B})$ is controllable. This implies that $\boldsymbol{v}\breve{\boldsymbol{B}} \neq 0$ and therefore the pair $(\breve{\boldsymbol{A}}, \breve{\boldsymbol{B}})$ will also be controllable for $\lambda = 0$. ■

The following corollary gives conditions for the stabilizability of the system in (4.61).

Corollary 4.2 (Stabilizability without ZOH) *Suppose that the pair $(\boldsymbol{A}, \boldsymbol{B})$ in (4.58) is stabilizable and* $\operatorname{rank}(\boldsymbol{B}) = n_u$. *Moreover, all n_u actuators are required for stabilizability of the plant. Under these conditions the following can be stated:*

(i) the pair $(\breve{\boldsymbol{A}}, \breve{\boldsymbol{B}})$ is stabilizable if condition (4.19) is satisfied and condition (4.21) is satisfied for any pair of eigenvalues, (λ_1, λ_2), $|\lambda_1| \geq 1$, $|\lambda_2| \geq 1$ of $\boldsymbol{A}$

(ii) the pair $(\breve{\boldsymbol{A}}, \breve{\boldsymbol{B}})$ is not stabilizable if (4.22) holds.

The following corollaries give conditions for the observability and detectability of the system in (4.61) using the concept of duality.

Corollary 4.3 (Observability without ZOH) *Suppose that the pair* $(\boldsymbol{C}, \boldsymbol{A})$ *in* (4.58) *is observable and* $\text{rank}(\boldsymbol{C}) = n_y$. *The system matrix* $\boldsymbol{A}$ *is invertible. Moreover, all* n_y *sensors are required for observability of the plant. Under these conditions the following can be stated:*

(i) the pair $(\breve{\boldsymbol{C}}, \breve{\boldsymbol{A}})$ *(the lifted system) is observable if*

$$\text{rank}\left(\sum_{k=0}^{p-1} \boldsymbol{S}_S(k)\right) = n_y, \tag{4.66}$$

and assuming that for any pair of eigenvalues, (λ_1, λ_2) *of* $\boldsymbol{A}$, *condition* (4.21) *holds*

(ii) the pair $(\breve{\boldsymbol{C}}, \breve{\boldsymbol{A}})$ *is not observable if*

$$\text{rank}\left(\sum_{k=0}^{p-1} \boldsymbol{S}_S(k)\right) < n_y. \tag{4.67}$$

Corollary 4.4 (Observability without ZOH for non-invertible A) *Suppose that the pair* $(\boldsymbol{C}, \boldsymbol{A})$ *in* (4.58) *is observable. If there exist eigenvalues* $\lambda = 0$ *of* $\boldsymbol{A}$ *and a ZOH strategy is not used, then the pair* $(\breve{\boldsymbol{C}}, \breve{\boldsymbol{A}})$ *in* (4.61) *is observable for* $\lambda = 0$ *if* $\sigma(0)$ *(the first element of the sequence) selects all the sensors corresponding to the mode of the zero eigenvalue.*

Remark 4.3 *It may seem counterintuitive the fact that, for the case on non-invertible* $\boldsymbol{A}$ *(Lemma 4.3), the location of the elements in the sequence matters. Here is an explanation. The mode corresponding to the zero eigenvalue has a dead-beat behavior and a non-ZOH strategy will allow this state to return to zero when it is not controlled. As the lifted system of* (4.61) *is considered, controllability is required for the states at time instant* $p, 2p, 3p, \ldots$. *If this actuator is not controlled last in the sequence, then the corresponding state will be zero at* $p, 2p, 3p, \ldots$ *and therefore uncontrollable. A similar argument holds for the observability case (Corollary 4.4).* •

Corollary 4.5 (Detectability without ZOH) *Suppose that the pair* $(\boldsymbol{C}, \boldsymbol{A})$ *in* (4.58) *is detectable and* $\text{rank}(\boldsymbol{C}) = n_y$. *The system matrix* $\boldsymbol{A}$ *is invertible. Moreover, all* n_y *sensors are required for detectability of the plant. Under these conditions the following can be stated:*

(i) the pair $(\breve{\boldsymbol{C}}, \breve{\boldsymbol{A}})$ *is detectable if condition* (4.66) *is satisfied and condition* (4.21) *is satisfied for any pair of eigenvalues,* (λ_1, λ_2), $|\lambda_1| \geq 1$, $|\lambda_2| \geq 1$ *of* $\boldsymbol{A}$

(ii) the pair $(\breve{\boldsymbol{C}}, \breve{\boldsymbol{A}})$ *is not detectable if* (4.67) *holds.*

A similar argument as the one in Observation 4.1 also applies for the non-ZOH case.

Remark 4.4 *The concept of feasible and minimum feasible sequences defined for σ in Definitions 4.6 and 4.7 also applies for sensor scheduling. Hence, a p-periodic sequence σ will be called feasible if it generates sequences of matrices S_A and S_S such that*

$$\text{rank}\left(\sum_{k=0}^{p-1} \boldsymbol{S}_A(k)\right) = n_u \quad \textit{and} \quad \text{rank}\left(\sum_{k=0}^{p-1} \boldsymbol{S}_S(k)\right) = n_y, \tag{4.68}$$

and minimum feasible if the actuators or the sensor signals appear only once in the sequence. •

4.4 Sampled-data case

Consider the more realistic case where the distributed plant to be controlled is a continuous linear time-invariant system described by

$$\begin{aligned} \dot{\boldsymbol{x}}(t) &= \boldsymbol{A}^c\boldsymbol{x}(t) + \boldsymbol{B}^c\hat{\boldsymbol{u}}(t), \\ \boldsymbol{y}(t) &= \boldsymbol{C}\boldsymbol{x}(t), \end{aligned} \tag{4.69}$$

where $\boldsymbol{x}(t) \in \mathbb{R}^{n_x}$, $\hat{\boldsymbol{u}}(t) \in \mathbb{R}^{n_u}$ and $\boldsymbol{y}(t) \in \mathbb{R}^{n_y}$. The control input $\hat{\boldsymbol{u}}(t)$ is a discrete input signal created by a ZOH element for a constant sampling period h,

$$\hat{\boldsymbol{u}}(t) = \hat{\boldsymbol{u}}(jh) = \textit{constant} \quad \text{for} \quad t \in [jh, (j+1)h). \tag{4.70}$$

The model in (4.69) has to be sampled with a periodic sampling interval h giving the sampled-data time-invariant system for constant input $\hat{\boldsymbol{u}}(t)$ in (4.1) where

$$\boldsymbol{A} = e^{\boldsymbol{A}^c h}, \quad \boldsymbol{B} = \int_0^h e^{\boldsymbol{A}^c s} ds \boldsymbol{B}^c. \tag{4.71}$$

Remark 4.5 *If the sampling frequency is non-pathological (*i.e. *$\boldsymbol{A}^c$ does not have two eigenvalues with equal real and imaginary parts that differ by an integral multiple of $\frac{2\pi}{h}$, [20, Theorem 3.2.1]), then*

$$\begin{aligned} (\boldsymbol{A}^c, \boldsymbol{B}^c) \textit{ controllable} \quad &\Rightarrow \quad (\boldsymbol{A}, \boldsymbol{B}) \textit{ controllable} \\ (\boldsymbol{C}, \boldsymbol{A}^c) \textit{ observable} \quad &\Rightarrow \quad (\boldsymbol{C}, \boldsymbol{A}) \textit{ observable.} \end{aligned} \tag{4.72}$$

This is a sufficient condition for preservation of controllability and observability after discretization (necessary and sufficient conditions can be found in [67]). If this condition is applied only to the unstable eigenvalues of $\boldsymbol{A}^c$ then stabilizability and detectability are preserved after discretization. •

Note that $\boldsymbol{A} = e^{\boldsymbol{A}^c h}$ and therefore $\boldsymbol{A}$ is always an invertible matrix. Hence, for sampled-data systems, only Theorems 4.1 and 4.2 and Corollaries 4.1, 4.2, 4.3 and 4.5 apply.

Remark 4.6 *The conditions in* (4.20) *and (4.21), for the eigenvalues* λ *of* $\boldsymbol{A} = e^{\boldsymbol{A}^c h}$*, can be regarded as an extended pathological sampling frequency condition for NCSs. These additional spectral conditions on the system can be easily satisfied by appropriate selection of the sampling time. This readily allows to achieve sequences of length* $\lceil \frac{n_u}{c} \rceil$ *which*[7] *is usually lower than* $\lceil \frac{n_x}{c} \rceil n_x$ *or* n_x *achieved in [150, 56, 51, 52].* •

4.5 Examples

The following examples should help clarifying the application of the above conditions and how these results compare to existing ones.

Example 4.2 *For comparative reasons, let us consider the numerical example in [51, Section IV A] where a 2-input, 2-output, unstable plant with matrices*

$$\boldsymbol{A} = \begin{bmatrix} -1.05 & 0 & 0 & 0 \\ 0 & 0.75 & 0 & 0 \\ 0 & 1.05 & 1.05 & 0 \\ 0 & 0 & -2 & 0.5 \end{bmatrix}, \quad \boldsymbol{B} = \begin{bmatrix} 0.5 & 0 \\ 0 & 0 \\ 0 & 0.5 \\ 0 & 0 \end{bmatrix}, \quad \boldsymbol{C} = \begin{bmatrix} 0 & 0 & 0 & 1 \\ 1 & 0 & 0 & 0 \end{bmatrix}, \tag{4.73}$$

is controlled through a shared communication medium with both sensor and actuator scheduling. The system is completely observable and stabilizable (but not completely controllable). Only one sensor and only one actuator can be read/controlled at any time tick. Delays are ignored here. The original plant is stabilizable and detectable and the shortest communication sequence that preserves stabilizability and detectability found by the iterative algorithm of [150, 51] is $\sigma_A = \sigma_S = \{2, 2, 1\}$ *(3-periodic).* σ_A *and* σ_S *are the sensor and actuator scheduling sequences defined in the context of [51]. These sequences can be written as a single sequence* σ *(to fit the framework of this book) as* $\sigma = \{\Delta(2), \Delta(2), \Delta(1)\}$ *where* $\Delta(1) = \{actuator\ 1, sensor\ 1\}$ *and* $\Delta(2) = \{actuator\ 2, sensor\ 2\}$*. In any case, they are not minimum feasible sequences according to Definition 4.7.*

Because $\boldsymbol{A}$ *is invertible, the controllable subsystem is analyzed via Corollaries 4.1 and 4.2 and the whole system via Corollary 4.3. It turns out that any (2-periodic) minimum feasible communication sequence (*e.g. $\sigma_A = \sigma_S = \{2, 1\}$ *or* $\sigma = \{\Delta(2), \Delta(1)\}$ *to use the notation of this book) would have sufficed to*

[7] The constant c was defined in Remark 4.2.

preserve stabilizability and detectability. This is valid with and without ZOH strategy. ♦

Example 4.3 *In the example of [150, Section 5] the discrete-time plant with two actuators and with matrices*

$$\boldsymbol{A} = \begin{bmatrix} 0 & -1 \\ 1 & 0 \end{bmatrix}, \quad \boldsymbol{B} = \begin{bmatrix} 0 & 1 \\ 1 & 0 \end{bmatrix}, \tag{4.74}$$

*(the output matrix $\boldsymbol{C}$ does not matter) is considered. It is claimed by the authors that a round robin communication sequence (*i.e. $\sigma = \{\Delta(1), \Delta(2)\}$ *or* $\sigma = \{\Delta(2), \Delta(1)\}$ *where* $\Delta(1) = \{actuator\ 1\}$ *and* $\Delta(2) = \{actuator\ 2\}$*) would fail to preserve the plant's controllability. The controllability preserving sequence calculated by the algorithms in [150] is* $\sigma = \{\Delta(2), \Delta(2), \Delta(1)\}$ *(3-periodic). This is consistent with the above presented results because of condition (4.21). In fact, the eigenvalues of* $\boldsymbol{A}$ *are* $\lambda_1 = \sqrt{-1}$ *and* $\lambda_2 = -\sqrt{-1}$ *and clearly*

$$\lambda_1 = \lambda_2 \exp\left(\frac{2\pi l\sqrt{-1}}{2}\right), \quad l = \ldots, -3, -1, 1, 3, \ldots, \tag{4.75}$$

which does not satisfy the condition in (4.21). However, if the sequence is 3-periodic

$$\lambda_1 \neq \lambda_2 \exp\left(\frac{2\pi l\sqrt{-1}}{3}\right), \quad l = \ldots, -1, 0, 1, 2, \ldots, \tag{4.76}$$

and therefore the system preserves controllability (of course, with condition (4.55) satisfied too). Notice that not all the actuators are required for controllability of the plant and consequently the trivial sequence $\sigma = \{\Delta(1)\}$ *or* $\sigma = \{\Delta(2)\}$ *can also preserve controllability. In general, for this example, the presented results show that any* $(2l-1)$*-periodic sequence,* $l = 1, 2, \ldots$*, could be used.* ♦

Example 4.4 *In the example of [52, Section IV] the following discrete-time plant is considered*

$$\boldsymbol{A} = \begin{bmatrix} -1.05 & 1.05 & -0.5 & 1.05 & -0.775 \\ 0 & 0 & 1.05 & 0 & 0 \\ 0 & 1.05 & 0 & 1.05 & 0 \\ 0 & 0 & -2.1 & 0 & 0 \\ 0 & -2.1 & 1 & -2.1 & 0.5 \end{bmatrix}, \quad \boldsymbol{B} = \begin{bmatrix} 1 & 0.5 \\ 1 & 0 \\ 1 & 0 \\ -1 & 0 \\ -2 & 0 \end{bmatrix},$$
$$\boldsymbol{C} = \begin{bmatrix} 0 & 0 & 0 & 1 & 0 \\ 1 & 0 & 0 & 0 & 1 \end{bmatrix}. \tag{4.77}$$

The eigenvalues of $\boldsymbol{A}$ *are* $\left\{1.05\sqrt{-1}, -1.05\sqrt{-1}, 0, 0.5, -1.05\right\}$*. Since* $\boldsymbol{A}$ *is non-invertible, by Lemma 4.2, controllability is lost if a ZOH strategy is* *adopted and therefore only the case without ZOH is considered. Similarly to the previous example, if a 2-periodic (minimum feasible) sequence is used, then condition (4.21) is not satisfied because of the eigenvalues* $1.05\sqrt{-1}$ *and* $-1.05\sqrt{-1}$. *Also in this case,* $(2l-1)$*-periodic sequences should be used. Furthermore, according to Lemma 4.3 and Corollary 4.4, the actuator associated to the mode of the zero eigenvalue (actuator 1) must appear as the last (for controllability) or the first (for observability) element of the sequence for controllability/observability of* $\lambda = 0$. *Therefore, the set of the shortest possible sequences, satisfying the constraints, that preserve the controllability structure is*

$$\{\{\Delta^A(1), \Delta^A(2), \Delta^A(1)\}, \{\Delta^A(2), \Delta^A(1), \Delta^A(1)\}, \{\Delta^A(2), \Delta^A(2), \Delta^A(1)\}\} \tag{4.78}$$

(3-periodic with actuator 1 controlled last) where $\Delta^A(1) = \{actuator\ 1\}$ *and* $\Delta^A(2) = \{actuator\ 2\}$. *The set of the shortest possible sequences, satisfying the constraints, that preserve the observability structure is*

$$\{\{\Delta^S(1), \Delta^S(1), \Delta^S(2)\}, \{\Delta^S(1), \Delta^S(2), \Delta^S(1)\}, \{\Delta^S(1), \Delta^S(2), \Delta^S(2)\}\} \tag{4.79}$$

(3-periodic with sensor 1 being read first) where $\Delta^S(1) = \{sensor\ 1\}$ *and* $\Delta^S(2) = \{sensor\ 2\}$. *Thus, other possible sequences for control communication are easily obtained by checking the conditions of Theorem 4.2, Lemma 4.3 or Corollaries 4.3 and 4.4. The iterative search algorithms of [52] are not required. Furthermore, it is only possible to preserve controllability and observability if* $\{actuator\ 1\}$ *is a subset of* $\sigma(p-1)$ *(the last scheduled node in the sequence) and* $\{sensor\ 2\}$ *is a subset of* $\sigma(0)$ *(the first scheduled node in the sequence) .* ♦

Example 4.5 *Another interesting example is the following. Consider the continuous-time system in (4.69) with* $\boldsymbol{A}^c = \text{diag}(\boldsymbol{A}_{c1}, \boldsymbol{A}_{c2})$, $\boldsymbol{B}^c = \text{diag}(\boldsymbol{B}_{c1}, \boldsymbol{B}_{c2})$, *where* $\boldsymbol{B}_{c1}$ *and* $\boldsymbol{B}_{c2}$ *are vectors. Suppose that the 2-periodic sequence* $\sigma = \{\Delta(1), \Delta(2)\}$ *is chosen where* $\Delta(1) = \{actuator\ 1\}$ *and* $\Delta(2) = \{actuator\ 2\}$ *(actuators 1 and 2 receive control every other step). In this case, it is required to check the stabilizability of the two discretized systems* $(\boldsymbol{A}_{c1}, \boldsymbol{B}_{c1})$ *and* $(\boldsymbol{A}_{c2}, \boldsymbol{B}_{c2})$ *with sampling period* $2h$, *separately. It would be possible that, for example,* $(\boldsymbol{A}_{c1}, \boldsymbol{B}_{c1})$ *is stabilizable with the shorter sampling period* h, *but not with the longer one* $2h$. *This can be only caused by* $\frac{1}{2h}$ *being a pathological sampling frequency. For the NCS case, condition (4.21) prevents this from happening. Assuming* p *is an even number, the choice* $l = \frac{p}{2}$ *in (4.21) will reject the system* $(\boldsymbol{A}^c, \boldsymbol{B}^c)$ *as unfeasible. For a sampled-data system, a solution to this problem would be to change the sampling time by a small amount. An odd value for* p *will prevent the pathological case, so that* $(\boldsymbol{A}^c, \boldsymbol{B}^c)$ *can be employed.* ♦

Example 4.6 *Finally, let us consider the more practical problem of stabilizing m inverted pendulums arranged in a circle controlled by torques T_i (see Figure 4.1). The rods are mechanically connected by ideal springs and the circle is*

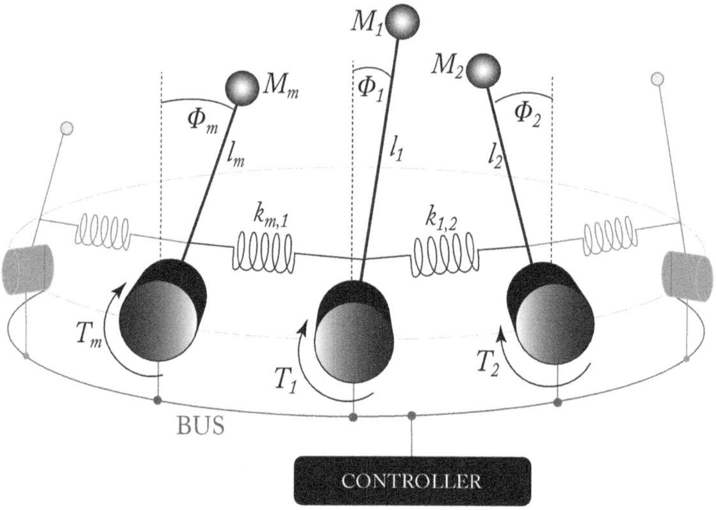

FIGURE 4.1: Inverted pendulums in a circle: an example of a distributed system.

large enough in radius so that the interaction due to the springs can be regarded as linear. ϕ_i is the rod's angle of the ith pendulum, M_i is the mass of the bob, l_i the pendulums' rod length, $k_{i,m}$ is the spring coefficient of the spring between pendulum i and m, and $g = 9.81$. The linearized dynamics of the system with $M_1 = M_2 = \ldots = M_m = 1$ can be described by (4.69) where

$$\boldsymbol{A}^c = \begin{bmatrix} 0 & 1 & 0 & 0 & \cdots & 0 & 0 \\ \frac{g}{l_1} - k_{1,2} - k_{m,1} & 0 & k_{1,2} & 0 & \cdots & k_{m,1} & 0 \\ 0 & 0 & 0 & 1 & \cdots & 0 & 0 \\ k_{1,2} & 0 & \frac{g}{l_2} - k_{2,3} - k_{1,2} & 0 & \cdots & 0 & 0 \\ \vdots & \vdots & \vdots & \vdots & \ddots & \vdots & \vdots \\ 0 & 0 & 0 & 0 & \cdots & 0 & 1 \\ k_{m,1} & 0 & 0 & 0 & \cdots & \frac{g}{l_m} - k_{m,1} - k_{m-1,m} & 0 \end{bmatrix},$$

$$\boldsymbol{B}^c = \begin{bmatrix} 0 & 0 & \cdots & 0 \\ \frac{g}{l_1} & 0 & \cdots & 0 \\ 0 & 0 & \cdots & 0 \\ 0 & \frac{g}{l_2} & \cdots & 0 \\ \vdots & \vdots & \ddots & \vdots \\ 0 & 0 & \cdots & 0 \\ 0 & 0 & \cdots & \frac{g}{l_m} \end{bmatrix}, \quad \boldsymbol{x}(t) = \begin{bmatrix} \phi_1(t) \\ \dot{\phi}_1(t) \\ \phi_2(t) \\ \dot{\phi}_2(t) \\ \vdots \\ \phi_m(t) \\ \dot{\phi}_m(t) \end{bmatrix}, \tag{4.80}$$

and $\boldsymbol{C} = \boldsymbol{I}$. The pendulums' rod length has been chosen to be $l_1 = l_5 = 1$,

$p=4$	$i=1$	$i=2$	$i=3$	$i=4$	$i=5$	$i=6$	$i=7$	$i=8$
$k_f^{(i)}$	2	0	3	1	3	2	2	3
$\bar{q}^{(i)}$	1	1	1	2	2	3	3	4
$\hat{q}^{(i)}(k_f^{(i)})$	0	0	0	1	0	2	2	3

TABLE 4.1: Values for controllability conditions of Example 4.6.

$l_2 = l_6 = 3$, $l_3 = l_7 = 3$, $l_4 = l_8 = 4$, and the spring coefficients $k_{1,2} = k_{2,3} = \ldots = k_{m,1} = 40$. This system is fully controllable and fully observable. It is required to check that these properties are preserved after it is implemented as an NCS. The sampling period has been set to $h = 0.1$. The controller and the actuators share the communication medium and up to three actuators can be controlled at any time tick. The chosen communication sequence is 4-periodic ($p = 4$) and it is

$$\sigma = \{\Delta(1,2,3), \Delta(1,4,5), \Delta(1,6,7), \Delta(3,5,8)\}, \tag{4.81}$$

where

$$\begin{aligned}
\Delta(1,2,3) &= \{actuator\ 1, actuator\ 2, actuator\ 3\},\\
\Delta(1,4,5) &= \{actuator\ 1, actuator\ 4, actuator\ 5\},\\
\Delta(1,6,7) &= \{actuator\ 1, actuator\ 6, actuator\ 7\},\\
\Delta(3,5,8) &= \{actuator\ 3, actuator\ 5, actuator\ 8\}.
\end{aligned} \tag{4.82}$$

For this particular sequence, a set of values for $k_f^{(i)}$, $\bar{q}^{(i)}$ and $\hat{q}^{(i)}(k_f^{(i)})$, given in Table 4.1, has been calculated according to Definitions 4.2 and 4.3.

The choice of nodes may not be very realistic if this set up is to be implemented in practice. However, their choice renders the example more interesting. The final step is to check that conditions (4.19), (4.20) and (4.21), are satisfied. Condition (4.19) can be easily verified by checking that every actuator is controlled at least once in the sequence. This is the case (see (4.81)). To verify the other two conditions, the system has to be discretized first as shown in Section 4.4. In (4.20), the expression $p - k_f^{(i)} + \hat{q}^{(i)}(k_f^{(i)})$ can take the values $\{2, 3, 4\}$ giving eleven distinct 'forbidden' eigenvalues (i.e. eigenvalues that must be avoided in order to guarantee controllability). These are shown in Figure 4.2 as squares. Also, the sixteen system eigenvalues (displayed as crosses) are shown in Figure 4.2. It can be seen that none of the eigenvalues are in the forbidden places, therefore condition (4.20) is satisfied. As already mentioned, this condition could be easily satisfied by a slight change in the sampling time. Finally, in condition (4.21), the expression $\exp\left(\frac{2\pi l\sqrt{-1}}{p}\right)$, for $p = 4$, can only take the values $\{-1, -\sqrt{-1}, 1, \sqrt{-1}\}$. It can be easily shown that (4.21) is also satisfied and therefore controllability is guaranteed, with and

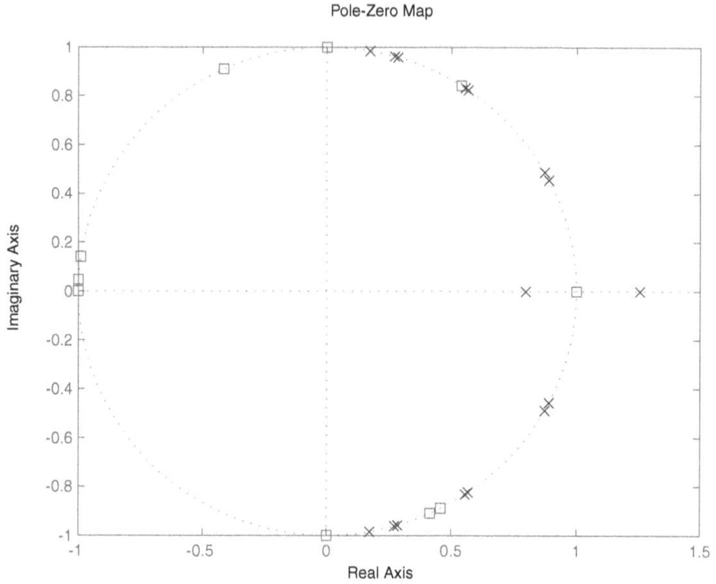

FIGURE 4.2: Pole-Zero map with forbidden (squares) eigenvalues.

without ZOH. A similar argument can be formulated for the dual observability problem. ♦

4.6 Summary

Because of the scheduled communication in NCSs, not only a stabilizing controller needs to be designed but also a communication sequence that preserves stability, controllability and observability when the plant is controlled through a network. In this chapter, it was proven that communication sequences that preserve controllability and observability (and the weaker stabilizability and detectability) properties of NCSs can be shorter than previously established and a tight lower bound was found. In particular, it was shown that for the sampled-data case, a communication sequence that avoids particularly defined pathological sampling rates and updates each signal only once in the sequence is sufficient to preserve the structural properties. These results apply to systems with and without ZOH strategy. The case of non-invertible plant models was addressed. Suitable examples showed that these results interestingly complement and improve existing ones.

5

Communication sequence optimization

CONTENTS

A universal optimization method able to efficiently solve any optimization problem does not yet exist. It is with our judgement that we select or develop the appropriate algorithm that best fits our problem. A communication sequence for an NCS has a finite discrete structure. Finding an optimal sequence is a complex combinatorial problem. However, our *a priory* knowledge about the search space can help simplifying this task.

5.1 Introduction

In Chapter 3, a framework for modeling NCSs was proposed. It was shown that the communication network, defined as $N = (\Delta, \sigma)$, where Δ is the set of nodes and σ the communication sequence, completely characterizes the limited communication between plant and controller. In Chapter 4, conditions were derived to guarantee preservation of the structural properties (controllability, observability and the weaker stabilizability and detectability) of the system. In other words, it was shown that, if a system with certain structural properties is implemented as an NCS and particular communication sequences are avoided, then the structural properties remain unchanged.

A model for an NCS can only be derived if a communication sequence and a set of nodes are given. For example, the system $\bar{\mathcal{P}}$ in (3.58) is an LTI model and therefore a controller can be designed for $\bar{\mathcal{P}}$ using standard techniques from control theory. Designing a controller for $\bar{\mathcal{P}}$ means designing a controller for the original plant $\mathcal{P}$ and for the communication policy defined by that particular communication network $N = (\Delta, \sigma)$. The set of nodes Δ is normally fixed because it emerges from the physical configuration of an NCS (*e.g.* the scenario in Example 3.2, p. 74). What is not normally fixed is the communication sequence σ which inevitably becomes an extra dimension in the design space. The way σ comes into play in the dynamics of the NCS is nontrivial and it will be one of the points of investigation in this work.

The modeling framework of Chapter 3 and the results from Chapter 4 will be used for the design of optimal and robust (in the sense of signal and parameter uncertainty) communication sequences. It turns out that finding these sequences is a complex combinatorial optimization problem. In this chapter, suitable optimization methods for finding optimal/robust offline communication sequences are discussed.

5.2 Optimization problem

The optimization problem is the design of a communication sequence σ such that the NCS retains certain fundamental structural properties (stability, controllability, observability, *etc.*) and a cost function is minimized. It will be shown that the cost is the result of either a controller and schedule codesign problem or system integration problem and it is subject to a given communication sequence σ. The cost functions proposed will be discussed in Chapters 6, 7 and 8. Whether this cost is quadratic or $\mathcal{H}_\infty$-based is not important for the description of the optimization procedure. All that is needed for the moment is to define a function of the form

$$J(\sigma) = \Phi(\mathcal{G}, N) \geq 0, \tag{5.1}$$

which associates a scalar $J(\sigma)$ to a communication network $N = (\Delta, \sigma)$ for a system $\mathcal{G}$. Note that $\mathcal{G}$ here can be any system (plant, closed-loop plant and controller, *etc.*). $J(\sigma)$ is a measure of performance that needs to be minimized by appropriate selection of a sequence σ (Δ is assumed to be fixed for practical applications but it does not need to be). The term 'feasible' is used to describe those sequences that preserve the structural properties of the system as if it was not implemented over a network (ideal feedback loops). The feasibility issue (in terms of controllability/observability) was discussed in detail in the previous chapter.

Finding an optimal feasible sequence that minimizes (5.1) is the optimization problem. This falls into the area of combinatorial optimization and can be described by the tuple $(\mathfrak{S}, \mathfrak{P}, \mathfrak{F}, J(\sigma), min)$ where $\mathfrak{S}$ is the solution space of size $|\mathfrak{S}|$ on which $\mathfrak{P}$ and $J(\sigma)$ are defined, $\mathfrak{P}$ is the feasibility predicate and $\mathfrak{F}$ is the set of feasible solutions of size $|\mathfrak{F}|$. The set of unfeasible solutions is $\mathfrak{U}$ and its size is $|\mathfrak{U}| = |\mathfrak{S}| - |\mathfrak{F}|$ since $\mathfrak{F} \cup \mathfrak{U} = \mathfrak{S}$.

Finally, the optimization problem is

$$\min_{s.t. \mathcal{G}, \Delta} J(\sigma), \quad \sigma \in \mathfrak{F}. \tag{5.2}$$

Remark 5.1 *If a system has d nodes and a sequence of length p is used, then the number of all possible sequences is* d^p. *The sequence which generates the lower* $J(\sigma)$ *can be found by calculating the* $J(\sigma)$ *for all* d^p *sequences and then choosing the one that minimizes it (exhaustive search). Although this is possible for systems with few nodes and short sequences, the problem becomes intractable very soon. For example, to calculate the cost of all possible permutations of the simple round robin sequence for a system with 10 nodes means that* $J(\sigma)$ *has to be calculated at least* 10! *times (just to guarantee feasibility). This is known as combinatorial explosion. The algorithms proposed here do not suffer from combinatorial explosion.* •

5.2.1 Analysis of properties

Some of the properties specific to this optimization problem will be analyzed here. The effectiveness of search techniques is dependent on the choices of parameters and operators. The behavior of an optimization algorithm can be predicted by analyzing the fitness landscape that the cost function $J(\sigma)$ gives rise to [104].

Definition 5.1 *The fitness landscape of the cost function* $J(\sigma)$ *is defined by* $\mathfrak{L} = (\mathfrak{S}, J(\sigma), \epsilon)$, *where* $\epsilon : \mathfrak{S} \times \mathfrak{S} \to \mathbb{R}^+ \cup \{\infty\}$ *is the distance measure such that for all* $\sigma_m, \sigma_n, \sigma_o$ *in the search space* $\mathfrak{S}$, *it follows that* $\epsilon(\sigma_m, \sigma_n) \geq 0$, $\epsilon(\sigma_m, \sigma_n) = 0 \Leftrightarrow \sigma_m = \sigma_n$ *and* $\epsilon(\sigma_m, \sigma_o) \leq \epsilon(\sigma_m, \sigma_n) + \epsilon(\sigma_n, \sigma_o)$. ▲

The distance measure is defined here by the number of unity changes needed for a sequence σ_m to become equal to a sequence σ_n. Once a distance measure on the search space is defined, the concept of neighborhood will follow.

Definition 5.2 *The neighborhood* $\mathfrak{N}(\sigma_m)$ *of a point* σ_m *in the search space is the set of points that can be reached from* σ_m *by a single application of an operator* ω. *Let* ϵ_ω *be the distance measure under the operator* ω *where* $t \in \mathfrak{N}(\sigma_m) \Leftrightarrow \epsilon_\omega(\sigma_m, \sigma_n) = 1$. *The distance between non-neighbors is the length of the shortest path between them.* ▲

The operator ω used here is the unity change as pointed out above. The number of local optima in the fitness landscape will be one important factor in the performance of the search algorithms. Even more important are the relative basins of attraction of the optima that will depend on the search strategy.

Definition 5.3 *The basin of attraction of a local optimum is the set of points in the search space from which that local optimum will be attained under some search strategy.* ▲

A simple example will clarify the importance of the fitness landscape analysis for this problem.

Example 5.1 *Assume that two identical plants are controlled by a unique controller via a network. A node is associated to each plant so that node $\Delta(1)$ includes all sensors, actuators and demands for plant 1 and node $\Delta(2)$ includes all sensors, actuators and demands for plant 2. A predefined sequence of length $p = 4$ is used. It is intuitive in this case that the best schedule will be the one that allocates equal 'attention' to the systems and also that will leave the systems 'unattended' for the shortest possible time. The optimal sequences will therefore be $\sigma_1 = \{\Delta(1), \Delta(2), \Delta(1), \Delta(2)\}$ and $\sigma_2 = \{\Delta(2), \Delta(1), \Delta(2), \Delta(1)\}$. Let us visualize all 2^4 possible sequence permutations as the vertices of a 4-dimensional hypercube (a tesseract) shown in Figure 5.1. It is clear that the two solutions σ_1 and σ_2 lie on two opposite*

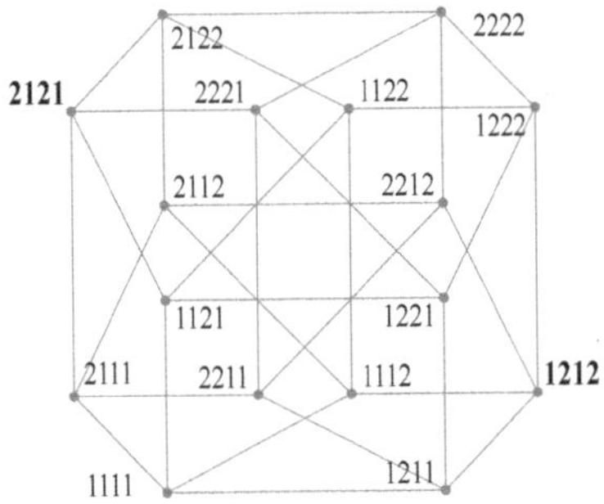

FIGURE 5.1: All 2^4 solution as vertices of a tesseract: the optimal solutions lie on opposite vertices.

vertices. Similarly, the two unfeasible solutions $\{\Delta(1), \Delta(1), \Delta(1), \Delta(1)\}$ and $\{\Delta(2), \Delta(2), \Delta(2), \Delta(2)\}$ (which are unfeasible because one of the systems is never controlled), lie on the other two opposite vertices. This situation creates two equivalent basins of attraction on the opposite sides of the search space. ♦

Remark 5.2 *It is intuitive to assume that given a periodic sequence $\sigma = \{\sigma(k)\}_{k=0}^{p-1}$ any k circular shifting of σ will result in another sequence σ_k with* *the same cost of σ. This is argued in [105] in what is called 'the set of equivalence classes'. Although this is true when the plant consists of several decoupled and identical systems, it may not be the case for other situations. For example, consider a plant which consists of different decoupled systems, of which some are stable and others unstable. For instance, consider the two systems of Example 5.1 and let system 1 be stable and system 2 be unstable. Also, assume that the performance measure is a quadratic cost function. The sequence $\sigma_1 = \{\Delta(1), \Delta(2), \Delta(1), \Delta(2)\}$ will leave system 2 'unattended' in the first control instance with a resulting higher cost when compared to the circularly shifted sequence $\sigma_2 = \{\Delta(2), \Delta(1), \Delta(2), \Delta(1)\}$.* •

A visual landscape analysis can be performed by inspecting the solution space as shown in the following example.

Example 5.2 *Figure 5.2 shows the quadratic cost for all the d^p sequence*

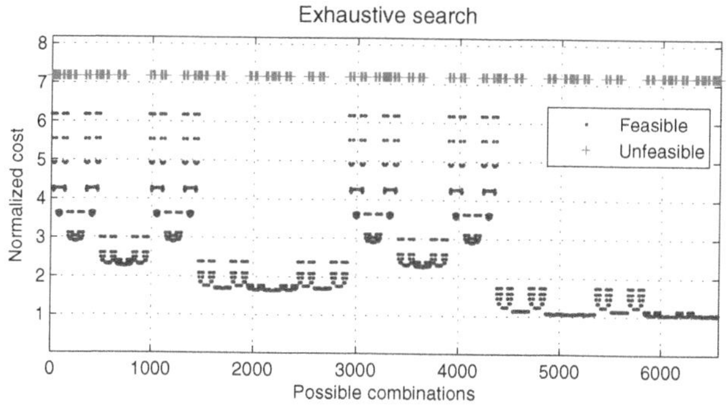

FIGURE 5.2: Typical quadratic costs for all the possible sequences applied to a system with 3 nodes and a communication sequence of length 8. The sequences are distributed according to the $(3, 8)$-Gray code.

permutations for a system with actuator scheduling only where $n_u = 3$, $\Delta = \{\Delta(1), \Delta(2), \Delta(3)\}$ (where 1, 2 and 3 identify the nodes corresponding to their respective actuators) and $p = 8$.[1] *All the possible combinations have been encoded using the (d, p)-Gray code algorithm of [43]. This code has the property that each pair of adjacent sequences differs in only one digit and the difference*

[1]Recall that d is the number of nodes, p is the communication sequence length, n_u is the number of actuators and Δ is the set of nodes $\Delta(i)$.

is either $+1$ *or* -1, i.e.

$$\begin{aligned} sequence\ 1 =& \{1,1,1,1,1,1,1,1\} \\ sequence\ 2 =& \{1,1,1,1,1,1,1,2\} \\ sequence\ 3 =& \{1,1,1,1,1,1,1,3\} \\ sequence\ 4 =& \{1,1,1,1,1,1,2,3\} \\ sequence\ 5 =& \{1,1,1,1,1,1,2,2\} \\ sequence\ 6 =& \{1,1,1,1,1,1,2,1\} \\ sequence\ 7 =& \{1,1,1,1,1,1,3,1\} \\ & \vdots \\ sequence\ 6560 =& \{3,3,3,3,3,3,3,2\} \\ sequence\ 6561 =& \{3,3,3,3,3,3,3,3\} \end{aligned} \tag{5.3}$$

(where for simplicity only the node identifier i *was written instead of the full node name* $\Delta(i)$*). This allows to see the changes in cost for neighbor solutions as a Hamiltonian path in a generalized hypercube network in* $(\mathbb{Z}_d)^p$ *(note that for an odd* d*, like in this case, the first and the last combinations cannot be neighbors). All the unfeasible solutions have been plotted with a symbolic high cost (crosses in the figure). Figure 5.2 is in fact the result of an exhaustive search.*

The set up for this example is discussed in more detail in Example 6.1, p. 138. It has been introduced here only to show the discontinuity of the search space in terms of neighbor solutions for landscape analysis. It should be appreciated that small changes in the sequence can cause large changes in cost or even lead to unfeasibility. Also, there are regions between unfeasible sequences that delimit a set of sequences of which cost is low compared to the rest. These sequences can be regarded as 'robust' in the sense that small permutations still lead to near-to-optimal solutions. ♦

Let us now consider the feasibility issue only for the actuator scheduling (the analysis is similar and can be extended to sensor and demand scheduling). Here the issue is modified in the following way.

Definition 5.4 *If every node from the set of nodes* Δ *appears in the sequence at least once, then the sequence is called admissible. Otherwise, it will be called inadmissible.* ▲

The admissibility, as defined in Definition 5.4, is a reasonable practical concept. This is because in practice a node often corresponds to a physical node as a microprocessor unit and, if a physical node does not need to be scheduled, then it probably does not need to exist in the network. Hence, admissible sequences are those sequences that schedule every physical node connected to the network at least once. Feasibility often corresponds to admissibility but it is not equivalent.

Recall that $\mathfrak{F}$ is the subspace of feasible solutions and $\mathfrak{U}$ the subspace of unfeasible ones. The relationship between $\mathfrak{F}$ and $\mathfrak{U}$ is analyzed in the following observation with the assumption that feasibility corresponds to admissibility (this assumption will made often from now on).

Observation 5.1 *Let us assume a communication sequence of length p, where only one of the nodes can be scheduled at any time step. If* $|\mathfrak{S}| = d^p$ *is the size of the solution space, the number of feasible and unfeasible permutations* ($|\mathfrak{F}|$ *and* $|\mathfrak{U}|$ *respectively) depends on d and p and can be found analytically as follows*

$$|\mathfrak{F}| = |\mathfrak{S}| - |\mathfrak{U}| \qquad \textit{and} \qquad |\mathfrak{U}| = \sum_{i=1}^{m-1} c(i)\binom{m}{i}, \tag{5.4}$$

where

$$c(i) = i^p - \sum_{j=1}^{i} c(i-j)\binom{i}{i-j}, \quad c(0) = 0, \tag{5.5}$$

and $\binom{n}{k} = \frac{n!}{k!(n-k)!}$. ▼

Proof. The size of the solution space $|\mathfrak{S}|$ is equal to the number of permutations with repetitions which is d^p. What needs to be counted now is the number of permutations $|\mathfrak{U}|$ when the order matters and a node can be chosen more than once but at least once. Let $0 \leq i < d$, then count the number of permutations $c(i)$ where only i nodes appear at least once. These are all unfeasible sequences as long as $i < d$:

$$\begin{aligned}
i = 0 \quad &\Rightarrow \quad c(0) = 0 \\
i = 1 \quad &\Rightarrow \quad c(1) = 1^p \\
i = 2 \quad &\Rightarrow \quad c(2) = 2^p - 1\binom{2}{1} = 2^p - 2 = 2^p - c(1)\binom{2}{1} \\
i = 3 \quad &\Rightarrow \quad c(3) = 3^p - (2^p - 2)\binom{3}{2} - 1\binom{3}{1} = 3^p - c(2)\binom{3}{2} - c(1)\binom{3}{1} \\
&\ \ \vdots
\end{aligned} \tag{5.6}$$

and the general recursion is (by inspection)

$$\begin{aligned}
c(i) &= i^p - c(i-1)\binom{i}{i-1} - c(i-2)\binom{i}{i-2} - \ldots - c(1)\binom{i}{1} \\
&= i^p - \sum_{j=1}^{i} c(i-j)\binom{i}{i-j}.
\end{aligned} \tag{5.7}$$

Note that the number of permutations $c(i-n)$ with $(i-n)$ nodes and $1 \leq n \leq i$ is subtracted from the number of permutations i^p and the result is multiplied

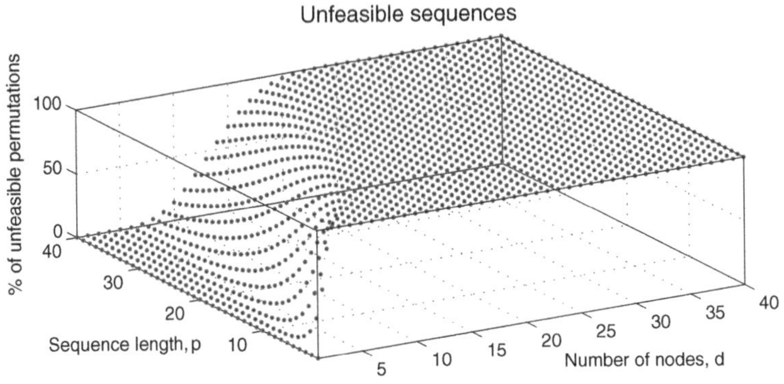

FIGURE 5.3: Percentage of unfeasible sequences in $\mathfrak{S}$ as a function of d and p.

by all possible combinations. Finally, the number of permutations where only $d-1$ objects appear at least once is calculated by summing all $c(i)$ multiplied by $\binom{d}{i}$ (the number of possible combinations in which it can appear). ■

Figure 5.3 shows the percentage of unfeasible sequences in $|\mathfrak{S}| = d^p$ as a function of the number of nodes d and the sequence length p for $1 \leq \{p, d\} \leq 40$. As the length of the sequence p gets closer to the number of nodes d, the number of unfeasible solutions $\mathfrak{U}$ increases until, when $d = p$, it becomes $|\mathfrak{U}| = d^p - d!$ (and therefore $|\mathfrak{F}| = d!$). Clearly, for $d > p \Rightarrow \mathfrak{F} = 0$. The ratio between $|\mathfrak{F}|$ and $|\mathfrak{U}|$ is important to consider when optimizing for sequences with fixed length. It gives a measure of the probability of the algorithm to end up in intermediate unfeasible solutions. The algorithms proposed here are implemented to cater for this.

5.3 Optimization algorithms

The analysis in the previous section will be used to select suitable parameters for the optimization algorithms and enhance their performance. Before the algorithms are presented, some definitions are needed.

The algorithm run time is expressed here as the number of evaluations of the cost function as this is the predominantly expensive computation. The solution space grows exponentially as the number of nodes and/or the sequence length increases. An exhaustive search will run in $O(d^p)$ time. This problem is similar to the symmetric Traveling Salesman Problem (TSP) with the difference that in this case the salesman (the scheduler) is allowed to return to a city (a node) already visited more than once. Also in this case it is difficult to specify the cost (like an Euclidian distance for example) of the path between two adjacent control actions as this is highly dependent on the previous and future actions. The TSP is strongly NP-hard.

The highly discontinuous and unstructured search space (see Figure 5.2) prevents any sort of gradient descent or local search algorithm to find satisfactory solutions. The results will be highly dependent on the initial states of the algorithm. The heuristics proposed by [105] greatly simplify the problem by breaking it into three disjoint parts but it is only valid for NCSs where $|\Delta(i)| = 1$, for all $i = 1, 2, \ldots, d$, (*i.e.* only one (actuator) signal per node [105]).

Definition 5.5 *For a periodic sequence σ, the elements of the set*

$$\mathcal{N}_\Delta = \{\mathcal{N}_{\Delta(1)}, \mathcal{N}_{\Delta(2)}, \cdots, \mathcal{N}_{\Delta(d)}\} = \{\mathcal{N}_{\Delta(i)}\}_{i=1}^{d} \tag{5.8}$$

are the number of times $\Delta(i)$ appears in σ, i.e.

$$\mathcal{N}_{\Delta(i)} = \sum_{k=1}^{p} n(k), \quad n(k) = \begin{cases} 1 & \textit{if } \sigma(k) = \Delta(i) \\ 0 & \textit{if } \sigma(k) \neq \Delta(i) \end{cases}, \qquad \sum_{i=1}^{d} \mathcal{N}_{\Delta(i)} = p. \tag{5.9}$$

▲

The algorithm of [105] consists of three nested loops that optimize over the sequence length p (outer loop) by an exhaustive search, the number of instances of control action $\mathcal{N}_{\Delta(i)}$ by a local search, and the control action distribution (inner loop), again by a local search. This partitioning allows a local search approach to be effective for small, single-channel problems but strong assumptions are made. In contrast, the optimization approach proposed here exploits the characteristics of two stochastic algorithms in the attempt to find a fast, near-to-optimal solution to avoid combinatorial explosion as the problem size grows. The potential large size of the search space makes it impractical to use an exhaustive search method. The element of randomness improves the chances to find a solution which is close to an optimum [42].

5.3.1 Genetic Algorithm optimization

The simple Genetic Algorithm (GA) is modified by using an integer-coded chromosome [30] instead of the traditional binary-string chromosome [127, 104]. This type of coding is the most natural way to represent the sequence $\sigma = \{\sigma(k)\}_{k=1}^{p}$, as each 'gene' $\varrho(k)$ in a chromosome $\boldsymbol{\varrho} = \{\varrho(k)\}_{k=1}^{p}$ will hold the identifier (*e.g.* an integer number) that directly corresponds to the node to be scheduled *i.e.* $\sigma \equiv \boldsymbol{\varrho}$. This eliminates the problems of generating binary unfeasible solutions due to the encoding when $d \neq 2^i, i \in \mathbb{N}_0$ and will also eliminate the need for decoding. The selection method used is the tournament which involves running several 'tournaments' among chromosomes randomly selected from the population. The steps of the algorithm are:

Step 1: Form an initial population $\mathfrak{A}(0)$ of feasible sequences of size $\mathcal{M}_{\mathfrak{A}} = |\mathfrak{A}(0)|$ *i.e.* $\mathfrak{A}(0) = \{\sigma_i : i = 1, 2, \ldots, \mathcal{M}_{\mathfrak{A}}\}$, $\mathfrak{A}(0) \subset \mathfrak{F}$
Step 2: Evaluate the cost $J(\sigma_i)$ of each sequence
Step 3: Perform selection in the population by tournament such that $|\mathfrak{A}(iter)| = \mathcal{M}_{\mathfrak{A}}$ (where *iter* is the iteration number)
Step 4: Perform crossover according to a parents selection probability $\Pr(P)$ inversely proportional to their cost and generate the set of new feasible sequences $\mathfrak{A}_{\mathfrak{n}}(iter)$
Step 5: Add new sequences into $\mathfrak{A}(iter)$ to form $\mathfrak{A}(iter + 1) = \mathfrak{A}(iter) \cup \mathfrak{A}_{\mathfrak{n}}(iter)$
Step 6: Perform mutation according to a mutation probability $\Pr(M)$ and ensure feasibility of mutated sequences
if end-test conditions are not met go to Step 2

For small $\Pr(M)$, this algorithm converges rapidly to a single (sub)optimal solution. Large $|\mathfrak{A}(iter)|$ will give more chances to find the global optimum but it will slow the search. A large $\Pr(P)$ will encourage a wider exploration of the solution space but will also deteriorate the convergence rate or convergence may never be achieved.

5.3.2 Particle Swarm Optimization algorithm

In the Particle Swarm Optimization (PSO) algorithm an optimal solution is found by a swarm of particles 'flying' through the solution space. Although the algorithm of [64] was originally proposed to find a solution in a continuous space, it can be readapted for discrete optimization problems by truncating the real values to integers which does not significantly affect the search performance [46]. The PSO algorithm is implemented in two substantially different ways and they will be referred as PSO1 and PSO2.

5.3.2.1 PSO1

In the PSO1 algorithm, the set of coordinates $\boldsymbol{\rho} = \{\rho(k)\}_{k=1}^{p}$ of the position of a particle directly corresponds to a candidate scheduling sequence *i.e.* $\sigma \equiv \boldsymbol{\rho}$. The advantages of this algorithm is its simplicity and that it requires only few parameters to be tuned. The core of the algorithm is in fact the velocity and position update equations for the ith particle given by

$$\begin{aligned} \boldsymbol{v}_i(iter+1) &= w\boldsymbol{v}_i(iter) + c^{ind}\boldsymbol{r}_i^{ind} \circ (\boldsymbol{\rho}_i^{best} - \boldsymbol{\rho}_i) + c^{col}\boldsymbol{r}_i^{col} \circ (\boldsymbol{g}^{best} - \boldsymbol{\rho}_i), \\ \boldsymbol{\rho}_i(iter+1) &= \boldsymbol{\rho}_i + \boldsymbol{v}_i(iter), \end{aligned} \tag{5.10}$$

where $\boldsymbol{v}_i$ and $\boldsymbol{\rho}_i$, $(\boldsymbol{v}_i, \boldsymbol{\rho}_i \in \mathbb{N}^p)$ are the velocity and position vectors respectively, $0 \leq w < 1$ is an inertial constant and $\boldsymbol{\rho}_i^{best}$ and $\boldsymbol{g}^{best}$ are the personal best and global best found solution of particle i respectively. The 'personal best' is the best solution found by a single particle while the 'global best' is the best solution among all the personal best solutions of all the particles.

The scalars c^{ind} and c^{col} are the individuality and collectivity constants that determine the 'desire' of a particle to move towards the personal or global best solution. $\boldsymbol{r}_i^{ind}$ and $\boldsymbol{r}_i^{col}$ are vectors with random elements in the range $[0,1]$ and $\circ$ is the Hadamard product. The steps of the algorithm are:

Step 1: Form an initial population $\mathfrak{A}(0)$ of feasible sequences of size $\mathcal{M}_{\mathfrak{A}} = |\mathfrak{A}(0)|$ *i.e.* $\mathfrak{A}(0) = \{\sigma_i : i = 1, 2, \ldots, \mathcal{M}_{\mathfrak{A}}\}$, $\mathfrak{A}(0) \subset \mathfrak{F}$
Step 2: Evaluate the cost $J(\sigma_i)$ of each sequence
Step 3: Store the best sequences found by individual particles, $\boldsymbol{\rho}_i^{best}$
Step 4: Store the global best sequence found, $\boldsymbol{g}^{best}$
Step 5: Calculate new sequences using (5.10)
Step 6: If the new sequence is not feasible, fly back to previous one
if end-test conditions are not met go to Step 2

A large inertial constant w will allow a wider search through the solution space but will slow the convergence rate. Also, the choice of the parameters c^{ind} and c^{col} is determinant for the convergence of the algorithm. With $c^{ind} > c^{col}$ the particles will have a strong attraction towards their personal best found solution $\boldsymbol{\rho}_i^{best}$. Because of the large number of local minima with similar size basins of attraction, the particles may never converge to a steady-state solution. With $c^{ind} < c^{col}$, the particles will be more attracted towards a global (sub)optimal solution and they will quickly converge to this.

5.3.2.2 PSO2

In the PSO2 algorithm, the search space is partitioned into two disjoint subproblems. The first subproblem is an optimization over $\mathcal{N}_\Delta$ (Definition 5.5) and it is solved in the following way: the set of coordinates $\boldsymbol{\rho}$ of a particle corresponds to the number of times a node appears in the sequence. In other words, the algorithm is used to optimize $\mathcal{N}_\Delta \equiv \boldsymbol{\rho}$ rather than σ directly. This gives another important advantage which is the fact that the sequence length p does not need to be specified (as for the GA and PSO1) but it automatically becomes an optimization parameter. The second subproblem is an optimization over the distribution of control actions. This can be indeed solved with the neighborhood search of [105] but numerical analysis showed that it is highly inefficient, especially for large problems. Instead, once an optimal $\mathcal{N}_\Delta$ is found, a sequence σ can be constructed by maximizing the distance between control actions of the same actuator. The distribution algorithm is:

Step 1: $p = \sum_{i=1}^{d} \mathcal{N}_{\Delta(i)}, \quad \sigma = \{0_1, 0_2, \ldots, 0_p\}$
Step 2: $c = \max_i\{\mathcal{N}_{\Delta(i)}\}, \quad k = i|_{\mathcal{N}_{\Delta(i)}=c}$
Step 3: $q_0 = \min_{j\in\{j:\ \sigma(j)=0\}} j$
Step 4: $q = \operatorname{argmin}_{j\in\{j:\ \sigma(j)=0\}} |q_0 - j|, \quad \sigma(q) = k, \quad c = c - 1$
Step 5: $q_0 = q + \lfloor (p-q)/c \rfloor$
if $c > 0$ go to Step 4
Step 6: $\mathcal{N}_{\Delta(k)} = 0$

if $\sum_{i=1}^{d} \mathcal{N}_{\Delta(i)} \neq 0$ go to Step 2

The distribution algorithm starts from an empty sequence (Step 1) and fills it up with the nodes (Step 4) by starting from the node that has to be scheduled most often (evaluated at Steps 2 and 3). In this way, the node that is scheduled most often is the first to be distributed along the sequence starting from the beginning ($\sigma(0)$). Because of Remark 5.2, this ensures that nodes that require more attention appear early in the sequence.

Observation 5.2 *Although the even distribution of the sequence is not proved to be optimal, experience suggests that it is often the case. When the solution is not optimal, however, the variation in the cost is small.* ▼

By constraining the particle position to $\mathcal{N}_{\Delta}^{min} \leq \mathcal{N}_{\Delta(i)}$ for all i, the sequence will always be within the feasibility region if $\mathcal{N}_{\Delta}^{min} \geq 1$. Also, it is easy to keep the period length p within desired limits, for example by introducing a penalty factor in the cost function proportional to the sequence length or by constraining the particle position to $\mathcal{N}_{\Delta(i)} \leq \mathcal{N}_{\Delta}^{max}$ where $\mathcal{N}_{\Delta}^{max}$ is the maximum specified number of times that a node can be scheduled in the sequence. The latter is the method that will be used. If a particle 'flies' outside the constraint boundary, it will be placed back on the constraint boundary and its velocity reset to zero.

5.3.3 Discussion on algorithm performance

The PSO2 algorithm has the advantage of simultaneously solving two subproblems (optimization over p and $\mathcal{N}_{\Delta}$) with an effective global search and it eliminates the need to solve the third subproblem. The search in the PSO2 algorithm is confined within $\mathfrak{F}$ (the feasibility region) only, ignoring the potentially large $\mathfrak{U}$ (the unfeasible region). Disadvantages of the PSO2 algorithm include the assumption that sequences obtained by the distribution algorithm of Section 5.3.2.2 are optimal.

The GA and PSO1 algorithm eliminate any assumption on the structural properties and solve the problem as a whole. The worst case run time will be $O(\mathcal{M}_{\mathfrak{A}}\mathcal{M}_{\mathfrak{C}})$ where $\mathcal{M}_{\mathfrak{A}}$ is the population size and $\mathcal{M}_{\mathfrak{C}}$ is the number of iterations until convergence. For the GA and PSO1 algorithm there is no optimization over the sequence length. Of course an exhaustive search over $\beta_l \leq p \leq p_{max}$ (where p_{max} is the maximum allowed sequence length) would be possible. If the ratio $|\mathfrak{U}|/|\mathfrak{F}|$ is high (see Figure 5.3), sequences as offsprings (for the GA) or position (for the PSO1 algorithm) have a high probability to evolve or move into an unfeasible one and, as unfeasible sequences are rejected, fast convergence is prevented.

The GA is less sensitive to parameter changes which suggests that, for this type of problems GAs are more suited. This can be explained by the fact that GAs naturally work with discrete variables while it is possible that the forced discretization of the PSO algorithms deteriorate their search characteristics.

5.4 Constraint handling

Constraint handling ability is an important issue in the optimization of systems for practical applications. The intrinsic search characteristics of the above mentioned algorithms easily allow constrained optimization. For the time-triggered communication scheduling problem common constraints are:

(i) ensure all deadlines are met

(ii) scheduling of nodes that are not actively contributing to the dynamics of the system (*e.g.* physical nodes only concerned with performance and status monitoring, data acquisition, *etc.*).

Both (i) and (ii) translate, in terms of communication, in producing sequences where some nodes appear with a predefined frequency or a predefined minimum frequency. For example, if the sampling time is 0.01 ms and node 1 has to be scheduled exactly every 2 ticks, node 2 has to be scheduled at least every 0.04 ms and it does not matter how often node 3 is scheduled, then the sequence $\sigma = \{1, 2, 1, 3, 1, 2, 1, 3, 1, 2\}$ will meet this requirement.

Before setting the constraints for the optimization, the following fact should be noted. Let $m_i \in \mathbb{N}$, be how often node $\Delta(i)$ must appear in the sequence (node $\Delta(i)$ must be scheduled every m_i ticks). Then, all nodes are schedulable under these constraints if and only if [140]

$$\sum_{i=1}^{d} \frac{1}{m_i} \leq 1 \qquad \text{and} \qquad GCD(m_i, m_\iota) = 1, \quad \forall i, \iota, \quad i \neq \iota, \tag{5.11}$$

where $GCD(\cdot, \cdot)$ is the greatest common divisor of the elements in brackets.

For the GA and PSO1 such constraints can be handled in the following way. Optimize the unconstrained sequence in the usual way, as described above. Then, before evaluating the cost, artificially augment the sequence by inserting those nodes which are constrained according to their requirement. In this way, as for the unconstrained process, the fittest solutions will emerge from the algorithm. The difference is that such solutions also satisfy the scheduling constraints. Since the unconstrained sequence is guaranteed to be feasible, the constrained one will also be feasible.

The PSO2 optimizes $\mathcal{N}_{\Delta(i)}$ (the number of times node i is scheduled in the sequence). If the lower limit, $\mathcal{N}_{\Delta(i)}^{min}$, for the dimension i of the search space is dynamically set to $\mathcal{N}_{\Delta(i)}^{min} \geq \left\lceil \frac{p}{m_i} \right\rceil$, then it is guaranteed that, after the equally spaced distribution, node i is scheduled at least every m_i nodes. Hence, constrained scheduling according to (i) and (ii) above can be easily satisfied.

5.5 Optimizing for Δ

So far, it was assumed that, for a communication network $N = (\Delta, \sigma)$, σ is the parameter to be optimized while Δ (the set of nodes) is given. This is acceptable since Δ depends on the physical configuration of the NCS and normally it cannot be designed. If it was possible to freely design Δ, then $\Delta = \{\Delta(all)\}$ where $\Delta(all) \stackrel{\text{def}}{=} \{all\ signals\}$, (*i.e.* the node with all the signals associated to it) would be the optimal solution. In this case the system is not an NCS anymore because the communication is ideal. However, if the size of the elements of Δ is appropriately constrained, then the flexibility of the algorithms proposed also allows for the optimization of Δ.

A candidate sequence in the population of possible solutions is a list of node identifiers. Nodes, like sequences, are a combination of objects (signals) that, if they had to be optimized by an exhaustive search, would suffer from combinatorial explosion. Instead, they could be appended at the end of a sequence to form an augmented sequence where the first p elements form the actual communication sequence and the rest of the elements, divided into d sets, are the nodes. For example,

$$\{\underbrace{\sigma(0), \sigma(1), \ldots, \sigma(p-1)}_{\sigma}, \underbrace{\underbrace{\star, \star, \ldots, \star}_{\Delta(1)}, \underbrace{\star, \star, \ldots, \star}_{\Delta(2)}, \cdots, \underbrace{\star, \star, \ldots, \star}_{\Delta(d)},}_{\Delta}\} \tag{5.12}$$

is an augmented sequence that contains the actual communication sequence σ and the signals, represented by $\star$, to be assigned to each node $\Delta(i)$, $i = 1, 2, \ldots d$ to form the optimized set of nodes Δ. From the practical viewpoint, node optimization is a way to decide what signals should be assigned to the same node for an improved performance. This brief discussion is included for completeness. Node optimization is only considered here as an example in Section 6.4.

5.6 Optimization of NCSs which are multirate systems

In a multirate system, the sampling rates of individual signals can be different. It was shown in Section 3.7 that NCSs generalize the concept of multirate systems and therefore any multirate system can be formulated under the framework discussed here.

In this context, the optimization of NCSs as multirate systems consists of finding optimal signal rates, *i.e.* finding periodic transmission times that minimize a cost function. This is equivalent to finding optimal communication sequences that, among other constraints, satisfy the one that scheduling of the same signal must be equidistant in time (hence the systems can be implemented as a multirate one). The approach here is to optimize a tuple of sampling times of each signal and then, from this tuple, construct the communication sequence.

Let h be the minimum possible sampling time (or the base sampling time or time tick) and signals can have sampling times that are integer multiples of h, *i.e.* permissible ones are mh, $m \in \mathbb{N}$. It could also be assumed, for practical reasons, that $m < m_{max}$ where m_{max} is the maximum periodicity.

Definition 5.6 *Let the tuple of sampling times be*

$$\zeta \stackrel{def}{=} (h_1, h_2, \ldots, h_n), \tag{5.13}$$

where $h_i = m_i h$ is the sampling time of signal i, n is the number of signals and $m_i \in \mathbb{N}$. ▲

Once a tuple of sampling times is given, the method in Section 3.7 can be used to construct a communication sequence σ from ζ which always exists.

Notice that σ and ζ are substantially different because, while the elements of σ indicate which signal (or set of signals) should be transmitted at each time tick, the elements of ζ are the sampling times of each signal. However, this difference does not affect the optimization procedure, as σ and ζ are in essence two ordered lists of numbers selected from finite sets. The optimization algorithms proposed, given an appropriate cost function, can optimize them regardless of their meaning.

5.6.1 Bus occupancy as a constraint

If no other constraints are introduced to the multirate system optimization problem stated above, the optimal solution would be the one with ζ having elements $h_i = h$ for all i, *i.e.* the one that gives the shortest possible sampling time for each signal. This solution allows the maximum exchange of data at the expense of 'bus occupancy'. Let us define the bus occupancy as

$$O = \sum_{i=1}^{n} \frac{h}{h_i}. \tag{5.14}$$

For example, if only two signals, *signal* 1 and *signal* 2, are scheduled with a periodicity of 2 and 4 respectively (*i.e.* with sampling times of $2h$ and $4h$), then, it will be said that this scheduling gives a bus occupancy of $O = \frac{h}{2h} + \frac{h}{4h} = 0.75$. The bus occupancy defined here does not take into account the actual transmission times as, since time-triggered communication is considered, it is assumed that the whole time slot (of length h) assigned to a signal is occupied. Hence, $O = 1$ means that each available time slot has been assigned to a single signal. It will be shown in Section 9.5 that, for the communication protocol FlexRay, it is possible to assign more than one signal per time slot. Hence, the bus occupancy can be a number larger than 1. O is a measure of how much data is sent through the communication bus.

The bus occupancy constraint can be handled in a similar way as for the other constraints (see Section 5.4). Hence, for the GA, if an offspring solution does not satisfy the bus occupancy constraint it will be rejected as an unfeasible one. For the PSO1 and PSO2, if a particle 'flies' outside the feasibility region, it will be brought back to its previous position and its velocity reset to zero. A limit on the maximum allowed bus occupancy, O_{max}, can be used as a constraint for the optimization.

It was mentioned in Section 5.3 that communication sequence optimization is analogous to the traveling salesman problem as each element in the sequence could be seen as a city the salesman has to visit. In a similar fashion, communication rate optimization is analogous to another famous problem called the knapsack problem. This is a problem in combinatorial optimization that states that, given a set of items, each with a value and a weight, determine which items to include in the 'knapsack' so that the total weight does not exceed a given limit and the total value is maximized [63]. The analogy for the communication rate problem is that the 'weight' is the sampling time h_i, the weight 'limit' is the limit on the bus occupancy O_{max} and the 'total value' is given by the cost function (that will be discussed in Chapters 6, 7 and 8).

5.7 Applying the optimization to the vehicle brake-by-wire control system

Let us consider the example of Section 3.9, p. 81. In that example, a brake-by-wire control system was modeled but the optimality of the communication sequence was not addressed. Finding the communication sequences that solve the problem in (5.2) is the optimization problem considered here. Let us say, for the sake of simplicity, that the optimization objective is to minimize the ℓ_2-gain (or the $\mathcal{H}_\infty$-norm) between *Brake demand* (the system input) and *Vehicle linear velocity error* (a system's artificial output which is an error between the demand signal and the vehicle actual velocity). The first action is to find the equations for the augmented closed-loop system $\hat{\mathcal{P}}_{cl}$ (as shown in Section 3.3.3 but with the due considerations) where, in this case

$$\hat{\mathcal{P}}_{cl}: \quad \textit{Brake demand} \mapsto \textit{Vehicle linear velocity error}. \tag{5.15}$$

As a consequence of the bus and task scheduling dynamics, $\hat{\mathcal{P}}_{cl}$ is periodic, with a periodicity of $p = \text{LCM}(p_E, p_t)$. Periodicity is eliminated by transforming $\hat{\mathcal{P}}_{cl}$ into the equivalent system $\bar{\mathcal{P}}_{cl}$. This is obtained by applying the discrete-time lifting operator

$$\bar{\mathcal{P}}_{cl} = \mathbb{L}_D\{\hat{\mathcal{P}}_{cl}, p\}, \tag{5.16}$$

(see Definitions 3.7, p. 69). $\bar{\mathcal{P}}_{cl}$ is an LTI discrete model.

The cost function (the ℓ_2-gain of this system) is defined as

$$J(\sigma^E, \sigma^t) = \sup_{\|Brake\ demand\|_{\ell_2} \neq 0} \frac{\|\,Vehicle\ linear\ velocity\ error\|_{\ell_2}}{\|Brake\ demand\|_{\ell_2}} \tag{5.17}$$

where the ℓ_2-norm of a discrete signal $\boldsymbol{\nu}(k)$ is defined in the usual way as

$$\|\boldsymbol{\nu}(k)\|_{\ell_2}^2 = \sum_{k=0}^{\infty} \boldsymbol{\nu}(k)^T \boldsymbol{\nu}(k), \tag{5.18}$$

and the supremum, sup, is taken over all the nonzero *brake demand* trajectories. The optimization problem is to find sequences σ^E and σ^t for the ECU and task scheduling respectively that minimize $J(\sigma^E, \sigma^t)$. Let us merge these two sequences together and define

$$\sigma \stackrel{\text{def}}{=} \{\sigma^E, \sigma^t\}. \tag{5.19}$$

This does not affect the optimization process as the resulting optimal σ will be split again into σ^E and σ^t. Similarly, define a unique set of nodes

$$\Delta \stackrel{\text{def}}{=} \Delta^E \cup \Delta^t. \tag{5.20}$$

Finally, the optimization problem is

$$\min_{s.t. \mathcal{P}_{cl}, \Delta} J(\sigma). \tag{5.21}$$

The result of the proposed algorithms (GA, PSO1 and PSO2) will be a sequence $\sigma_{opt} = \{\sigma^E_{opt}, \sigma^t_{opt}\}$ that is optimal (or suboptimal) in the sense that it minimizes the specified cost function.

5.8 Summary

Finding optimal communication sequences for NCSs is a complex combinatorial optimization problem. This chapter has shown that it is theoretically possible to solve this problem and fast stochastic algorithms, that naturally handle constraints (an important feature for practical applications), were proposed. A method, given by some analysis of the fitness landscape that the cost function gave rise to, was given for tuning the parameters of the optimization algorithms.

6

Optimal controller and schedule codesign

CONTENTS

Optimal control design is an old and well established discipline. It gives us the confidence to say that those controllers are the 'best' possible controllers according to some assumptions. One of these assumptions is that controllers and plants that are controlled can communicate freely like friends on a conference call on Skype. Because in reality the connection is via a shared network, the communication becomes more like the one of a push-to-talk radio system where many users share the same frequency and only one person can talk at a time. Under these circumstances, the 'optimal' controller is not optimal anymore....

But we also have the freedom to design the communication policy (schedule). And if we find a way to cast the controller and schedule design into a single optimization problem, then we will regain the confidence to say that the controller and the schedule are the 'best' possible controller-schedule combination.

6.1 Introduction

In Chapter 4, it was shown how, in an NCS with scheduled communication between the system components, the choice of particular communication policies can alter the structural properties of the system. In other words, controllability and observability can be lost if the communication sequence is not chosen appropriately. This is an important result that allows partitioning of the set of all possible communication sequences into feasible (those sequences that preserve the structural properties) and unfeasible (those sequences that do not). In the previous chapter, three algorithms were introduced to efficiently solve the combinatorial problem of finding optimal solutions from the set of the feasible ones. However, no particular cost functions were specified.

The aim of this chapter is to define a cost function for finding sequences (in the subset of the feasible ones) that are optimal under a quadratic cost. Furthermore, the aim is to design a controller, together with the communication sequence, such that the overall codesign is optimal.

In the process of development of real-time systems, there is often a gap between control theory and computer science [122]. Limitations on the practical implementation lead to an overall suboptimal design. This is equivalent to designing a controller for a plant model with important unmodeled dynamics. If the bandwidth constraints of the communication medium are considered at the controller design stage, the performance of the controller significantly increases [80, 105, 14, 41].

The focus is on contention-free protocols. The main problem to be solved is the following: given a continuous-time infinite horizon LQR problem for a distributed system, find a scheduler with a fixed periodic communication sequence and the corresponding sampled-data controller based on the continuous-time LQR cost that takes into account the limited communication medium and inter-sampling behavior. This problem was introduced in [105] and reconsidered in [14]. A subset of the problem has been investigated in [41] where only a fully decentralized control system with decoupled plant is discussed. In [150, 56, 51], it was proved that communication sequences that preserve reachability and observability exist and have an upper bound on their period but the optimality of such sequences is not addressed.

The joint optimization problem of finding an optimal communication sequence with an optimal controller has been treated in [105, 80, 14, 75]. The solution to the complex combinatorial optimization problem proposed by [105] is a heuristic algorithm based on a partition of the problem into three subproblems. The subproblems are solved within nested loops with a combination of exhaustive and local searches. Only the single-channel case is considered *i.e.*, only one control signal can be updated at any time tick. In [80], the (sub)optimal sequence for an $\mathcal{H}_\infty$ control problem is solved by using another heuristic; the sequence period is gradually increased and the related optimal sequence for this period is computed by a (presumably exhaustive) search algorithm until the cost converges. In [14], the LQ problem is translated into a mixed integer quadratic programming formulation and solved using a branch and bound based method. It is extremely dependent on the chosen initial states and developed for finite horizon problems only. Also, an appropriate theoretical justification for the branch and bound algorithm remains limited. In [75], a tree pruning technique is used and optimality for finite horizon problems is proven if the number of pruned branches is kept small. This is in fact a compromise between an almost exhaustive search with a near-to-optimal solution or a faster search with a suboptimal solution. The result depends on the bounding (or pruning) parameter.

The approach here is to start from a standard continuous-time LQR problem and obtain the equivalent sampled-data representation. Then, model the limited communication channel as shown in Section 3.3 (p. 53) and merge the two models to obtain an augmented model which includes the dynamics of the scheduler. Since the pre-planned schedule is periodic, the resulting LQ problem will also be periodic. By using the discrete-time lifting technique (Section 3.5, p. 63) for a given sequence, this periodicity can be eliminated by creating a higher dimensional system and the solution to the problem will be given by solving a Discrete Algebraic Riccati Equation (DARE).

The optimization is performed by the stochastic algorithms presented in the previous chapter.

6.2 Problem formulation

The formulation of the problem of plants connected to a controller via a limited communication network with time-triggered communication was extensively discussed in Section 3.3, p. 53. It was then reiterated in Chapter 4 for the controllability/observability problem. Hence, the modeling process is only summarized here and the focus is on the formulation of the quadratic cost function needed for the codesign of an optimal controller and schedule.

6.2.1 NCS model

Consider, as an example, the NCS shown in Figure 6.1. The plant actuators are spatially distributed and the limited communication medium used for the actuator signals is represented by a shared bus. A scheduler acts in the form of switches between the end-points and the bus. For simplicity, and direct comparison to [105], sensor signals are not subject to bus communication constraints; in that case, sensor signals can also be scheduled and an observer-based controller can be implemented [150]. Only a limited number of actuators (via a physical node) can be controlled and the assumption is that the actuator inputs latch so that when communication stops the actuator holds its signal value (ZOH). The assumption is that the spatially distributed plant $\mathcal{P}^c$ is an LTI system described by

$$\mathcal{P}^c: \quad \dot{\boldsymbol{x}}(t) = \boldsymbol{A}^c\boldsymbol{x}(t) + \boldsymbol{B}^c\hat{\boldsymbol{u}}(t), \tag{6.1}$$

where $\boldsymbol{x}(t) \in \mathbb{R}^{n_x}$, $\hat{\boldsymbol{u}}(t) \in \mathbb{R}^{n_u}$ and $(\boldsymbol{A}^c, \boldsymbol{B}^c)$ is a controllable pair. The system described by (6.1) may be a decoupled system communicating the control signals over a single bus. The control input $\hat{\boldsymbol{u}}(t)$ is a discrete input signal

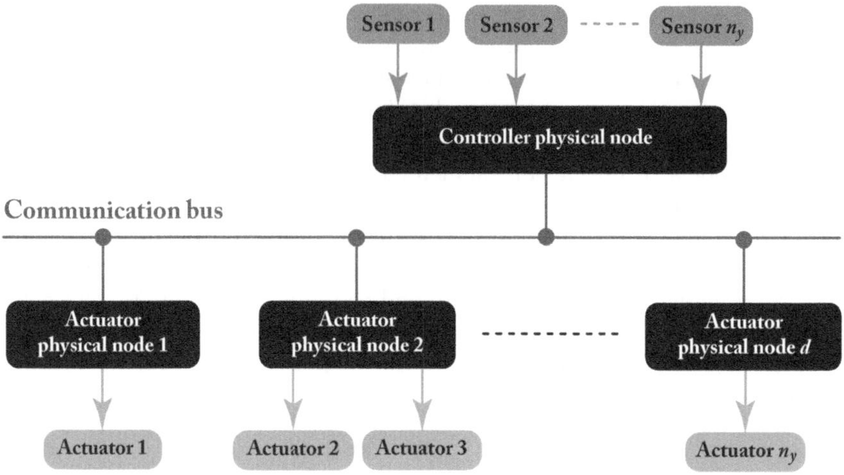

FIGURE 6.1: Architecture of an NCS with limited communication: the actuators communicate through a shared bus and a reduced number of control signals can be updated at any time tick.

created by a sample-and-hold element for a constant sampling period h, *i.e.*

$$\hat{\boldsymbol{u}}(t) = \hat{\boldsymbol{u}}(jh) = constant \quad \text{for} \quad t \in [jh, (j+1)h). \tag{6.2}$$

The model in (6.1) has to be sampled with a periodic sampling interval h, giving the sampled-data system

$$\mathcal{P}: \quad \boldsymbol{x}(jh+h) = \boldsymbol{A}\boldsymbol{x}(jh) + \boldsymbol{B}\hat{\boldsymbol{u}}(jh), \tag{6.3}$$

where $\boldsymbol{A}$ and $\boldsymbol{B}$ are given in (2.6) (and can be computed using the matrix exponential method described in [125]) and j is the sampling instant. For simplicity, the sampling period h will be omitted when obvious.

The next step is to obtain a model for the scheduler that emulates the limited communication channel (Section 3.3, p. 53). This is characterized by the communication network $N = (\Delta, \sigma)$ where $\sigma = \{\sigma(k)\}_{k=0}^{p-1}$ is the p-periodic communication sequence (Definition 3.6, p. 64) and $\Delta = \{\Delta(i)\}$ is the set of nodes (Definition 3.1, p. 50) that in this case only contains actuator nodes. The model for the actuator scheduler is $\mathcal{S}_A$ and it is given, for periodic sequences, in (4.2).

The plant $\mathcal{P}$ state evolution, with the limited communication (*i.e.* augmented with $\mathcal{S}_A$), can be described by the *augmented plant*

$$\hat{\mathcal{P}}: \quad \hat{\boldsymbol{x}}(j+1) = \hat{\boldsymbol{A}}(k)\hat{\boldsymbol{x}}(j) + \hat{\boldsymbol{B}}(k)\boldsymbol{u}(j), \tag{6.4}$$

where the matrices are given in (4.4). $\hat{\mathcal{P}}$ is periodic in p (see Section 3.3.1, p. 58 for more detail). Periodicity is removed by applying the discrete-time lifting operator $\bar{\mathcal{P}} = \mathbb{L}_D\{\hat{\mathcal{P}}, p\}$ to obtain an LTI equivalent system, $\bar{\mathcal{P}}$, that models the original plant plus the actuator's limited communication. It is given by

$$\bar{\mathcal{P}}: \quad \hat{\boldsymbol{x}}(pl+p) = \tilde{\boldsymbol{A}}_p \hat{\boldsymbol{x}}(pl) + \tilde{\boldsymbol{B}}_p \bar{\boldsymbol{u}}(pl), \quad l = 0, 1, \ldots, \tag{6.5}$$

where

$$\tilde{\boldsymbol{A}}_p = (\boldsymbol{e}_p^T \otimes \boldsymbol{I})\bar{\boldsymbol{A}}, \quad \tilde{\boldsymbol{B}}_p = (\boldsymbol{e}_p^T \otimes \boldsymbol{I})\bar{\boldsymbol{B}},$$
$$\bar{\boldsymbol{u}}(pl) = \begin{bmatrix} \boldsymbol{u}(pl) \\ \boldsymbol{u}(pl+1) \\ \vdots \\ \boldsymbol{u}(pl+p-1) \end{bmatrix}, \tag{6.6}$$

and

$$\bar{\boldsymbol{A}} = \begin{bmatrix} \prod_{i=1}^{1} \hat{\boldsymbol{A}}(1-i) \\ \prod_{i=1}^{2} \hat{\boldsymbol{A}}(2-i) \\ \vdots \\ \prod_{i=1}^{p} \hat{\boldsymbol{A}}(p-i) \end{bmatrix},$$
$$\bar{\boldsymbol{B}} = \begin{bmatrix} \hat{\boldsymbol{B}}(0) & \boldsymbol{0} & \cdots & \boldsymbol{0} \\ \left(\prod_{i=1}^{1} \hat{\boldsymbol{A}}(2-i)\right)\hat{\boldsymbol{B}}(0) & \hat{\boldsymbol{B}}(1) & \cdots & \boldsymbol{0} \\ \left(\prod_{i=1}^{2} \hat{\boldsymbol{A}}(3-i)\right)\hat{\boldsymbol{B}}(0) & \left(\prod_{i=1}^{1} \hat{\boldsymbol{A}}(3-i)\right)\hat{\boldsymbol{B}}(1) & \cdots & \boldsymbol{0} \\ \vdots & \vdots & \ddots & \vdots \\ \left(\prod_{i=1}^{p-1} \hat{\boldsymbol{A}}(p-i)\right)\hat{\boldsymbol{B}}(0) & \left(\prod_{i=1}^{p-2} \hat{\boldsymbol{A}}(p-i)\right)\hat{\boldsymbol{B}}(1) & \cdots & \hat{\boldsymbol{B}}(p-1) \end{bmatrix}. \tag{6.7}$$

Column vector $\boldsymbol{e}_k^T$, $1 \le k \le p$, is the p-dimensional standard basis vector (for more detail see Section 3.5, p. 63). Notice that $\tilde{\boldsymbol{A}}_p$ and $\tilde{\boldsymbol{B}}_p$ are the p-lifted system matrices but in general, for $1 \le k \le p$, the k-lifted matrices can be written as

$$\tilde{\boldsymbol{A}}_k = (\boldsymbol{e}_k^T \otimes \boldsymbol{I})\bar{\boldsymbol{A}}, \quad \tilde{\boldsymbol{B}}_k = (\boldsymbol{e}_k^T \otimes \boldsymbol{I})\bar{\boldsymbol{B}}, \tag{6.8}$$

which gives

$$\hat{\boldsymbol{x}}(pl+k) = \tilde{\boldsymbol{A}}_k \hat{\boldsymbol{x}}(pl) + \tilde{\boldsymbol{B}}_k \bar{\boldsymbol{u}}(pl). \tag{6.9}$$

6.2.2 Quadratic cost function

Consider the given continuous-time, infinite horizon LQ problem:

$$\min_{\hat{\boldsymbol{u}}(t)} J \quad \text{subject to (6.1)}, \tag{6.10}$$

where

$$J = \int_0^\infty \boldsymbol{x}(t)^T \boldsymbol{Q}_{c1} \boldsymbol{x}(t) + \hat{\boldsymbol{u}}(t)^T \boldsymbol{Q}_{c2} \hat{\boldsymbol{u}}(t) dt. \tag{6.11}$$

Assume that $\boldsymbol{Q}_{c1} \geq \boldsymbol{0}$ and $\boldsymbol{Q}_{c2} > \boldsymbol{0}$ are given as weights for a desirable ideal closed-loop response and that $(\boldsymbol{A}^c, \boldsymbol{Q}_{c1})$ is an observable pair. The solution to the continuous-time control problem is given by the Continuous-time Algebraic Riccati Equation (CARE)

$$(\boldsymbol{A}^c)^T \boldsymbol{P}_c + \boldsymbol{P}_c \boldsymbol{A}^c - \boldsymbol{P}_c \boldsymbol{B}^c \boldsymbol{Q}_{c2}^{-1} (\boldsymbol{B}^c)^T \boldsymbol{P}_c + \boldsymbol{Q}_{c1} = \boldsymbol{0}. \tag{6.12}$$

However, as the system is subject to communication constraints in the actuators, this has to be accounted for in the optimal solution approach. The equivalent discrete problem for the sampled system (6.3) is:

$$\min_{\hat{\boldsymbol{u}}(j)} J \quad \text{subject to (6.3)}, \tag{6.13}$$

where

$$J = \sum_{j=0}^{\infty} \begin{bmatrix} \boldsymbol{x}(j) \\ \hat{\boldsymbol{u}}(j) \end{bmatrix}^T \underbrace{\begin{bmatrix} \boldsymbol{Q}_1 & \boldsymbol{Q}_{12} \\ \boldsymbol{Q}_{12}^T & \boldsymbol{Q}_2 \end{bmatrix}}_{\boldsymbol{Q}} \begin{bmatrix} \boldsymbol{x}(j) \\ \hat{\boldsymbol{u}}(j) \end{bmatrix}, \tag{6.14}$$

and

$$\begin{gathered} \begin{bmatrix} \boldsymbol{Q}_1 & \boldsymbol{Q}_{12} \\ \boldsymbol{Q}_{12}^T & \boldsymbol{Q}_2 \end{bmatrix} = \int_0^h \begin{bmatrix} \acute{\boldsymbol{A}}(\tau)^T & \boldsymbol{0} \\ \acute{\boldsymbol{B}}(\tau)^T & \boldsymbol{I} \end{bmatrix} \begin{bmatrix} \boldsymbol{Q}_{c1} & \boldsymbol{0} \\ \boldsymbol{0} & \boldsymbol{Q}_{c2} \end{bmatrix} \begin{bmatrix} \acute{\boldsymbol{A}}(\tau) & \acute{\boldsymbol{B}}(\tau) \\ \boldsymbol{0} & \boldsymbol{I} \end{bmatrix} d\tau, \\ \acute{\boldsymbol{A}}(\tau) = e^{\boldsymbol{A}^c \tau}, \quad \acute{\boldsymbol{B}}(\tau) = \int_0^\tau e^{\boldsymbol{A}^c s} ds \boldsymbol{B}^c, \end{gathered} \tag{6.15}$$

($\boldsymbol{Q}$ can be computed using the matrix exponential method described in [125]). With the assumption that the input $\hat{\boldsymbol{u}}(t)$ is constant over the interval $t \in [jh, (j+1)h)$ and the states $\boldsymbol{x}(jh)$ are sampled at the time instant $t = jh$, there is no approximation between (6.11) and (6.14) [1]. The sampling process is fully accounted for.

After the plant has been augmented with the scheduler dynamics (6.4), the optimization problem becomes

$$\min_{\boldsymbol{u}(j)} J(\sigma) \quad \text{subject to (6.4)}, \tag{6.16}$$

where, using the substitution $j = pl + k$, $l = 0, 1, \ldots$,

$$J(\sigma) = \sum_{l=0}^{\infty} \sum_{k=0}^{p-1} \begin{bmatrix} \hat{\boldsymbol{x}}(pl+k) \\ \boldsymbol{u}(pl+k) \end{bmatrix}^T \underbrace{\begin{bmatrix} \hat{\boldsymbol{Q}}_1(k) & \hat{\boldsymbol{Q}}_{12}(k) \\ \hat{\boldsymbol{Q}}_{12}(k)^T & \hat{\boldsymbol{Q}}_2(k) \end{bmatrix}}_{\hat{\boldsymbol{Q}}(k)} \begin{bmatrix} \hat{\boldsymbol{x}}(pl+k) \\ \boldsymbol{u}(pl+k) \end{bmatrix}, \tag{6.17}$$

and

$$\hat{\boldsymbol{Q}}_1(k) = \begin{bmatrix} \boldsymbol{Q}_1 & \boldsymbol{Q}_{12}\bar{\boldsymbol{S}}_A(k) \\ \bar{\boldsymbol{S}}_A(k)^T\boldsymbol{Q}_{12}^T & \bar{\boldsymbol{S}}_A(k)^T\boldsymbol{Q}_2\bar{\boldsymbol{S}}_A(k) \end{bmatrix}, \quad \hat{\boldsymbol{Q}}_2(k) = \boldsymbol{S}_A(k)^T\boldsymbol{Q}_2\boldsymbol{S}_A(k),$$
$$\hat{\boldsymbol{Q}}_{12}(k) = \begin{bmatrix} \boldsymbol{Q}_{12}\boldsymbol{S}_A(k) \\ \bar{\boldsymbol{S}}_A(k)^T\boldsymbol{Q}_2\boldsymbol{S}_A(k) \end{bmatrix}. \tag{6.18}$$

The cost function in (6.17) comes directly from the one in (6.14) where $\hat{\boldsymbol{u}}(j)$ has been substituted with the expression (from (3.15))

$$\hat{\boldsymbol{u}}(j) = \bar{\boldsymbol{S}}_A(j)\boldsymbol{x}_A(j) + \boldsymbol{S}_A(j)\boldsymbol{u}(j), \quad \boldsymbol{x}_A(j) = \hat{\boldsymbol{u}}(j-1), \tag{6.19}$$

and the substitution $j = pl + k$ allows for partitioning of the summation into an infinite one and a finite one for the periodic matrix $\hat{\boldsymbol{Q}}(k)$.

The dependence of $J(\sigma)$ in σ is to remind us that this cost is a function of a given sequence. The dynamics of the time-varying scheduler will inevitably introduce time-varying cost function matrices $\hat{\boldsymbol{Q}}_1(k)$, $\hat{\boldsymbol{Q}}_2(k)$ and $\hat{\boldsymbol{Q}}_{12}(k)$ (p-periodic). The original cost function (6.14) has been divided into two summations (6.17) to appreciate the existence of the scheduler periodicity.

By considering the set of lifted systems given by (6.7) and (6.9), it can be implied that

$$\begin{bmatrix} \hat{\boldsymbol{x}}(pl+k) \\ \boldsymbol{u}(pl+k) \end{bmatrix} = \begin{bmatrix} \tilde{\boldsymbol{A}}_k\hat{\boldsymbol{x}}(pl) + \tilde{\boldsymbol{B}}_k\bar{\boldsymbol{u}}(pl) \\ (\boldsymbol{e}_k^T \otimes \boldsymbol{I})\bar{\boldsymbol{u}}(pl) \end{bmatrix}, \tag{6.20}$$

where the matrix $(\boldsymbol{e}_k^T \otimes \boldsymbol{I})$ extracts the particular control signal $\boldsymbol{u}(pl+k)$. Equation (6.20) will be now used in (6.17) to obtain the equivalent cost function. For the discrete-time lifted plant in (6.5), the optimization problem becomes

$$\min_{\bar{\boldsymbol{u}}(pl)} J(\sigma) \quad \text{subject to (6.5)}, \tag{6.21}$$

where

$$J(\sigma) = \sum_{l=0}^{\infty} \begin{bmatrix} \hat{\boldsymbol{x}}(pl) \\ \bar{\boldsymbol{u}}(pl) \end{bmatrix}^T \underbrace{\left(\sum_{k=0}^{p-1} \begin{bmatrix} \bar{\boldsymbol{Q}}_1(k) & \bar{\boldsymbol{Q}}_{12}(k) \\ \bar{\boldsymbol{Q}}_{12}(k)^T & \bar{\boldsymbol{Q}}_2(k) \end{bmatrix} \right)}_{\overset{\text{def}}{=} \begin{bmatrix} \tilde{\boldsymbol{Q}}_1 & \tilde{\boldsymbol{Q}}_{12} \\ \tilde{\boldsymbol{Q}}_{12}^T & \tilde{\boldsymbol{Q}}_2 \end{bmatrix}} \begin{bmatrix} \hat{\boldsymbol{x}}(pl) \\ \bar{\boldsymbol{u}}(pl) \end{bmatrix}, \tag{6.22}$$

and

$$\begin{aligned} \bar{\boldsymbol{Q}}_1(k) &= \tilde{\boldsymbol{A}}_k^T\hat{\boldsymbol{Q}}_1(k)\tilde{\boldsymbol{A}}_k, \\ \bar{\boldsymbol{Q}}_2(k) &= \tilde{\boldsymbol{B}}_k^T\hat{\boldsymbol{Q}}_1(k)\tilde{\boldsymbol{B}}_k + (\boldsymbol{e}_k^T \otimes \boldsymbol{I})^T\hat{\boldsymbol{Q}}_{12}(k)^T\tilde{\boldsymbol{B}}_k + \tilde{\boldsymbol{B}}_k^T\hat{\boldsymbol{Q}}_{12}(k)(\boldsymbol{e}_k^T \otimes \boldsymbol{I}) \\ &\quad + (\boldsymbol{e}_k^T \otimes \boldsymbol{I})^T\hat{\boldsymbol{Q}}_2(k)(\boldsymbol{e}_k^T \otimes \boldsymbol{I}), \\ \bar{\boldsymbol{Q}}_{12}(k) &= \tilde{\boldsymbol{A}}_k^T\hat{\boldsymbol{Q}}_1(k)\tilde{\boldsymbol{B}}_k + \tilde{\boldsymbol{A}}_k^T\hat{\boldsymbol{Q}}_{12}(k)(\boldsymbol{e}_k^T \otimes \boldsymbol{I}), \\ \tilde{\boldsymbol{A}}_0 &= \boldsymbol{I}, \quad \tilde{\boldsymbol{B}}_0 = \boldsymbol{0}. \end{aligned} \tag{6.23}$$

Before setting the final LQR time-invariant problem, it should be noticed that the control input dimension of the matrix $\tilde{\boldsymbol{B}}_p$ has increased by a factor of p to $n_u p$. Only $\sum_{k=0}^{p-1} |\sigma(k)|$ columns[1] via n_u actuators affect the augmented dynamics via $\tilde{\boldsymbol{B}}_p$ and the weighting matrices $\tilde{\boldsymbol{Q}}_2 \in \mathbb{R}^{n_u p \times n_u p}$ and $\tilde{\boldsymbol{Q}}_{12} \in \mathbb{R}^{n_x \times n_u p}$. Hence, $\text{rank}(\tilde{\boldsymbol{B}}_p) = n_u$ and $\text{rank}(\tilde{\boldsymbol{Q}}_2) = \sum_{k=0}^{p-1} |\sigma(k)|$, *i.e.* $\tilde{\boldsymbol{Q}}_2$ is singular. It is easily shown that there exists a binary matrix $\boldsymbol{F} \in \mathbb{R}^{n_u p \times \sum_{k=0}^{p-1} |\sigma(k)|}$ with elements 0 and 1 only such that $\text{rank}(\tilde{\boldsymbol{B}}_p \boldsymbol{F}) = n_u$, $\tilde{\boldsymbol{B}}_p \boldsymbol{F} \in \mathbb{R}^{(n_u+p)\times p}$ and $\text{rank}(\boldsymbol{F}^T \tilde{\boldsymbol{Q}}_2 \boldsymbol{F}) = \sum_{k=0}^{p-1} |\sigma(k)|$, $\boldsymbol{F}^T \tilde{\boldsymbol{Q}}_2 \boldsymbol{F} \in \mathbb{R}^{p \times p}$, *i.e.* $\boldsymbol{F}^T \tilde{\boldsymbol{Q}}_2 \boldsymbol{F}$ is non-singular. Thus, $\boldsymbol{F}$ removes all zero columns (and rows) in $\tilde{\boldsymbol{Q}}_2$ and $\tilde{\boldsymbol{Q}}_{12}$. The final LQR time-invariant problem is

$$\min_{\check{\boldsymbol{u}}} J(\sigma) \quad \text{subject to (6.5)}, \tag{6.24}$$

where

$$J(\sigma) = \sum_{l=0}^{\infty} \begin{bmatrix} \hat{\boldsymbol{x}}(pl) \\ \check{\boldsymbol{u}}(pl) \end{bmatrix}^T \begin{bmatrix} \tilde{\boldsymbol{Q}}_1 & \tilde{\boldsymbol{Q}}_{12}\boldsymbol{F} \\ \boldsymbol{F}^T \tilde{\boldsymbol{Q}}_{12}^T & \boldsymbol{F}^T \tilde{\boldsymbol{Q}}_2 \boldsymbol{F} \end{bmatrix} \begin{bmatrix} \hat{\boldsymbol{x}}(pl) \\ \check{\boldsymbol{u}}(pl) \end{bmatrix}, \tag{6.25}$$

and $\bar{\boldsymbol{u}}(pl) = \boldsymbol{F}\check{\boldsymbol{u}}(pl)$. With a given feasible sequence of length p the solution to (6.24) is given by the feedback controller

$$\begin{aligned} \bar{\boldsymbol{u}}(pl) &= -\boldsymbol{F}\bar{\boldsymbol{K}}\hat{\boldsymbol{x}}(pl) \\ &= -\boldsymbol{F}(\boldsymbol{F}^T \tilde{\boldsymbol{B}}_p^T \bar{\boldsymbol{P}} \tilde{\boldsymbol{B}}_p \boldsymbol{F} + \boldsymbol{F}^T \tilde{\boldsymbol{Q}}_2 \boldsymbol{F})^{-1} (\boldsymbol{F}^T \tilde{\boldsymbol{B}}_p^T \bar{\boldsymbol{P}} \tilde{\boldsymbol{A}}_p + \boldsymbol{F}^T \tilde{\boldsymbol{Q}}_{12}^T) \hat{\boldsymbol{x}}(pl), \end{aligned} \tag{6.26}$$

where $\bar{\boldsymbol{P}}$ is the solution of the DARE

$$\begin{aligned} \bar{\boldsymbol{P}} = \tilde{\boldsymbol{A}}_p^T \bar{\boldsymbol{P}} \tilde{\boldsymbol{A}}_p - (\tilde{\boldsymbol{A}}_p^T \bar{\boldsymbol{P}} \tilde{\boldsymbol{B}}_p \boldsymbol{F} + \tilde{\boldsymbol{Q}}_{12}\boldsymbol{F})(\boldsymbol{F}\tilde{\boldsymbol{B}}_p^T \bar{\boldsymbol{P}} \tilde{\boldsymbol{B}}_p \boldsymbol{F} + \boldsymbol{F}^T \tilde{\boldsymbol{Q}}_2 \boldsymbol{F})^{-1} \\ (\tilde{\boldsymbol{A}}_p^T \bar{\boldsymbol{P}} \tilde{\boldsymbol{B}}_p \boldsymbol{F} + \tilde{\boldsymbol{Q}}_{12}\boldsymbol{F})^T + \tilde{\boldsymbol{Q}}_1. \end{aligned} \tag{6.27}$$

Since the lifted system is considered, only one Riccati equation needs to be solved.

Remark 6.1 *The solution of* (6.27) *can be computed for given feasible sequences satisfying controllability as in Theorem 4.1, p. 92. This is also guaranteed by the fact that*

$$\begin{bmatrix} \tilde{\boldsymbol{Q}}_1 & \tilde{\boldsymbol{Q}}_{12}\boldsymbol{F} \\ \boldsymbol{F}^T \tilde{\boldsymbol{Q}}_{12}^T & \boldsymbol{F}^T \tilde{\boldsymbol{Q}}_2 \boldsymbol{F} \end{bmatrix} \geq \boldsymbol{0}, \quad \boldsymbol{F}^T \tilde{\boldsymbol{Q}}_2 \boldsymbol{F} > \boldsymbol{0}. \tag{6.28}$$

•

[1]Recall that $|\cdot|$ for a set denotes its cardinality, therefore, $|\sigma(k)|$ is the number of actuators (in this case) in the node scheduled at $\sigma(k)$.

Assuming the initial states $\hat{\boldsymbol{x}}(0)$ are random following a Gaussian process with unity variance, the expectation of the cost $J(\sigma)$ is

$$J_E(\sigma) = \mathbb{E}\{J(\sigma)\} = \mathrm{tr}(\bar{\boldsymbol{P}}). \tag{6.29}$$

Moreover, the cost equivalent to the optimality function used in [105] is

$$J_O(\sigma) = \max\left\{\lambda(\boldsymbol{P}_c^{-1}\bar{\boldsymbol{P}}_{11})\right\}, \tag{6.30}$$

where $\boldsymbol{P}_c$ is the solution of the CARE (6.12) and $\bar{\boldsymbol{P}}_{11}$ is the top left submatrix of $\bar{\boldsymbol{P}}$, *i.e.*

$$\bar{\boldsymbol{P}} = \begin{bmatrix} \bar{\boldsymbol{P}}_{11} & \star \\ \star & \star \end{bmatrix}. \tag{6.31}$$

The cost in (6.30), which is a worst case normalized cost, will be used for direct comparison with the results in [105].

Finally, the optimization problem is: find an optimal feasible sequence that minimizes the cost in (6.29) or (6.30). The associated feedback controller will be given by (6.26).

Remark 6.2 *The solvability of the LQR problem is implied from controllability*[2] *of the lifted augmented system that also depends on the communication sequence and the observability of* $(\boldsymbol{A}^{\mathrm{c}}, \boldsymbol{Q}_{\mathrm{c}1})$. *Sufficient conditions involving 'feasible' scheduling sequences (Definition 4.6, p. 99) and a more general form of non-pathological sampling frequencies are given in Chapter 4.* •

Remark 6.3 *Although the cost function here is only defined for the LQR problem, the approach is readily extended to LQG (*e.g. *[149]),* $\mathcal{H}_2$ *and* $\mathcal{H}_\infty$ *(*e.g. *[80]) design.* •

6.3 Optimal codesign

In the previous section it was shown how an LQR controller, $\boldsymbol{F}\bar{\boldsymbol{K}}$, given in (6.26), can be designed for $\bar{\mathcal{P}}$. $\bar{\mathcal{P}}$ is the plant augmented with the states of the actuator scheduler and lifted to remove its periodicity. In other words, $\boldsymbol{F}\bar{\boldsymbol{K}}$ is the controller that minimizes the quadratic cost in (6.11) for a given communication network $N = (\Delta, \sigma)$. Since the aim here is in the codesign problem, (*i.e.* in the design of an optimal controller together with an optimal communication policy) a σ needs to be found such that the combination of σ together with $\boldsymbol{F}\bar{\boldsymbol{K}}$ is optimal. The set of nodes Δ is normally given. However,

[2] The more relaxed assumption of stabilizability could be used instead.

the design of an optimal set of nodes is also possible (Section 5.5) and it will be considered later in one of the examples.

The optimization of σ, for a generic cost function $J(\sigma)$, was discussed in Chapter 5. The cost functions utilized here are $J_E(\sigma)$ (6.29) and $J_O(\sigma)$ (6.30). Notice that either $J_E(\sigma)$ or $J_O(\sigma)$ could be used for the optimization. However, $J_O(\sigma)$ gives a direct comparison with the results of [105].

The proposed search algorithms start from a population of feasible sequences where the feasibility predicate $\mathfrak{P}$ is given by the conditions in Theorem 4.1, p. 92. $\mathfrak{P}$ is used to divide the search space into $\mathfrak{F}$ and $\mathfrak{U}$, the set of feasible and unfeasible solutions respectively. As the iteration number increases, the population of candidate sequences 'evolves' (in the case of the GA) or 'flies' through the feasible search space (in the case of PSO1 and PSO2) with the aim of minimizing $J_E(\sigma)$ (or $J_O(\sigma)$). The algorithms stop when a maximum number of iterations has been reached. Parameters like population size, maximum number of iterations, *etc.* are chosen heuristically with the help of the landscape analysis described in Section 5.2.1, p. 113.

6.4 Examples

The effectiveness of the optimization approaches proposed will be demonstrated by some numerical results. The distributed system to be controlled is the inverted pendulums arranged in a circle of Example 4.6, p. 107 (Figure 4.1). The sampling period has been set to $h = 0.01$ and $K \stackrel{\text{def}}{=} k_{1,2} = k_{2,3} = \ldots = k_{m,1}$. When $K = 0$ (decoupled systems case), the time constants of the pendulum systems will depend on the length of the rod l_i according to $\tau_i = \sqrt{l_i/g}$. By varying l_i, the system's dynamics can be altered and this, together with different choices of weight matrices, will be used to show how different sequences are assigned to different system's requirements. Furthermore, the actuators share the communication medium used to control the pendulum's torque.

Example 6.1 *This example shows how the proposed algorithms are used to find optimal sequences. The pendulums were set with $K = 0$, $n_u = 3$, $\tau_1 = \tau_2 = 0.32$, $\tau_3 = 3.2$, and the cost with $Q_{c1} = Q_{c2} = \mathrm{diag}(10, \cdots, 10)$. The set of nodes is $\Delta = \{\Delta(1), \Delta(2), \Delta(3)\}$ which are nodes connected to actuators 1,2 and 3 respectively. This is the case where two systems (pendulums 1 and 2) have smaller time constants than the other one (pendulum 3). (This setup was used in Example 5.2, p. 115, to produce Figure 5.2 where the 'normalized cost' is $J_O(\sigma)$.) First, let us run the PSO2 algorithm with $\mathcal{N}_\Delta^{min} = 1$ and $\mathcal{N}_\Delta^{max} = 4$. The optimal sequence was found for $p = 5$. The reason for keeping p small is to allow a direct comparison with the results from an exhaustive search. The GA and PSO1 algorithm have been used to find an optimal sequence* *by setting* $p = 5$. *The sequences returned by the three algorithms have been shown in Figure 6.2 in comparison to their location with the entire feasible*

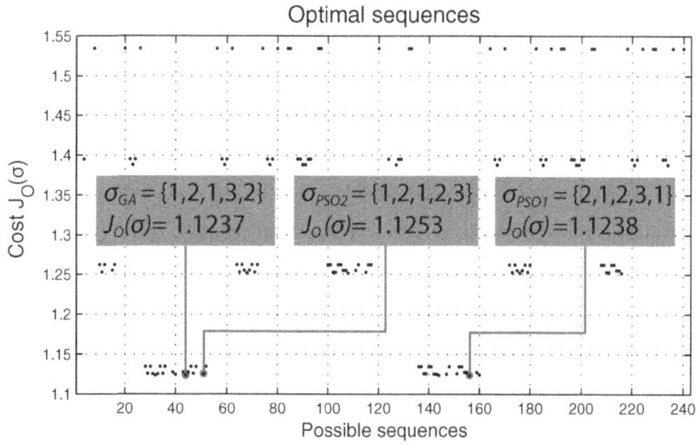

FIGURE 6.2: Optimal sequences found by the three proposed algorithms and their relation to the entire solution space.

solution space. This is represented by the dots in Figure 6.2 where the x*-axis is all the* $(3,5)$*-Gray encoded solution (see Example 5.2, p. 115) and the* y*-axis is the cost* $J_O(\sigma)$. *The three algorithms returned different solutions, with the GA having found the actual optimal one. However, the sequences found by the PSO1 and PSO2 algorithms are close to the optimal. More time slots are allocated to the systems with higher requirement. Notice that* $\mathcal{N}^{GA}_{\Delta(i)} = \mathcal{N}^{PSO1}_{\Delta(i)} = \mathcal{N}^{PSO2}_{\Delta(i)}$ *for all* i*, and therefore different costs are to be attributed to different distribution of control actions*[3]. *This result is in accordance with Remark 5.2, p. 114.*

The same set up was used but this time it was decided to also optimize the combination of actuators connected to a node as explained in Section 5.5. Let $|\Delta(i)| = 2$ *for all* i, i.e. *each node can be connected to (any) two actuators (this is the multi-channel case). The optimal sequences found with the corresponding costs are*

$$\begin{aligned} \sigma_{GA} = \sigma_{PSO1} &= \{\Delta(1,2), \Delta(1,2), \Delta(2,3)\}, \quad \textit{with} \quad J_O(\sigma) = 1.003 \\ \sigma_{PSO2} &= \{\Delta(1,2), \Delta(1,2), \Delta(1,3)\}, \quad \textit{with} \quad J_O(\sigma) = 1.003. \end{aligned} \tag{6.32}$$

where $\Delta(1,2)$ *is the node that is connected to actuators 1 and 2 and similarly for the others. Both sequences actually correspond to the two equal minima and therefore produce the same cost. The lower cost achieved compared to the previous (single-channel) example is due to higher channel capacity* i.e. *the possibility to schedule two instead of one actuator at each time step. Obviously,*

[3]Recall that $\mathcal{N}_{\Delta(i)}$ is the number of times node $\Delta(i)$ appears in the sequence (see Definition 5.5, p. 119).

letting $|\Delta(i)| = 3$ *for all* i*, means that all the actuators can be scheduled at the same time (no network). The optimal solution would be* $\Delta = \{\Delta(1,2,3)\}$, $\sigma = \{\Delta(1,2,3)\}$. ♦

Example 6.2 *This example demonstrates the algorithm performance. The algorithms are first compared with the one of [105] using their exact set up. Then, a larger scale, more realistic example is presented. For the first example the algorithms have been tested for the two identical decoupled pendulum systems with weight matrices*

$$Q_{c1} = \begin{bmatrix} 1.72 & 0 \\ 0 & 0 \end{bmatrix}, \quad Q_{c2} = \begin{bmatrix} 2000000 & 0 \\ 0 & 0 \end{bmatrix}, \tag{6.33}$$

for a desired time constant for pendulum 1 of 0.5 *seconds and pendulum 2 of* 0.015 *seconds. The left plot of Figure 6.3 shows the run time*[4] *between*

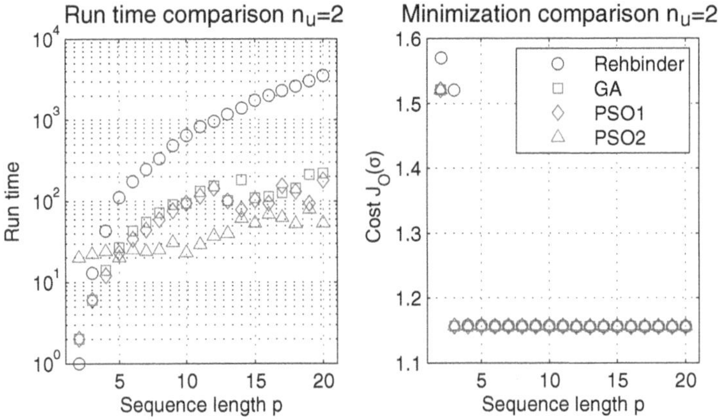

FIGURE 6.3: Run time comparison between the algorithms (on the left) and respective minimum cost (on the right) for $n_u = 2$.

the algorithm in [105] (called 'Rehbinder') and the three algorithms proposed here. Apart from the trivial sequence of length 2, for any increasing length, the time required to find an optimal solution for the Rehbinder algorithm is much larger compared to the one of the other algorithms. The three algorithms proposed here on the other hand, after an initial increase in run time, are not as affected by the length of the sequence. The right plot of Figure 6.3 gives the measure $J_O(\sigma)$ *of the quality of the solution found. Although it is not clear from the plot, the PSO1 algorithm and GA found equally or slightly better solutions than the Rehbinder algorithm for any sequence length. The results do not exactly match the ones in [105] because of Remark 5.2, p. 114.*

[4]The algorithm run time is expressed here as the number of evaluations of the cost function as this is the predominantly expensive computation (see Section 5.3, p. 118)

For the more realistic example, the system was set up with $n_u = 8$, $K = 0.1$ and

$$Q_{c1} = \text{diag}\,(1.72, 0, 250, 0, 250, 0, 1.72, 0, 1.72, 0, 250, 0, 250, 0, 1.72, 0)\,,$$
$$Q_{c2} = \text{diag}\,(1, \ldots, 1)\,, \tag{6.34}$$

(eight coupled pendulums with different control specifications). The cost function used this time is $J_E(\sigma)$ in (6.29). The results are shown in Figure 6.4. The run time of the Rehbinder algorithm becomes very large for $p > 15$. The

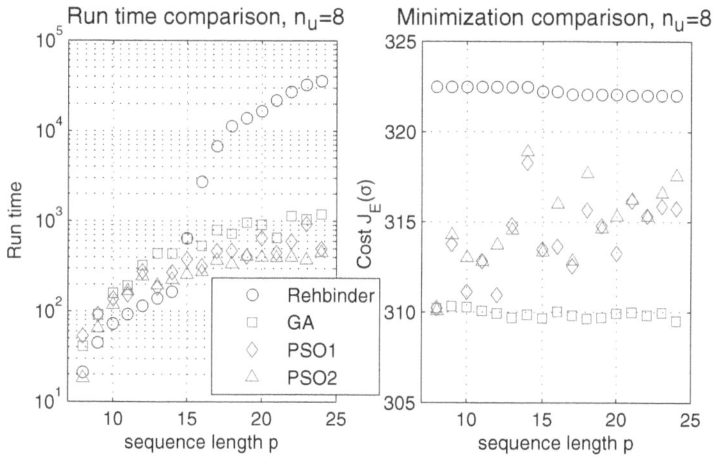

FIGURE 6.4: Run time comparison between the algorithms (on the left) and the respective minimum cost (on the right) for $n_u = 8$.

algorithms proposed here showed to solve the problem in reasonable time, where the PSO2 algorithm is still the fastest. The GA, in this case, for all p, returns better solutions than the other algorithms. ♦

6.5 Summary

An overall optimal design can be achieved if a controller and a communication schedule are optimized simultaneously. In this chapter, the joint controller and communication scheduling optimization problem was considered. The NCS was modeled using the theoretical framework of Chapter 3. A cost function for the sampled-data problem was derived starting from a standard LQR problem. Various singularity issues were catered for throughout the process. The communication sequence optimization problem was solved using the algorithms of Chapter 5. Their performance seemed to be superior when comparing the results to existing ones, especially for large scale problems. In general, the use of stochastic search methods to solve scheduling problems was effective in terms of quality of solutions found and computational time.

7

Optimal schedule design

CONTENTS

Modern control systems are becoming more and more modular. This means that systems may be enhanced by adding new controllers that share the already existing components like sensors and actuators. In this scenario, the challenge for the control engineer has shifted from the classical 'controller design' to the more modern 'controller integration'.

The integration challenge is the design of policies that orchestrate the communication among the networked control system components. In other words, the design of communication schedulers. In computer science, scheduling is often seen as a way to ensure that tasks meet their deadlines. In real-time control systems, the scheduler becomes part of the dynamics of the system and therefore, rather than merely meeting deadlines, scheduling must cater for performance. If we find a way to predict the performance of a system for a given scheduling policy, then we can implement the policy that maximizes this performance.

7.1 Introduction

In the previous chapter, the codesign of an optimal controller and communication sequence for an NCS with a contention-free (or time-triggered) communication protocol was shown. The advantage of such a codesign is the fact that the optimal controller takes into account the communication scheduling which is also a design parameter optimized simultaneously. Although this is theoretically appealing, the freedom to design a controller in real NCSs might not always be available.

Manufacturers tend to provide systems with their own sophisticated and optimized controllers which need to be integrated in an NCS. For example, Robert Bosch GmbH[1], a large supplier of automobile components, has developed an intelligent Adaptive Cruise Controller (ACC) system which is sold to many car manufacturers. The ACC system needs information from sensors (like the vehicle velocity) to regulate the vehicle's relative position and speed via actuators (like throttle and brakes). Car manufacturers have to wire the ACC system into their cars which already have most of the sensors and actuators (connected to a communication network) required by the ACC system. Since it would be highly inefficient to replicate existing sensors and actuators, the ACC system must be integrated to the other components of the car by connecting it to the car communication network.

The integration challenge is the design of a particular scheduling policy that not only guarantees schedulability of the overall NCS but also minimizes the performance degradation caused by the network. Often, existing scheduling algorithms only consider the maximum allowed delay for communication scheduling (*e.g.* [109, 130]). Such schedules ensure that messages (exchanged between nodes in the network) meet their deadlines but the dynamics (and therefore the performance) of the control systems are somehow ignored.

In this chapter, methods for optimal static communication scheduling in an NCS are proposed. Starting from a continuous-time LTI system model with a given controller (and control cost), the equivalent sampled-data representation of the plant is obtained. Then, the framework of Chapter 3 is used to obtain an LTI model of the closed-loop system which includes the scheduled communication. A quadratic cost function for the NCS is constructed and a cost is given by solving a Lyapunov function. This cost permits subsequent optimization of the schedule which is a combinatorial optimization problem. The algorithms of Chapter 5 are used to solve this problem. In this formulation, optimal scheduling of demand signals is not possible. However, demands can be regarded as a scheduling constraint.

Theoretical results are validated via practically relevant examples in simulation. Results from these tests will show how the overall performance of the system is improved by the optimal scheduling of the nodes' communication even when communication constraints are present.

[1] `www.bosch.com`.

7.2 Problem formulation

The procedures of Section 3.3 (p. 53) are used here to model the limited communication between a plant in feedback with a controller. Contrary to the previous chapter, the more general case of sensor and actuator scheduling is considered here.

7.2.1 NCS model

Consider the NCS in Figure 1.5, p. 6, (but for simplicity, without demand signals). As usual, controllers, sensors and actuators are spatially distributed and the limited communication medium is represented by a shared bus. Only a limited number of data can be transmitted at every time tick. The assumption is that the spatially distributed plant is an LTI system described by

$$\mathcal{P}^c: \quad \begin{cases} \dot{\boldsymbol{x}}(t) = \boldsymbol{A}^c\boldsymbol{x}(t) + \boldsymbol{B}^c\hat{\boldsymbol{u}}(t) \\ \boldsymbol{y}(t) = \boldsymbol{C}\boldsymbol{x}(t) + \boldsymbol{D}\hat{\boldsymbol{u}}(t) \end{cases}, \tag{7.1}$$

where $\boldsymbol{x}(t) \in \mathbb{R}^{n_x}$, $\hat{\boldsymbol{u}}(t) \in \mathbb{R}^{n_u}$, $\boldsymbol{y}(t) \in \mathbb{R}^{n_y}$. $\boldsymbol{A}^c, \boldsymbol{B}^c, \boldsymbol{C}$ and $\boldsymbol{D}$ are matrices of appropriate dimensions and, without loss of generality, $\boldsymbol{D} = \boldsymbol{0}$. The control input $\hat{\boldsymbol{u}}(t)$ is a discrete input signal created by the ZOH element. The model in (7.1) is sampled with a periodic sampling interval h giving the sampled-data plant model

$$\mathcal{P}: \quad \begin{cases} \boldsymbol{x}(jh+h) = \boldsymbol{A}\boldsymbol{x}(jh) + \boldsymbol{B}\hat{\boldsymbol{u}}(jh) \\ \boldsymbol{y}(jh) = \boldsymbol{C}\boldsymbol{x}(jh) \end{cases}, \tag{7.2}$$

where $\boldsymbol{A}$ and $\boldsymbol{B}$ are given in (2.6) and j is the sampling instant.

The next step is to obtain a model for the scheduler that emulates the limited communication channel between the plant and the controller (see Section 3.3). This is given by the communication network $N = (\Delta, \sigma)$ where $\Delta = \{\Delta(i)\}$ is the set of nodes (Definition 3.1, p. 50) and $\sigma = \{\sigma(k)\}_{k=0}^{p-1}$ is the periodic communication sequence (Definition 3.6, p. 64). Since both the sensor and the actuator signals need to be scheduled (via their respective nodes) models for the sensor scheduler $\mathcal{S}_S$ and the actuator scheduler $\mathcal{S}_A$ are required. Equations for $\mathcal{S}_A$ and $\mathcal{S}_S$ are given in (3.15) and (3.16) respectively. However, since periodic sequences are considered here, the matrices of $\mathcal{S}_A$ and $\mathcal{S}_S$ become k-dependent where $k \stackrel{\text{def}}{=} \text{mod}(j, p)$.

The discrete controller is described by

$$\mathcal{K}: \quad \begin{cases} \boldsymbol{x}_K(j+1) = \boldsymbol{A}_K\boldsymbol{x}_K(j) + \boldsymbol{B}_K\hat{\boldsymbol{y}}(j) \\ \boldsymbol{u}(j) = \boldsymbol{C}_K\boldsymbol{x}_K(j) + \boldsymbol{D}_K\hat{\boldsymbol{y}}(j) \end{cases}, \tag{7.3}$$

where $\boldsymbol{x}_K(j) \in \mathbb{R}^{n_k}$, $\hat{\boldsymbol{y}}(t) \in \mathbb{R}^{n_y}$ and $\boldsymbol{u}(j) \in \mathbb{R}^{n_u}$. $\boldsymbol{A}_K, \boldsymbol{B}_K, \boldsymbol{C}_K$ and $\boldsymbol{D}_K$ are matrices of appropriate dimensions. The closed-loop NCS is depicted in

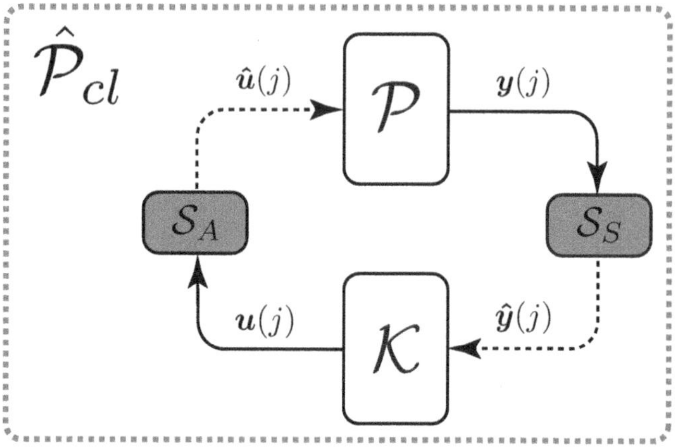

FIGURE 7.1: Closed-loop system with schedulers in the loop without demand.

Figure 7.1. The plant $\mathcal{P}$ with schedulers $\mathcal{S}_A$ and $\mathcal{S}_S$ and controller $\mathcal{K}$ form an *augmented closed-loop* autonomous system that is described by

$$\hat{\mathcal{P}}_{cl}: \quad \hat{\boldsymbol{x}}(j+1) = \hat{\boldsymbol{A}}_{cl}(k)\hat{\boldsymbol{x}}(j), \tag{7.4}$$

where

$$\hat{\boldsymbol{x}}(j) = \begin{bmatrix} \boldsymbol{x}_A(j) \\ \boldsymbol{x}(j) \\ \boldsymbol{x}_S(j) \\ \boldsymbol{x}_K(j) \end{bmatrix} = \begin{bmatrix} \hat{\boldsymbol{u}}(j-1) \\ \boldsymbol{x}(j) \\ \hat{\boldsymbol{y}}(j-1) \\ \boldsymbol{x}_K(j) \end{bmatrix},$$

$$\hat{\boldsymbol{A}}_{cl}(k) = \begin{bmatrix} \bar{\boldsymbol{S}}_A(k) & \boldsymbol{S}_A(k)\boldsymbol{D}_K\boldsymbol{S}_S(k)\boldsymbol{C} & \boldsymbol{S}_A(k)\boldsymbol{D}_K\bar{\boldsymbol{S}}_S(k) & \boldsymbol{S}_A(k)\boldsymbol{C}_K \\ \boldsymbol{B}\bar{\boldsymbol{S}}_A(k) & \boldsymbol{A}+\boldsymbol{B}\boldsymbol{S}_A(k)\boldsymbol{D}_K\boldsymbol{S}_S(k)\boldsymbol{C} & \boldsymbol{B}\boldsymbol{S}_A(k)\boldsymbol{D}_K\bar{\boldsymbol{S}}_S(k) & \boldsymbol{B}\boldsymbol{S}_A(k)\boldsymbol{C}_K \\ \boldsymbol{0} & \boldsymbol{S}_S(k)\boldsymbol{C} & \bar{\boldsymbol{S}}_S(k) & \boldsymbol{0} \\ \boldsymbol{0} & \boldsymbol{B}_K\boldsymbol{S}_S(k)\boldsymbol{C} & \boldsymbol{B}_K\bar{\boldsymbol{S}}_S(k) & \boldsymbol{A}_K \end{bmatrix}. \tag{7.5}$$

(This was shown in more detail and for a generalized plant in feedback with a two degrees of freedom controller in Section 3.3.3, p. 61.) Because of the scheduler in the feedback loop, $\hat{\mathcal{P}}_{cl}$ is periodic in p.

Periodicity is removed by applying the discrete-time lifting operator, $\bar{\mathcal{P}}_{cl} = \mathbb{L}_D\{\hat{\mathcal{P}}_{cl}, p\}$ (Definition 3.7, p. 69), to obtain an LTI equivalent system, $\bar{\mathcal{P}}_{cl}$, given by

$$\bar{\mathcal{P}}_{cl}: \quad \hat{\boldsymbol{x}}(pl+p) = \tilde{\boldsymbol{A}}_{cl,p}\hat{\boldsymbol{x}}(pl), \quad l = 0, 1, \ldots, \tag{7.6}$$

where

$$\tilde{\boldsymbol{A}}_{cl,p} = (\boldsymbol{e}_p^T \otimes \boldsymbol{I})\bar{\boldsymbol{A}}_{cl}, \tag{7.7}$$

and

$$\bar{\boldsymbol{A}}_{cl} = \begin{bmatrix} \prod_{i=1}^{1} \hat{\boldsymbol{A}}_{cl}(1-i) \\ \prod_{i=1}^{2} \hat{\boldsymbol{A}}_{cl}(2-i) \\ \vdots \\ \prod_{i=1}^{p} \hat{\boldsymbol{A}}_{cl}(p-i) \end{bmatrix}. \tag{7.8}$$

Column vector $\boldsymbol{e}_k^T$, $1 \leq k \leq p$, is the p-dimensional standard basis vector (for more detail see Section 3.5, p. 63). Notice that $\tilde{\boldsymbol{A}}_{cl,p}$ is the p-lifted system matrix but in general, for $1 \leq k \leq p$, the k-lifted matrices can be written as

$$\tilde{\boldsymbol{A}}_{cl,k} = (\boldsymbol{e}_k^T \otimes \boldsymbol{I})\bar{\boldsymbol{A}}_{cl}, \tag{7.9}$$

which gives

$$\hat{\boldsymbol{x}}(pl+k) = \tilde{\boldsymbol{A}}_{cl,k}\hat{\boldsymbol{x}}(pl). \tag{7.10}$$

7.2.2 Quadratic cost function

Let us assume that a performance measure $\boldsymbol{Q}_c$ for the original continuous-time system in (7.1), with the full control input $\boldsymbol{u}(t)$ (where $\boldsymbol{u}(t) = \hat{\boldsymbol{u}}(t)$ *i.e.* not subject to scheduling) and the associated controller are given. For example, if the controller is the result of an LQR design, then the matrix $\boldsymbol{Q}_c$ is the one of the quadratic cost

$$J = \int_0^\infty \begin{bmatrix} \boldsymbol{x}(t) \\ \boldsymbol{u}(t) \end{bmatrix}^T \boldsymbol{Q}_c \begin{bmatrix} \boldsymbol{x}(t) \\ \boldsymbol{u}(t) \end{bmatrix} dt. \tag{7.11}$$

The equivalent discrete cost function of (7.11) for the sampled system in (7.2) is:

$$J = \sum_{j=0}^{\infty} \begin{bmatrix} \boldsymbol{x}(j) \\ \boldsymbol{u}(j) \end{bmatrix}^T \boldsymbol{Q} \begin{bmatrix} \boldsymbol{x}(j) \\ \boldsymbol{u}(j) \end{bmatrix}, \tag{7.12}$$

where an expression for $\boldsymbol{Q}$ is given in (6.14)-(6.15). Now consider that

$$\begin{bmatrix} \boldsymbol{x}(j) \\ \boldsymbol{u}(j) \end{bmatrix} = \boldsymbol{E}(k)\hat{\boldsymbol{x}}(j), \tag{7.13}$$

where

$$\boldsymbol{E}(k) = \begin{bmatrix} \boldsymbol{0} & \boldsymbol{I} & \boldsymbol{0} & \boldsymbol{0} \\ \bar{\boldsymbol{S}}_A(k) & \boldsymbol{S}_A(k)\boldsymbol{D}_K\boldsymbol{S}_S(k)\boldsymbol{C} & \boldsymbol{S}_A(k)\boldsymbol{D}_K\bar{\boldsymbol{S}}_S(k) & \boldsymbol{S}_A(k)\boldsymbol{C}_K \end{bmatrix}. \tag{7.14}$$

Notice that the bottom block matrix row of $\boldsymbol{E}(k)$ is the top block matrix row of $\hat{\boldsymbol{A}}_{cl}(k)$ (top n_u rows). It follows from (7.12)-(7.13) that

$$J(\sigma) = \sum_{j=0}^{\infty} (\boldsymbol{E}(k)\hat{\boldsymbol{x}}(j))^T \boldsymbol{Q}(\boldsymbol{E}(k)\hat{\boldsymbol{x}}(j)) = \sum_{l=0}^{\infty}\sum_{k=0}^{p-1} \hat{\boldsymbol{x}}(pl+k)^T \hat{\boldsymbol{Q}}(k)\hat{\boldsymbol{x}}(pl+k), \tag{7.15}$$

where $\hat{\boldsymbol{Q}}(k) = \boldsymbol{E}(k)^T\boldsymbol{Q}\boldsymbol{E}(k)$ and the substitution $j = pl + k$, $l = 0, 1, \ldots$ has been used to partition the summation into an infinite one and a finite one for the periodic matrix $\hat{\boldsymbol{Q}}(k)$. The dependence of $J(\sigma)$ in σ is to highlight the fact that this cost is a function of the sequence σ. Because the lifted system considered is the one of (7.10), the cost of (7.15) can be written as

$$J(\sigma) = \sum_{l=0}^{\infty}\sum_{k=0}^{p-1} (\tilde{\boldsymbol{A}}_{cl,k}\hat{\boldsymbol{x}}(pl))^T \hat{\boldsymbol{Q}}(k)(\tilde{\boldsymbol{A}}_{cl,k}\hat{\boldsymbol{x}}(pl)) = \sum_{l=0}^{\infty} \hat{\boldsymbol{x}}(pl)^T \bar{\boldsymbol{Q}}\hat{\boldsymbol{x}}(pl), \tag{7.16}$$

where $\bar{\boldsymbol{Q}} = \sum_{k=0}^{p-1} \tilde{\boldsymbol{A}}_{cl,k}^T \hat{\boldsymbol{Q}}(k)\tilde{\boldsymbol{A}}_{cl,k}$. Equation (7.16) is the lifted cost criterion corresponding to the lifted system in (7.6). At this point, the discrete Lyapunov equation

$$\tilde{\boldsymbol{A}}_{cl,p}\boldsymbol{P}\tilde{\boldsymbol{A}}_{cl,p}^T - \boldsymbol{P} = -\bar{\boldsymbol{Q}}, \tag{7.17}$$

can be solved. This can be easily solved with the MATLAB® command `dlyap`. Assuming the initial states $\hat{\boldsymbol{x}}(0)$ are random following a Gaussian process with unity variance, the expectation of the cost $J(\sigma)$ is

$$J_E(\sigma) = \mathbb{E}\{J(\sigma)\} = \mathrm{tr}(\boldsymbol{P}). \tag{7.18}$$

Finally, the optimization problem is: find an optimal feasible sequence that minimizes the cost in (7.18).

Remark 7.1 *The solvability of this problem is implied from controllability/observability (or stabilizability/detectability) of the lifted system that also depends on the communication sequence (see Chapter 4).* •

7.2.3 A model reference approach for the performance matrix

The cost in (7.18) can be calculated only if the performance measure $\boldsymbol{Q}_c$ is given. This can indeed be the performance measure for an LQR controller design, but in practice it is unlikely that such matrix is given and, in any case, the design of optimal sequences should not be limited to plants with LQR controllers. A method for constructing a performance matrix $\boldsymbol{Q}_c$ for any LTI controller is shown next.

Consider the plant $\mathcal{P}^c$ in (7.1) with input $\boldsymbol{u}(t)$ instead of $\hat{\boldsymbol{u}}(t)$ (*i.e.* there is no actuator scheduler). $\mathcal{P}^c$ is connected to the LTI continuous-time controller $\mathcal{K}^c$ described by

$$\mathcal{K}^c : \quad \begin{cases} \dot{\boldsymbol{x}}_K(t) = \boldsymbol{A}_K^c \boldsymbol{x}_K(t) + \boldsymbol{B}_K^c \boldsymbol{y}(t) \\ \boldsymbol{u}(t) = \boldsymbol{C}_K \boldsymbol{x}_K(t) + \boldsymbol{D}_K \boldsymbol{y}(t) = \boldsymbol{C}_K \boldsymbol{x}_K(t) + \boldsymbol{D}_K \boldsymbol{C} \boldsymbol{x}(t) \end{cases}, \tag{7.19}$$

(notice that $\mathcal{K}^c$ takes $\boldsymbol{y}(t)$ as an input instead of $\hat{\boldsymbol{y}}(t)$ because there is no sensor scheduler). $\mathcal{P}^c$ in feedback with $\mathcal{K}^c$ is the nominal closed-loop system as perfect communication takes place via the signal $\boldsymbol{u}(t)$ and $\boldsymbol{y}(t)$ rather than the scheduled signals $\hat{\boldsymbol{u}}(t)$ and $\hat{\boldsymbol{y}}(t)$ (see Figure 7.1). The nominal closed-loop system, denoted by $\mathcal{P}_{cl}^c$ is described by

$$\mathcal{P}_{cl}^c : \quad \begin{cases} \dot{\boldsymbol{x}}_{cl}(t) = \boldsymbol{A}_{cl}^c \boldsymbol{x}_{cl}(t) \\ \boldsymbol{y}(t) = \boldsymbol{C}_{cl} \boldsymbol{x}_{cl}(t) \end{cases}, \tag{7.20}$$

where

$$\boldsymbol{x}_{cl}(t) = \begin{bmatrix} \boldsymbol{x}(t) \\ \boldsymbol{x}_K(t) \end{bmatrix}, \quad \boldsymbol{A}_{cl}^c = \begin{bmatrix} \boldsymbol{A}^c + \boldsymbol{B}^c \boldsymbol{D}_K \boldsymbol{C} & \boldsymbol{B}^c \boldsymbol{C}_K \\ \boldsymbol{B}_K^c \boldsymbol{C} & \boldsymbol{A}_K^c \end{bmatrix}, \quad \boldsymbol{C}_{cl} = \begin{bmatrix} \boldsymbol{C} & \boldsymbol{0} \end{bmatrix}. \tag{7.21}$$

Let us form an augmented model made up of the original system with the closed-loop nominal system. The state equation of the augmented model is

$$\dot{\boldsymbol{x}}_a(t) = \boldsymbol{A}_a^c \boldsymbol{x}_a(t) + \boldsymbol{B}_a^c \boldsymbol{u}(t), \tag{7.22}$$

where

$$\boldsymbol{x}_a(t) = \begin{bmatrix} \boldsymbol{x}(t) \\ \boldsymbol{x}_{cl}(t) \end{bmatrix}, \quad \boldsymbol{A}_a^c = \begin{bmatrix} \boldsymbol{A}^c & \boldsymbol{0} \\ \boldsymbol{0} & \boldsymbol{A}_{cl}^c \end{bmatrix}, \quad \boldsymbol{B}_a^c = \begin{bmatrix} \boldsymbol{B}^c \\ \boldsymbol{0} \end{bmatrix}. \tag{7.23}$$

From the augmented model, an expression for the newly defined nominal controller signal, $\acute{\boldsymbol{u}}(t)$, can be derived. This is, from (7.19),

$$\acute{\boldsymbol{u}}(t) = \underbrace{\begin{bmatrix} \boldsymbol{0} & \boldsymbol{D}_K \boldsymbol{C} & \boldsymbol{C}_K \end{bmatrix}}_{\stackrel{\text{def}}{=} \boldsymbol{F}_u} \underbrace{\begin{bmatrix} \boldsymbol{x}(t) \\ \boldsymbol{x}(t) \\ \boldsymbol{x}_K(t) \end{bmatrix}}_{\boldsymbol{x}_a(t)}. \tag{7.24}$$

The performance index for the control signal is defined as the difference between the actual control signal $\boldsymbol{u}(t)$ and the nominal one, $\acute{\boldsymbol{u}}(t)$, *i.e.*

$$\tilde{\boldsymbol{u}}(t) = \boldsymbol{u}(t) - \acute{\boldsymbol{u}}(t) = \boldsymbol{u}(t) - \boldsymbol{F}_u \boldsymbol{x}_a(t). \tag{7.25}$$

The actual system output, from the augmented system state, is

$$\boldsymbol{y}(t) = \begin{bmatrix} \boldsymbol{C} & \boldsymbol{0} \end{bmatrix} \begin{bmatrix} \boldsymbol{x}(t) \\ \boldsymbol{x}_{cl}(t) \end{bmatrix} \tag{7.26}$$

and the newly defined nominal output, $\acute{\boldsymbol{y}}(t)$, is

$$\acute{\boldsymbol{y}}(t) = \begin{bmatrix} \mathbf{0} & \boldsymbol{C}_{cl} \end{bmatrix} \begin{bmatrix} \boldsymbol{x}(t) \\ \boldsymbol{x}_{cl}(t) \end{bmatrix}. \tag{7.27}$$

The performance index for the output is defined as the difference between the actual output $\boldsymbol{y}(t)$ and the nominal one, $\acute{\boldsymbol{y}}(t)$, *i.e.*

$$\tilde{\boldsymbol{y}}(t) = \boldsymbol{y}(t) - \acute{\boldsymbol{y}}(t) = \underbrace{\begin{bmatrix} \boldsymbol{C} & -\boldsymbol{C}_{cl} \end{bmatrix}}_{\stackrel{\text{def}}{=} \boldsymbol{F}_y} \underbrace{\begin{bmatrix} \boldsymbol{x}(t) \\ \boldsymbol{x}_{cl}(t) \end{bmatrix}}_{\boldsymbol{x}_a(t)}. \tag{7.28}$$

Finally, the quadratic cost to be minimized is

$$\begin{aligned}
&\int_0^\infty \left(\tilde{\boldsymbol{y}}(t)^T \tilde{\boldsymbol{y}}(t) + \tilde{\boldsymbol{u}}(t)^T \boldsymbol{R} \tilde{\boldsymbol{u}}(t)\right) dt \\
&= \int_0^\infty \boldsymbol{x}_a(t)^T \boldsymbol{F}_y(t)^T \boldsymbol{F}_y(t) \boldsymbol{x}_a(t) + \left(\boldsymbol{u}(t)^T - \boldsymbol{x}_a(t)^T \boldsymbol{F}_u^T\right) \boldsymbol{R} \left(\boldsymbol{u}(t) - \boldsymbol{F}_u \boldsymbol{x}_a(t)\right) dt \\
&= \int_0^\infty \begin{bmatrix} \boldsymbol{x}_a(t) \\ \boldsymbol{u}(t) \end{bmatrix}^T \underbrace{\begin{bmatrix} \boldsymbol{F}_y^T \boldsymbol{F}_y + \boldsymbol{F}_u^T \boldsymbol{R} \boldsymbol{F}_u & -\boldsymbol{F}_u^T \boldsymbol{R} \\ -\boldsymbol{R} \boldsymbol{F}_u & \boldsymbol{R} \end{bmatrix}}_{\boldsymbol{Q}_c} \begin{bmatrix} \boldsymbol{x}_a(t) \\ \boldsymbol{u}(t) \end{bmatrix} dt,
\end{aligned} \tag{7.29}$$

where $\boldsymbol{R} > \mathbf{0}$ is a weighting matrix.

In summary, it was shown here how a performance matrix $\boldsymbol{Q}_c$ in (7.29) can be derived for any plant/controller by creating a squared error between the plant output (and control effort) and the nominal continuous-time closed-loop output (and control effort). A practically well chosen weight, $\boldsymbol{R}$, on the control signal error needs to be included.

7.3 Optimal design

Finding a communication sequence that minimizes the cost in (7.18) is the combinatorial optimization problem. The algorithms discussed in Chapter 5 will be used. The proposed search algorithms start from a population of feasible sequences where the feasibility predicate $\mathfrak{P}$ is given by the conditions in Chapter 4. $\mathfrak{P}$ is used to divide the search space into $\mathfrak{F}$ and $\mathfrak{U}$, the set of feasible and unfeasible solutions respectively. As the iteration number increases, the population of candidate sequences 'evolves' (in the case of the GA) or 'flies' through the feasible search space (in the case of PSO1 and PSO2) with the aim of minimizing $J_E(\sigma)$. The algorithms stop when a maximum number of iterations has been reached. Parameters like population size, maximum number of iterations, *etc.* are chosen heuristically following the landscape analysis described in Section 5.2.1, p. 113.

Since, to the best of the authors' knowledge, this particular optimization problem has never been solved before, the performance of the proposed algorithms cannot be compared to previous ones (in contrast to the codesign problem in the previous chapter). However, the use of stochastic algorithms can be motivated by showing a visual example of landscape for this problem and by inspecting the solution space. Figure 7.2 shows typical costs $J_E(\sigma)$ for

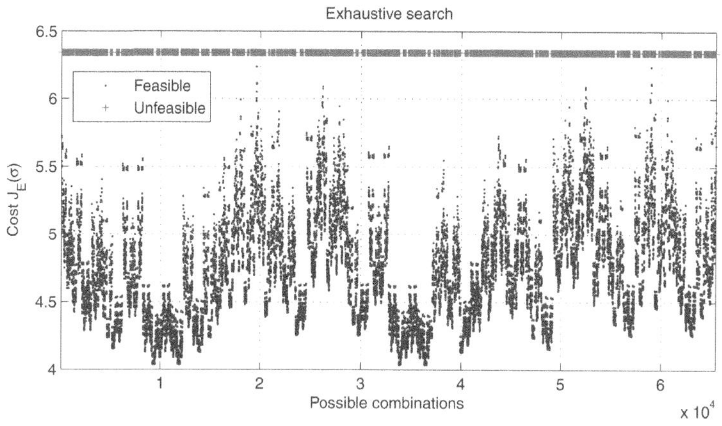

FIGURE 7.2: Typical LG cost of every (4, 8)-Gray encoded solution.

all the d^p permutations of output scheduling for the usual inverted pendulum system (see Example 4.6) made, in this case, of four pendulums, $d = 4$ (one for each pendulum) and $p = 8$ (recall, d is the number of nodes in the NCS and p is the sequence length). The four nodes are two sensor nodes and two actuator nodes for the two pendulums. All the possible combinations are shown in the x-axis ordered using the (d, p)-Gray code. All the unfeasible solutions (those sequences that do not preserve the structural properties, see Chapter 4) have been plotted with a symbolic high cost (on top). The example shows the discontinuity of the search space in terms of neighbor solutions. This search space, to a visual analysis, seems to be even more unstructured than the typical space for the codesign problem (see Figure 5.2, p. 115). Gradient descent or local search algorithms would not be suited for finding satisfactory solutions. The optimization approach suggested here exploits the characteristics of stochastic search in the attempt to find a fast, near-to-optimal solution and avoid combinatorial explosion as the problem size grows. Stochastic algorithms are often regarded as 'slow'. It will be shown later with examples that convergence speed is not an issue, especially because optimization is performed offline.

7.4 Examples

The effectiveness of the optimization approach will be demonstrated by a numerical example. The optimization is performed on a desktop PC with a 1.86 GHz processor and 3.24 GB of RAM.

The distributed system to be controlled is formed by two inverted pendulums (see Example 4.6, p. 107) without spring coupling ($K = k_{1,2} = k_{2,1} = 0$). The time constants of the pendulum systems depend on the length of the rod l_i according to $\tau_i = \sqrt{l_i/g}$. To investigate on systems with different bandwidth requirements, the rod lengths were chosen to be $l_1 = 100$ and $l_2 = 1$. Therefore $\tau_1 = 3.19$ and $\tau_2 = 0.319$ (pendulum 2 is expected to be 'more unstable'). An LQR controller was designed with a weight matrix $\boldsymbol{Q}_c = \text{diag}(1, 1, 1)$ for both pendulums. Furthermore, it is assumed that each pendulum has two sensors (angular velocity and angular position) and an actuator located on three separate physical nodes connected by a communication bus. The total number of physical nodes to be scheduled is therefore six. The communication network is defined as usual as $N = (\Delta, \sigma)$. The set of nodes is

$$\Delta = \{\Delta(1p), \Delta(1v), \Delta(1a), \Delta(2p), \Delta(2v), \Delta(2a)\}, \tag{7.30}$$

where

$$\begin{aligned}
\Delta(1v) &= \{\textit{Pendulum } 1 \textit{ angular velocity}\},\\
\Delta(2v) &= \{\textit{Pendulum } 2 \textit{ angular velocity}\},\\
\Delta(1p) &= \{\textit{Pendulum } 1 \textit{ angular position}\},\\
\Delta(2p) &= \{\textit{Pendulum } 2 \textit{ angular position}\},\\
\Delta(1a) &= \{\textit{Pendulum } 1 \textit{ actuator}\},\\
\Delta(2a) &= \{\textit{Pendulum } 2 \textit{ actuator}\},
\end{aligned} \tag{7.31}$$

are the nodes that correspond to each physical node. The control algorithm is implemented on a controller node. The sampling period is set to $h = 0.03$.

Example 7.1 *The simplest scheduling sequence for controlling such a system is a Round Robin (RR) schedule like*

$$\sigma_{RR} = \{\Delta(1p), \Delta(1v), \Delta(1a), \Delta(2p), \Delta(2v), \Delta(2a)\}. \tag{7.32}$$

Using the RR schedule is very common in practical applications due to its simplicity. The GA was used in this case to find an optimal sequence of length $p = 20$. The returned optimized sequence was

$$\begin{aligned}
\sigma_{opt} = \{&\Delta(2p), \Delta(2v), \Delta(2a), \Delta(1p), \Delta(1v), \Delta(1a), \Delta(1p), \Delta(2v), \Delta(2a), \Delta(2p),\\
&\Delta(2a), \Delta(1v), \Delta(2v), \Delta(2a), \Delta(1a), \Delta(1p), \Delta(1v), \Delta(1a), \Delta(2a), \Delta(1p)\},
\end{aligned} \tag{7.33}$$

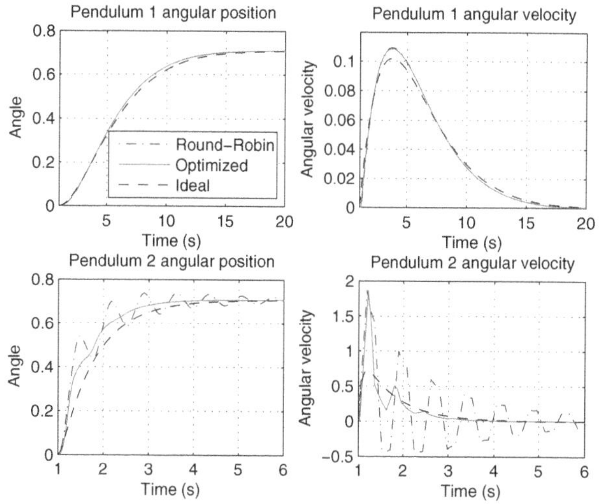

FIGURE 7.3: Comparison between ideal network, RR schedule, and optimized schedule. The optimized schedule shows improved performance.

and it was found in 48 seconds. Figure 7.3 shows the unit step response comparison between an NCS adopting an RR schedule, an NCS adopting an optimized schedule, and an 'ideal' system. The ideal system is the one with a dedicated connection between sensors, actuators and controller and no need for scheduling, i.e. *a system which is an NCS with a communication sequence of* $\sigma = \{\Delta(all)\}$ *where* $\Delta(all) = \{all\ sensors\ ,\ all\ actuators\}$.

Although the theoretical ideal performance cannot be achieved by any sequence for the given set of nodes Δ*, Pendulum 1's transient response for both RR and GA-optimized sequences is similar. However, for pendulum 2 (the more unstable one), the performance of the GA-optimized schedule considerably improves its transient response.* ♦

Example 7.2 *In this other example, the case with* $l_1 = l_2 = 100$ *and* $h = 0.1$ *is considered. This time the assumption is that the physical node corresponding to the actuator of Pendulum 1 is a critical node and has to be scheduled at least every 300 ms. This means that the corresponding node (called* $\Delta(1a)$*) must appear in a sequence at least every third element. Also, an extra node, called* $\Delta(x)$*, needs to be scheduled every 600 ms.* $\Delta(x)$ *could be a performance monitor node that needs to communicate with the controller for safety reasons but does not influence the dynamics. The PSO2 algorithm was used and the sequence returned (with constraints) was*

$$\sigma_{PSO2} = \{\Delta(x), \Delta(1a), \Delta(1v), \Delta(2v), \Delta(1a), \Delta(2a), \Delta(x), \Delta(1a), \Delta(1p), \Delta(1v), \Delta(1a), \Delta(2v), \Delta(x), \Delta(1a), \Delta(2p), \Delta(2a), \Delta(1a), \Delta(1v).\Delta(x), \Delta(1a), \Delta(2v)\}. \tag{7.34}$$

It was found in 57 seconds. It is clear that, while the algorithm was trying to find a sequence to minimize the performance cost, it also considered the constraints for $\Delta(1a)$ *and* $\Delta(x)$*. A simulation result is shown in Figure 7.4. The*

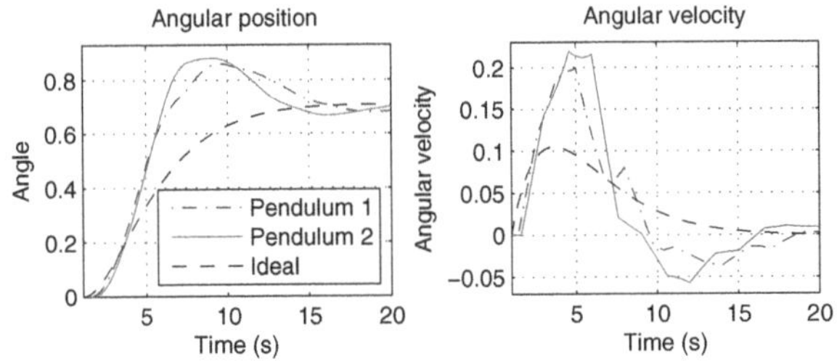

FIGURE 7.4: Constrained simulation: the performance is degraded due to the constrained optimization and the introduction of a 'passive' node.

overall performance is of course degraded due to the constrained optimization and the introduction of a 'passive' node. Note that any of the three algorithms proposed could have been used for these examples. ♦

7.5 Summary

Integrating dynamical systems in an optimal way is the design of optimal schedules to orchestrate the communication between the systems' components. In this chapter, the communication scheduling optimization problem for a given plant and controller was considered. The NCS was modeled using the theoretical framework of Chapter 3. A quadratic Lyapunov-based cost function for the sampled-data problem was derived. The communication sequence optimization problem was solved using the algorithms of Chapter 5. Examples showed the performance improvement due to optimal communication scheduling and the ability of the algorithms to handle constraints.

8

Robust schedule design

CONTENTS

Almost any real system has nonlinearities that, in the modeling process, are often ignored for the sake of simplicity. So far, we have designed communication sequences that are optimal for a particular system described by a given mathematical model. What happens if that mathematical model does not truly describe the dynamics of the real system? Can we reformulate the design procedure in a way that the uncertain unmodeled dynamics are also taken into account? If we could do so, then the optimal communication schedule will become robust. A robust design is not optimal for just the given mathematical model, but it is optimal for a whole family of possible models.

8.1 Introduction

In the previous chapter, methods for NCS optimal communication scheduling were proposed. Starting from a continuous-time LTI system model with a given controller, the equivalent sampled-data representation of the plant was obtained. Then, the framework of Chapter 3 was used to model the resulting NCS which is the closed-loop system including the model of the scheduled communication. A quadratic cost function was constructed and a cost was given by solving a Lyapunov function. This cost permitted subsequent optimization of the schedule achieved by the algorithms of Chapter 5. Although the theoretically 'optimal' sequences performed well in the numerical simulation examples, practical applications showed that they were lacking in robustness, *i.e.* they were sensitive to unmodeled dynamics and parameter uncertainties. This will be shown in the next chapter when such sequences are applied to a real system.

In this chapter, the design process is revisited so that robustness (in terms of model uncertainties) can be incorporated. The procedure (NCS modeling and communication sequence optimization) remains the same but the cost function is reformulated. The cost function is a measure of 'how good' a particular schedule is, *i.e.* it maps a communication sequence to a scalar. An optimal sequence is the one that minimizes that scalar. Hence, an appropriate formulation of this cost is crucial for the optimization process.

Tools from robust control theory will be used to construct $\mathcal{H}_\infty$-based cost functions. Three methods are proposed.

- In the first method, a quadratic cost similar to the one presented in the previous chapter is used but, instead of solving a Lyapunov function, the induced ℓ_2-gain of the system is calculated. The advantage of this approach, with respect to the one in the previous chapter, is that scheduling of demand signals can also be optimized. However, robustness is not yet introduced.

- In the second method, the cost is formulated as the ℓ_2-gain of a generalized discrete plant in feedback with a controller and with schedulers in the loop. The advantage of this method is that robustness can be included by an appropriate formulation of the generalized plant. However, a discretization (which is an approximation) of the continuous-time plant is needed.

- In the third method, the cost is formulated as the ℓ_2-gain of a sampled-data system where the plant is in continuous-time and it is connected to a discrete controller via a Zero-Order-Hold (ZOH) and sampler. The advantage of this method compared to the previous one is that an exact $\mathcal{H}_\infty$-norm of the NCS can be calculated (there is no need for discretization).

8.2 Formulation of an $\mathcal{H}_\infty$-based cost for performance

The following formulation of an $\mathcal{H}_\infty$-based cost function does not include robustness but the demand signals, which are the inputs of the closed-loop system, can now be optimally scheduled. In the Lyapunov-based formulation of the previous chapter, the demand was regarded as a disturbance and a method for including it as an optimization object was not available.

8.2.1 NCS model

Consider the NCS in Figure 1.5 (p. 6) which includes demand signals. As usual, controllers, sensors and actuators are spatially distributed and the limited communication medium is represented by a shared bus. Only limited data can be transmitted at every time tick. The assumption is that the spatially distributed plant is an LTI system described by

$$\mathcal{P}^c: \quad \begin{cases} \dot{\boldsymbol{x}}(t) = \boldsymbol{A}^c\boldsymbol{x}(t) + \boldsymbol{B}^c\hat{\boldsymbol{u}}(t) \\ \boldsymbol{y}(t) = \boldsymbol{C}\boldsymbol{x}(t) + \boldsymbol{D}\hat{\boldsymbol{u}}(t) \end{cases}, \tag{8.1}$$

where $\boldsymbol{x}(t) \in \mathbb{R}^{n_x}$, $\hat{\boldsymbol{u}}(t) \in \mathbb{R}^{n_u}$, $\boldsymbol{y}(t) \in \mathbb{R}^{n_y}$. $\boldsymbol{A}^c, \boldsymbol{B}^c, \boldsymbol{C}$ and $\boldsymbol{D}$ are matrices of appropriate dimensions and, without loss of generality, $\boldsymbol{D} = \boldsymbol{0}$. The control input $\hat{\boldsymbol{u}}(t)$ is a discrete input signal created by the ZOH element. The model in (8.1) is sampled with a periodic sampling interval h giving the sampled-data plant model

$$\mathcal{P}: \quad \begin{cases} \boldsymbol{x}(j+1) = \boldsymbol{A}\boldsymbol{x}(j) + \boldsymbol{B}\hat{\boldsymbol{u}}(j) \\ \boldsymbol{y}(j) = \boldsymbol{C}\boldsymbol{x}(j) \end{cases}, \tag{8.2}$$

where $\boldsymbol{A}$ and $\boldsymbol{B}$ are given in (2.6) and j is the sampling instant.

The next step is to obtain a model for the scheduler that emulates the limited communication channel between the plant and the controller (see Section 3.3). This is given by the communication network $N = (\Delta, \sigma)$ where $\Delta = \{\Delta(i)\}$ is the set of nodes (Definition 3.1, p. 50) and $\sigma = \{\sigma(k)\}_{k=0}^{p-1}$ is the periodic communication sequence (Definition 3.6, p. 64). Since demand, sensor and the actuator signals need to be scheduled (via their respective nodes) models for the demand scheduler $\mathcal{S}_D$, sensor scheduler $\mathcal{S}_S$ and actuator scheduler $\mathcal{S}_A$ are required. Their equations are given in (3.17), (3.16) and (3.15) respectively. However, since periodic sequences are considered here, the matrices of $\mathcal{S}_D$, $\mathcal{S}_S$ and $\mathcal{S}_A$ become k-dependent where $k \stackrel{\text{def}}{=} \text{mod}(j, p)$.

The discrete controller, denoted by $\mathcal{K}$, is a stabilizing controller given in (3.8). The closed-loop NCS is depicted in Figure 8.1. The plant $\mathcal{P}$ with schedulers $\mathcal{S}_D$, $\mathcal{S}_S$ and $\mathcal{S}_A$ and controller $\mathcal{K}$ form an *augmented closed-loop* system that is described by

$$\hat{\mathcal{P}}_{cl}: \quad \hat{\boldsymbol{x}}_{cl}(j+1) = \hat{\boldsymbol{A}}_{cl}(k)\hat{\boldsymbol{x}}_{cl}(j) + \hat{\boldsymbol{B}}_{cl}(k)\boldsymbol{d}(j), \tag{8.3}$$

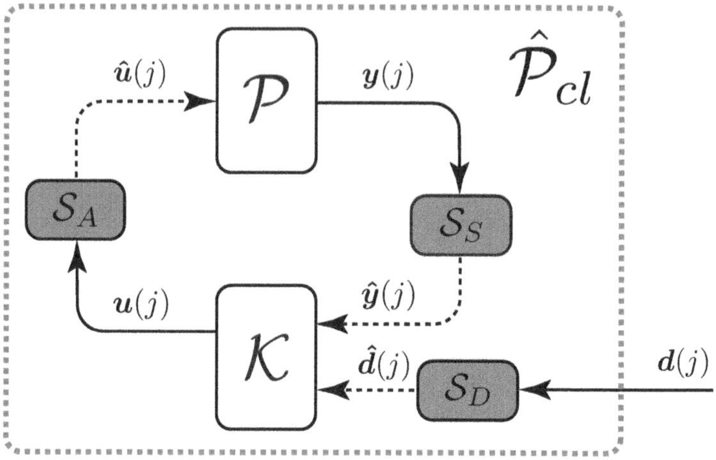

FIGURE 8.1: Closed-loop system with schedulers in the loop with demand.

where the augmented state vector is

$$\hat{\boldsymbol{x}}_{cl}(j) = \begin{bmatrix} \boldsymbol{x}(j) \\ \boldsymbol{x}_S(j) \\ \boldsymbol{x}_D(j) \\ \boldsymbol{x}_K(j) \\ \boldsymbol{x}_A(j) \end{bmatrix} = \begin{bmatrix} \boldsymbol{x}(j) \\ \hat{\boldsymbol{y}}(j-1) \\ \hat{\boldsymbol{d}}(j-1) \\ \boldsymbol{x}_K(j) \\ \hat{\boldsymbol{u}}(j-1) \end{bmatrix}, \tag{8.4}$$

the system matrices are

$$\begin{aligned} \hat{\boldsymbol{A}}_{cl}(k) &= \begin{bmatrix} \boldsymbol{A} + \boldsymbol{B}\hat{\boldsymbol{D}}_{K1}(k)\boldsymbol{C} & \boldsymbol{B}\hat{\boldsymbol{C}}_K(k) \\ \hat{\boldsymbol{B}}_{K1}(k)\boldsymbol{C} & \hat{\boldsymbol{A}}_K(k) \end{bmatrix}, \\ \hat{\boldsymbol{B}}_{cl}(k) &= \begin{bmatrix} \boldsymbol{0} \\ \hat{\boldsymbol{B}}_{K2}(k) \end{bmatrix}, \end{aligned} \tag{8.5}$$

and $\hat{\boldsymbol{A}}_K(k)$, $\hat{\boldsymbol{B}}_{K1}(k)$, $\hat{\boldsymbol{B}}_{K2}(k)$, $\hat{\boldsymbol{C}}_K(k)$ and $\hat{\boldsymbol{D}}_{K1}(k)$ are given in (3.29) for $k = j$. (This was also shown for a generalized plant and non-periodic schedulers in Section 3.3.3, p. 61.) Because of the scheduler in the feedback loop, $\hat{\mathcal{P}}_{cl}$ is periodic in p.

Periodicity is removed by applying the discrete-time lifting operator, $\bar{\mathcal{P}}_{cl} = \mathbb{L}_D\{\hat{\mathcal{P}}_{cl}, p\}$ (Definition 3.7, p. 69) to obtain an LTI equivalent system given by

$$\bar{\mathcal{P}}_{cl}: \quad \hat{\boldsymbol{x}}_{cl}(pl+p) = \tilde{\boldsymbol{A}}_{cl,p}\hat{\boldsymbol{x}}_{cl}(pl) + \tilde{\boldsymbol{B}}_{cl,p}\bar{\boldsymbol{d}}(pl), \quad l = 0, 1, \ldots, \tag{8.6}$$

where

$$\tilde{\boldsymbol{A}}_{cl,p} = (\boldsymbol{e}_p^T \otimes \boldsymbol{I})\bar{\boldsymbol{A}}_{cl}, \quad \tilde{\boldsymbol{B}}_{cl,p} = (\boldsymbol{e}_p^T \otimes \boldsymbol{I})\bar{\boldsymbol{B}}_{cl},$$
$$\bar{\boldsymbol{d}}(pl) = \begin{bmatrix} \boldsymbol{d}(pl) \\ \boldsymbol{d}(pl+1) \\ \vdots \\ \boldsymbol{d}(pl+p-1) \end{bmatrix}, \tag{8.7}$$

and

$$\bar{\boldsymbol{A}}_{cl} = \begin{bmatrix} \prod_{i=1}^{1} \hat{\boldsymbol{A}}_{cl}(1-i) \\ \prod_{i=1}^{2} \hat{\boldsymbol{A}}_{cl}(2-i) \\ \vdots \\ \prod_{i=1}^{p} \hat{\boldsymbol{A}}_{cl}(p-i) \end{bmatrix},$$
$$\bar{\boldsymbol{B}}_{cl} = \begin{bmatrix} \hat{\boldsymbol{B}}_{cl}(0) & \boldsymbol{0} & \cdots & \boldsymbol{0} \\ \left(\prod_{i=1}^{1} \hat{\boldsymbol{A}}_{cl}(2-i)\right) \hat{\boldsymbol{B}}_{cl}(0) & \hat{\boldsymbol{B}}_{cl}(1) & \cdots & \boldsymbol{0} \\ \left(\prod_{i=1}^{2} \hat{\boldsymbol{A}}_{cl}(3-i)\right) \hat{\boldsymbol{B}}_{cl}(0) & \left(\prod_{i=1}^{1} \hat{\boldsymbol{A}}_{cl}(3-i)\right) \hat{\boldsymbol{B}}_{cl}(1) & \cdots & \boldsymbol{0} \\ \vdots & \vdots & \ddots & \vdots \\ \left(\prod_{i=1}^{p-1} \hat{\boldsymbol{A}}_{cl}(p-i)\right) \hat{\boldsymbol{B}}_{cl}(0) & \left(\prod_{i=1}^{p-2} \hat{\boldsymbol{A}}_{cl}(p-i)\right) \hat{\boldsymbol{B}}_{cl}(1) & \cdots & \hat{\boldsymbol{B}}_{cl}(p-1) \end{bmatrix}. \tag{8.8}$$

Column vector $\boldsymbol{e}_k^T$, $1 \leq k \leq p$, is the p-dimensional standard basis vector (for more detail see Section 3.5, p. 63). Notice that $\tilde{\boldsymbol{A}}_{cl,p}$ and $\tilde{\boldsymbol{B}}_{cl,p}$ are the p-lifted system matrices but in general, for $1 \leq k \leq p$, the k-lifted matrices can be written as

$$\tilde{\boldsymbol{A}}_{cl,k} = (\boldsymbol{e}_k^T \otimes \boldsymbol{I})\bar{\boldsymbol{A}}_{cl} \quad \text{and} \quad \tilde{\boldsymbol{B}}_{cl,k} = (\boldsymbol{e}_k^T \otimes \boldsymbol{I})\bar{\boldsymbol{B}}_{cl}, \tag{8.9}$$

which gives

$$\hat{\boldsymbol{x}}_{cl}(pl+k) = \tilde{\boldsymbol{A}}_{cl,k}\hat{\boldsymbol{x}}_{cl}(pl) + \tilde{\boldsymbol{B}}_{cl,k}\bar{\boldsymbol{d}}(pl). \tag{8.10}$$

8.2.2 Cost function

The construction of this cost function follows very closely the one of Section 7.2.2, p. 147. Let us assume that a performance measure, $\boldsymbol{Q}_c$, for the original continuous-time system in (8.1), with the full control input $\boldsymbol{u}(t)$ (where

$\boldsymbol{u}(t) = \hat{\boldsymbol{u}}(t)$ *i.e.* not subject to scheduling), and the associated controller is given. For example, if the controller is the result of an LQR design, then the matrix $\boldsymbol{Q}_c$ is the one of the quadratic cost in (7.11). Otherwise, $\boldsymbol{Q}_c$ can be derived as shown in Section 7.2.3, p. 148. The associated discrete-time cost function for the system in (8.2) is given in (7.12). Let us define this cost function, from (7.12), as

$$J = \sum_{j=0}^{\infty} \begin{bmatrix} \boldsymbol{x}(j) \\ \boldsymbol{u}(j) \end{bmatrix}^T \boldsymbol{Q} \begin{bmatrix} \boldsymbol{x}(j) \\ \boldsymbol{u}(j) \end{bmatrix} \overset{\text{def}}{=} \sum_{j=0}^{\infty} \boldsymbol{z}(j)^T \boldsymbol{z}(j), \tag{8.11}$$

where $\boldsymbol{z}(j)$ is a generalized output. Now consider that

$$\begin{bmatrix} \boldsymbol{x}(j) \\ \boldsymbol{u}(j) \end{bmatrix} = \boldsymbol{E}(k) \begin{bmatrix} \hat{\boldsymbol{x}}_{cl}(j) \\ \boldsymbol{d}(j) \end{bmatrix}, \tag{8.12}$$

where

$$\boldsymbol{E}(k) = \left[\begin{matrix} \boldsymbol{I} & \boldsymbol{0} & \boldsymbol{0} \\ \boldsymbol{S}_A(k)\boldsymbol{D}_{K1}\boldsymbol{S}_S(k)\boldsymbol{C} & \boldsymbol{S}_A(k)\boldsymbol{D}_{K1}\bar{\boldsymbol{S}}_S(k) & \boldsymbol{S}_A(k)\boldsymbol{D}_{K2}\bar{\boldsymbol{S}}_D(k) \end{matrix} \right. \\ \left. \begin{matrix} \boldsymbol{0} & \boldsymbol{0} & \boldsymbol{0} \\ \boldsymbol{S}_A(k)\boldsymbol{C}_K & \bar{\boldsymbol{S}}_A(k) & \boldsymbol{S}_A(k)\boldsymbol{D}_{K2}\boldsymbol{S}_D(k) \end{matrix} \right]., \tag{8.13}$$

and $\hat{\boldsymbol{x}}_{cl}(j)$ is given in (8.4). Hence, the bottom block matrix row of $\boldsymbol{E}(k)$ is the top block matrix row of $\hat{\boldsymbol{A}}_{cl}(k)$ and $\hat{\boldsymbol{B}}_{cl}(k)$ in (8.5). The cost J in (8.11) can now be written as

$$\begin{aligned} J(\sigma) &= \sum_{j=0}^{\infty} \left(\boldsymbol{E}(k) \begin{bmatrix} \hat{\boldsymbol{x}}_{cl}(j) \\ \boldsymbol{d}(j) \end{bmatrix} \right)^T \boldsymbol{Q} \left(\boldsymbol{E}(k) \begin{bmatrix} \hat{\boldsymbol{x}}_{cl}(j) \\ \boldsymbol{d}(j) \end{bmatrix} \right) \\ &= \sum_{l=0}^{\infty} \sum_{k=0}^{p-1} \begin{bmatrix} \hat{\boldsymbol{x}}_{cl}(pl+k) \\ \boldsymbol{d}(pl+k) \end{bmatrix}^T \underbrace{\begin{bmatrix} \hat{\boldsymbol{Q}}_1(k) & \hat{\boldsymbol{Q}}_{12}(k) \\ \hat{\boldsymbol{Q}}_{12}^T(k) & \hat{\boldsymbol{Q}}_2(k) \end{bmatrix}}_{\hat{\boldsymbol{Q}}(k)} \begin{bmatrix} \hat{\boldsymbol{x}}_{cl}(pl+k) \\ \boldsymbol{d}(pl+k) \end{bmatrix}, \end{aligned} \tag{8.14}$$

where $\hat{\boldsymbol{Q}}(k) = \boldsymbol{E}(k)^T \boldsymbol{Q} \boldsymbol{E}(k)$ and the substitution $j = pl + k$, $l = 0, 1, \ldots$ has been used to partition the summation into an infinite one and a finite one for the periodic matrix $\hat{\boldsymbol{Q}}(k)$. Because of (8.10) and the definition of $\bar{\boldsymbol{d}}(pl)$ in (8.7),

$$\begin{bmatrix} \hat{\boldsymbol{x}}_{cl}(pl+k) \\ \boldsymbol{d}(pl+k) \end{bmatrix} = \begin{bmatrix} \tilde{\boldsymbol{A}}_{cl,k} \hat{\boldsymbol{x}}_{cl}(pl) + \tilde{\boldsymbol{B}}_{cl,k} \bar{\boldsymbol{d}}(pl) \\ (\boldsymbol{e}_k^T \otimes \boldsymbol{I}) \bar{\boldsymbol{d}}(pl) \end{bmatrix}, \tag{8.15}$$

hence, Equation (8.14) (after some algebraic manipulation) becomes

$$J(\sigma) = \sum_{l=0}^{\infty} \begin{bmatrix} \hat{\boldsymbol{x}}_{cl}(pl) \\ \bar{\boldsymbol{d}}(pl) \end{bmatrix}^T \underbrace{\left(\sum_{k=0}^{p-1} \begin{bmatrix} \bar{\boldsymbol{Q}}_1(k) & \bar{\boldsymbol{Q}}_{12}(k) \\ \bar{\boldsymbol{Q}}_{12}(k)^T & \bar{\boldsymbol{Q}}_2(k) \end{bmatrix} \right)}_{\overset{\text{def}}{=} \tilde{\boldsymbol{Q}}} \begin{bmatrix} \hat{\boldsymbol{x}}_{cl}(pl) \\ \bar{\boldsymbol{d}}(pl) \end{bmatrix}, \tag{8.16}$$

where

$$\begin{aligned}
\bar{\boldsymbol{Q}}_1(k) &= \tilde{\boldsymbol{A}}_{cl,k}^T \hat{\boldsymbol{Q}}_1(k) \tilde{\boldsymbol{A}}_{cl,k}, \\
\bar{\boldsymbol{Q}}_2(k) &= \tilde{\boldsymbol{B}}_{cl,k}^T \hat{\boldsymbol{Q}}_1(k) \tilde{\boldsymbol{B}}_{cl,k} + (\boldsymbol{e}_k^T \otimes \boldsymbol{I})^T \hat{\boldsymbol{Q}}_{12}(k)^T \tilde{\boldsymbol{B}}_{cl,k} + \tilde{\boldsymbol{B}}_{cl,k}^T \hat{\boldsymbol{Q}}_{12}(k)(\boldsymbol{e}_k^T \otimes \boldsymbol{I}) \\
&\quad + (\boldsymbol{e}_k^T \otimes \boldsymbol{I})^T \hat{\boldsymbol{Q}}_2(k)(\boldsymbol{e}_k^T \otimes \boldsymbol{I}), \\
\bar{\boldsymbol{Q}}_{12}(k) &= \tilde{\boldsymbol{A}}_{cl,k}^T \hat{\boldsymbol{Q}}_1(k) \tilde{\boldsymbol{B}}_{cl,k} + \tilde{\boldsymbol{A}}_{cl,k}^T \hat{\boldsymbol{Q}}_{12}(k)(\boldsymbol{e}_k^T \otimes \boldsymbol{I}), \\
\tilde{\boldsymbol{A}}_{cl,0} &= \boldsymbol{I}, \quad \tilde{\boldsymbol{B}}_{cl,0} = \boldsymbol{0}.
\end{aligned} \tag{8.17}$$

For $\boldsymbol{d}(pl) = \boldsymbol{0}$, an LQR cost is obtained while $\boldsymbol{d}(pl) \neq \boldsymbol{0}$ allows the investigation of the $\mathcal{H}_\infty$ problem that follows.

Let $\tilde{\boldsymbol{Q}} \stackrel{\text{def}}{=} \tilde{\boldsymbol{Q}}^{\frac{1}{2}} \tilde{\boldsymbol{Q}}^{\frac{1}{2}}$, and $\tilde{\boldsymbol{Q}}^{\frac{1}{2}} \stackrel{\text{def}}{=} \begin{bmatrix} \bar{\boldsymbol{C}}_{cl} & \bar{\boldsymbol{D}}_{cl} \end{bmatrix}$. Define the generalized output signal as

$$\bar{\boldsymbol{z}}(pl) \stackrel{\text{def}}{=} \begin{bmatrix} \bar{\boldsymbol{C}}_{cl} & \bar{\boldsymbol{D}}_{cl} \end{bmatrix} \begin{bmatrix} \hat{\boldsymbol{x}}_{cl}(pl) \\ \bar{\boldsymbol{d}}(pl) \end{bmatrix}. \tag{8.18}$$

The system of which $\mathcal{H}_\infty$-norm has to be minimized is

$$\mathcal{C}: \quad \begin{cases} \hat{\boldsymbol{x}}_{cl}(pl+p) = \tilde{\boldsymbol{A}}_{cl,p} \hat{\boldsymbol{x}}_{cl}(pl) + \tilde{\boldsymbol{B}}_{cl,p} \bar{\boldsymbol{d}}(pl) \\ \bar{\boldsymbol{z}}(pl) = \bar{\boldsymbol{C}}_{cl} \hat{\boldsymbol{x}}_{cl}(pl) + \bar{\boldsymbol{D}}_{cl} \bar{\boldsymbol{d}}(pl) \end{cases}. \tag{8.19}$$

Finally, the cost function, which is the system's induced ℓ_2-gain, is defined as

$$\gamma_s = \sup_{\|\bar{\boldsymbol{d}}(pl)\|_{\ell_2} \neq 0} \frac{\|\bar{\boldsymbol{z}}(pl)\|_{\ell_2}}{\|\bar{\boldsymbol{d}}(pl)\|_{\ell_2}}, \tag{8.20}$$

where the ℓ_2-norm of a discrete signal $\boldsymbol{\nu}(k)$ is defined in the usual way as

$$\|\boldsymbol{\nu}(k)\|_{\ell_2}^2 = \sum_{k=0}^{\infty} \boldsymbol{\nu}(k)^T \boldsymbol{\nu}(k), \tag{8.21}$$

and the supremum, sup, is taken over all the nonzero trajectories starting from $\hat{\boldsymbol{x}}_{cl}(0)$. This can be easily calculated with the MATLAB® command `dhfnorm` once the matrices of $\mathcal{C}$ are obtained.

Remark 8.1 *For the optimization, the set of unfeasible solutions $\mathfrak{U}$ was defined as the set of sequences which do not preserve the structural properties when a system is implemented as an NCS (this was discussed in detail in Chapter 4). Here, for calculation of the ℓ_2-gain, the closed-loop system, must be stable. Hence, the set $\mathfrak{U}$ must also include those sequences that do not stabilize the system.* •

8.3 Formulation of a discrete $\mathcal{H}_\infty$-based cost for robustness and performance

In the following formulation, both robustness and performance can be included into the design.

8.3.1 NCS model

The model used here is the one of a generalized plant in feedback with a two Degrees Of Freedom (DOF) controller. The network is modeled as schedulers. This model was discussed in Section 3.3, p. 53. The plant model, $\mathcal{P}$, in (3.7), can be a discretized version of a continuous-time plant and, without loss of generality, assume that $\boldsymbol{D}_{22} = \boldsymbol{0}$. The controller, $\mathcal{K}$, in (3.8) is a stabilizing discrete controller. Equations for the demand scheduler $\mathcal{S}_D$, sensor scheduler $\mathcal{S}_S$ and actuator scheduler $\mathcal{S}_A$ are given in (3.17), (3.16) and (3.15) respectively. The diagram of the closed-loop system is shown in Figure 3.2.

The approach here is to obtain the equations for the lifted augmented controller and the plant separately and then form the closed-loop system.

8.3.1.1 Discrete-time lifted controller

First, obtain the equations for $\hat{\mathcal{K}}$, the *augmented controller*. This is explained in Section 3.3.2 (p. 60) and the equations are given in (3.27)-(3.29). $\hat{\mathcal{K}}$ is periodic in p (because of the schedulers). Periodicity is removed by applying the discrete-time lifting operator, $\bar{\mathcal{K}} = \mathbb{L}_D\{\hat{\mathcal{K}}, p\}$ (Definition 3.7, p. 69), to obtain the LTI equivalent system given by

$$\bar{\mathcal{K}}: \quad \begin{cases} \hat{\boldsymbol{x}}_K(pl+p) = \tilde{\boldsymbol{A}}_{K,p}\hat{\boldsymbol{x}}_K(pl) + \tilde{\boldsymbol{B}}_{K1,p}\bar{\boldsymbol{y}}(pl) + \tilde{\boldsymbol{B}}_{K2,p}\bar{\boldsymbol{d}}(pl) \\ \bar{\boldsymbol{u}}(pl) = \bar{\boldsymbol{C}}_K\hat{\boldsymbol{x}}_K(pl) + \bar{\boldsymbol{D}}_{K1}\bar{\boldsymbol{y}}(pl) + \bar{\boldsymbol{D}}_{K2}\bar{\boldsymbol{d}}(pl) \end{cases}, \tag{8.22}$$

where $\hat{\boldsymbol{x}}_K(pl)$ (for $pl = j$) is given in (3.28),

$$\bar{\boldsymbol{y}}(pl) = \begin{bmatrix} \boldsymbol{y}(pl) \\ \boldsymbol{y}(pl+1) \\ \vdots \\ \boldsymbol{y}(pl+p-1) \end{bmatrix}, \quad \bar{\boldsymbol{d}}(pl) = \begin{bmatrix} \boldsymbol{d}(pl) \\ \boldsymbol{d}(pl+1) \\ \vdots \\ \boldsymbol{d}(pl+p-1) \end{bmatrix},$$
$$\bar{\boldsymbol{u}}(pl) = \begin{bmatrix} \hat{\boldsymbol{u}}(pl) \\ \hat{\boldsymbol{u}}(pl+1) \\ \vdots \\ \hat{\boldsymbol{u}}(pl+p-1) \end{bmatrix}, \tag{8.23}$$

and the lifted system matrices $\tilde{\boldsymbol{A}}_{K,p}$, $\tilde{\boldsymbol{B}}_{K1,p}$, $\tilde{\boldsymbol{B}}_{K2,p}$, $\bar{\boldsymbol{C}}_K$, $\bar{\boldsymbol{D}}_{K1}$ and $\bar{\boldsymbol{D}}_{K2}$ can be derived easily as their structure is the same as the one of the lifted matrices in (3.49) and (3.55) (Section 3.5, p. 63). $\bar{\mathcal{K}}$ is the equivalent higher dimensional system. The lifting of $\hat{\mathcal{K}}$ is shown diagrammatically in Figure 8.2.

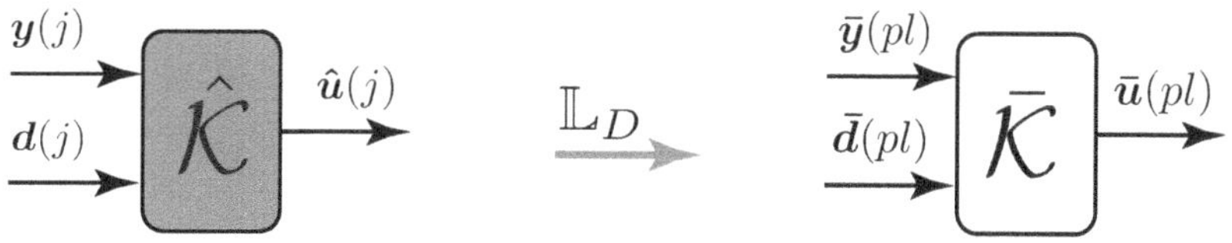

FIGURE 8.2: Discrete-time lifting applied to the augmented (periodic) controller.

8.3.1.2 Discrete-time lifted closed-loop system

After the discrete-time lifting, the dimensions of the augmented controller input and output signals have increased by a factor of p. To match the input and output signals of the plant with the discrete-time lifted controller's input and output (see (8.22)), the generalized plant

$$\mathcal{P}: \quad \begin{cases} \boldsymbol{x}(j+1) = \boldsymbol{A}\boldsymbol{x}(j) + \boldsymbol{B}_1\boldsymbol{w}(j) + \boldsymbol{B}_2\hat{\boldsymbol{u}}(j) \\ \boldsymbol{z}(j) = \boldsymbol{C}_1\boldsymbol{x}(j) + \boldsymbol{D}_{11}\boldsymbol{w}(j) + \boldsymbol{D}_{12}\hat{\boldsymbol{u}}(j) \\ \boldsymbol{y}(j) = \boldsymbol{C}_2\boldsymbol{x}(j) + \boldsymbol{D}_{21}\boldsymbol{w}(j) \end{cases}, \tag{8.24}$$

also needs to be lifted. The discrete-time lifting operator is thus applied to $\mathcal{P}$ to obtain $\bar{\mathcal{P}} = \mathbb{L}_D\{\mathcal{P}, p\}$. $\bar{\mathcal{P}}$ is the equivalent higher dimensional system described by

$$\bar{\mathcal{P}}: \quad \begin{cases} \boldsymbol{x}(pl+p) = \tilde{\boldsymbol{A}}_p\boldsymbol{x}(pl) + \tilde{\boldsymbol{B}}_{1,p}\bar{\boldsymbol{w}}(pl) + \tilde{\boldsymbol{B}}_{2,p}\bar{\boldsymbol{u}}(j) \\ \bar{\boldsymbol{z}}(pl) = \bar{\boldsymbol{C}}_1\boldsymbol{x}(pl) + \bar{\boldsymbol{D}}_{11}\bar{\boldsymbol{w}}(pl) + \bar{\boldsymbol{D}}_{12}\bar{\boldsymbol{u}}(j) \\ \bar{\boldsymbol{y}}(pl) = \bar{\boldsymbol{C}}_2\boldsymbol{x}(pl) + \bar{\boldsymbol{D}}_{21}\bar{\boldsymbol{w}}(pl) \end{cases}, \tag{8.25}$$

where the vectors and matrices in (8.25) can be easily derived by inspecting the matrix structure of previously lifted systems (they are also given in Remark 3.4, p. 69). The lifting of $\mathcal{P}$ is shown diagrammatically in Figure 8.3.

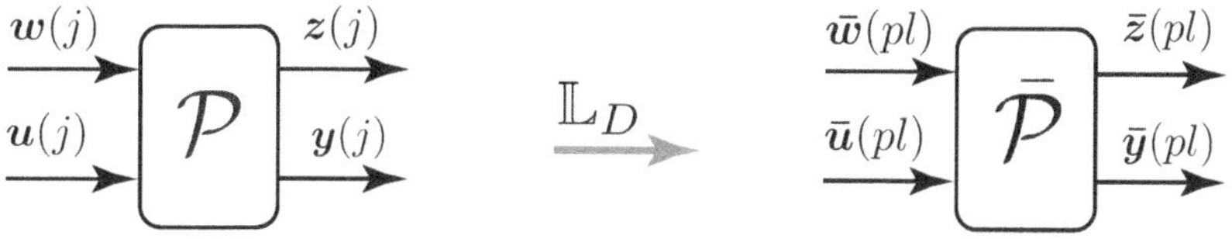

FIGURE 8.3: Discrete-time lifting applied to the discretized plant.

Finally, let us use the notation $\mathbb{F}(\mathcal{G}, \mathcal{H})$ to denote the closed-loop mapping from the exogenous input to the regulated output of the generalized plant $\mathcal{G}$ in feedback with controller $\mathcal{H}$. The discrete LTI closed-loop system is, as

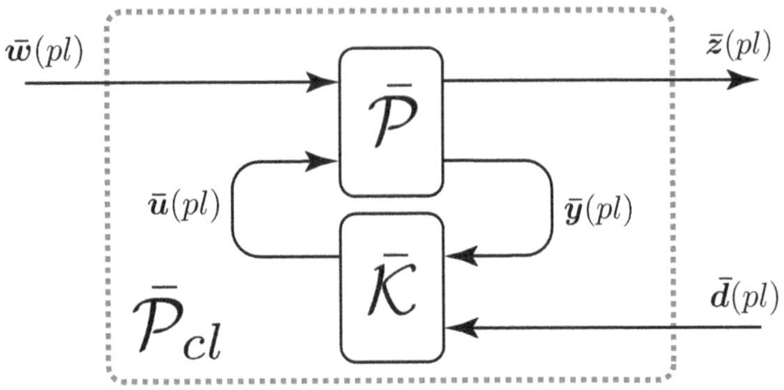

FIGURE 8.4: The closed-loop equivalent system.

shown in Figure 8.4, $\bar{\mathcal{P}}_{cl} = \mathbb{F}(\bar{\mathcal{P}}, \bar{\mathcal{K}})$. Hence, the discrete-time lifted version of the augmented closed-loop system is the map

$$\bar{\mathcal{P}}_{cl}: \quad \bar{v}(pl) \mapsto \bar{z}(pl), \tag{8.26}$$

where

$$\bar{v}(pl) = \begin{bmatrix} \bar{d}(pl) \\ \bar{w}(pl) \end{bmatrix}. \tag{8.27}$$

8.3.2 Cost function

Also in this case, the cost function is the system's induced ℓ_2-gain. This is defined for $\bar{\mathcal{P}}_{cl}$ as

$$\gamma_d = \sup_{\|\bar{v}(pl)\|_{\ell_2} \neq 0} \frac{\|\bar{z}(pl)\|_{\ell_2}}{\|\bar{v}(pl)\|_{\ell_2}}. \tag{8.28}$$

This can be easily calculated with the MATLAB® command `dhfnorm` once the matrices of $\bar{\mathcal{P}}_{cl}$ are obtained. In contrast to the approach of Section 8.2, robustness can be included here by appropriate construction of the generalized plant $\mathcal{P}$. This will be discussed in more detail for a practical application in Section 9.4. For the feasibility issue see Remark 8.1.

8.4 Formulation of a sampled-data $\mathcal{H}_\infty$-based cost for robustness and performance

The problem of designing $\mathcal{H}_\infty$ controllers for NCSs has been widely tackled (*e.g.* recently in [16]). However, the design of optimal and/or robust schedules presented so far in this work and also in [80, 105, 14] is possible only for a discrete-time model of the plant. Since the plant is generally a continuous-time system, it must be discretized and the communication sequence is then designed for the discretized plant. Clearly, the inter-sample behavior is lost during discretization. In robust control problems, it is more natural to consider a disturbance or a perturbation as a continuous-time signal since the controlled system evolves in continuous time. Hence, it is preferable to calculate the norms of the continuous-time systems. On the other hand, the scheduling mechanism is inherently a discrete-time process because the connections to sensor, actuator and demand signals switch according to the communication sequence. Also, the controller is likely to be implemented in a microprocessor resulting in discrete-time dynamics. The overall closed-loop dynamics are therefore those of a hybrid system made up of a continuous plant and discrete scheduler and controller.

In this section, the robust scheduling problem of the sampled-data system is considered. The plant to be controlled is described by a linear continuous-time system and it is connected to a discrete-time controller via a network with communication constraints modeled as discrete switchings. The interface between the continuous and discrete dynamics are synchronized, ideal ZOH and sampler elements. The continuous plant with ZOH and sampler form a periodic system with period equal to the sampling time. The lifting technique for continuous-time periodic systems [11, 20] is used to 'lift' continuous-time signals into discrete-time signals. This will be called the *continuous-time lifting* to distinguish it from the *discrete-time lifting* used for periodic discrete systems (Section 3.5, p. 63). The continuous-time lifting allows to convert the continuous periodic system with specific induced $\mathcal{L}_2$-norm into an equivalent discrete system with equal induced ℓ_2-norm. The lifted plant is a discrete-time, time-invariant system with infinite-dimensional input and output vectors but with finite-dimensional states. Since the state space is finite-dimensional, then some infinite-dimensional operators are in fact finite rank operators [11] and the problem has been shown to reduce to a finite dimensional one.

The discrete-time lifting technique is used to eliminate the periodicity caused by the scheduling mechanism. The result is a linear discrete-time non-periodic equivalent representation of the hybrid system with scheduled communication. Hence, the exact $\mathcal{H}_\infty$-norm of the NCS can be calculated. This cost permits subsequent optimization of the schedule (which is the communication sequence).

8.4.1 Continuous-time lifting

In this section, the continuous-time lifting and loop-transfer procedures derived in [11] are discussed in the context of a linear discrete periodic controller which contrasts [11]. Periodicity occurs because a discrete controller, used in feedback with a continuous plant, is interfaced to the plant by a sample device and a (usually zero-order) hold device. The lifting of a continuous signal, defined on the real line, can be visualized as the action of slicing up this signal into segments of finite duration, h (see Figure 8.5); a discrete sequence

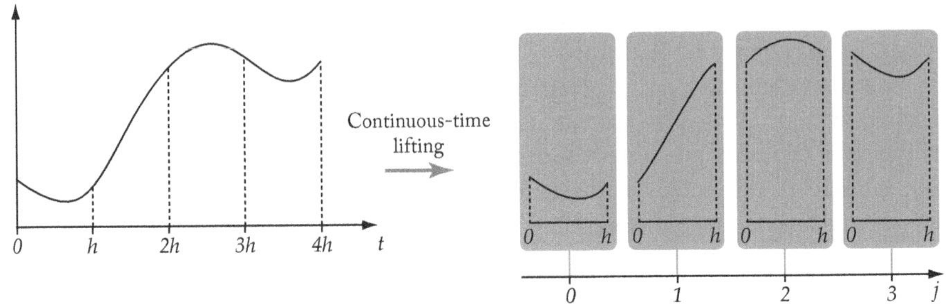

FIGURE 8.5: Lifting of a continuous-time signal.

of functions in the interval $[0, h)$ is obtained. The lifting of continuous-time signals can be used to convert a hybrid system (*i.e.* the plant with a ZOH element at its input and sampler element at its output) into an equivalent discrete system with equal induced norms. Let[1] $\mathcal{L}_p^n[0,\infty)$, $1 \le p \le \infty$ be the space of signals (that can be n-dimensional vectors of scalar signals) defined as

$$\mathcal{L}_p^n[0,\infty) \overset{\text{def}}{=} \left\{ \boldsymbol{f} : [0,\infty) \to \mathbb{R}, \sqrt[p]{\int_0^\infty \|f(s)\|^p ds} < \infty \right\}, \tag{8.29}$$

and $\mathcal{L}_p^{n,e}[0,\infty)$ be the extended signal space. For a Banach space X, let $\{\breve{\boldsymbol{f}}(j)\} : \mathbb{N} \to X$ be the sequences that take values in X. The space of sequences $\{\breve{\boldsymbol{f}}(j)\}$, where $\breve{\boldsymbol{f}}(j)$ are the elements of the sequence, is

$$l_X = \left\{ \{\breve{\boldsymbol{f}}(j)\} : \breve{\boldsymbol{f}}(j) \in X \right\}. \tag{8.30}$$

Hence, a signal in $l_{\mathcal{L}_p^n[0,h]}$ are sequences $\{\boldsymbol{f}(j)\}$ (or discrete-time signals) whose sequence elements are functions in $\mathcal{L}_p^n[0,h]$. The continuous-time lifting of $\boldsymbol{f}(t)$ for fixed h is defined as

$$\breve{\boldsymbol{f}} = W_h \boldsymbol{f}, \quad \breve{\boldsymbol{f}}(j,t) = \boldsymbol{f}(jh+t), \quad 0 \le t \le h, \tag{8.31}$$

where

$$W_h : \quad \mathcal{L}_p^{n,e}[0,\infty) \to l_{\mathcal{L}_p^n[0,h]}. \tag{8.32}$$

$\breve{\boldsymbol{f}}$ is a sequence, given by (8.32), where each of its elements is a function of t.

[1]Notation: When discussing $\mathcal{L}_p$ spaces, p is not the same p as the one used to indicate the periodicity of the communication sequence. The difference should be clear from the context.

Now that the lifting of continuous-time signals has been defined, it can be used to convert a hybrid system (*i.e.* the plant with a ZOH element at its input and sampler element at its output) into an equivalent discrete system with equal induced norms. Consider the strictly proper continuous-time plant

$$\mathcal{P}^c : \quad \begin{bmatrix} \dot{\tilde{\boldsymbol{x}}}(t) \\ \tilde{\boldsymbol{z}}(t) \\ \tilde{\boldsymbol{y}}(t) \end{bmatrix} = \begin{bmatrix} \boldsymbol{A}^c & \boldsymbol{B}_1^c & \boldsymbol{B}_2^c \\ \boldsymbol{C}_1 & \boldsymbol{D}_{11} & \boldsymbol{D}_{12} \\ \boldsymbol{C}_2 & \boldsymbol{0} & \boldsymbol{0} \end{bmatrix} \begin{bmatrix} \tilde{\boldsymbol{x}}(t) \\ \tilde{\boldsymbol{w}}(t) \\ \tilde{\boldsymbol{u}}(t) \end{bmatrix}, \tag{8.33}$$

where $\tilde{\boldsymbol{x}}(t) \in \mathbb{R}^{n_x}$, $\tilde{\boldsymbol{w}}(t) \in \mathbb{R}^{n_w}$, $\tilde{\boldsymbol{u}}(t) \in \mathbb{R}^{n_u}$, $\tilde{\boldsymbol{z}}(t) \in \mathbb{R}^{n_z}$, $\tilde{\boldsymbol{y}}(t) \in \mathbb{R}^{n_y}$, and $\boldsymbol{A}^c, \boldsymbol{B}_1^c$, *etc.* are matrices with appropriate dimensions. The state equation that describes the behavior of the state in between samples (*i.e.* for $0 \leq \check{t} \leq h$) is given by

$$\begin{aligned} \tilde{\boldsymbol{x}}(jh+\check{t}) &= e^{\boldsymbol{A}^c(jh+\check{t})}\tilde{\boldsymbol{x}}(0) \\ &\quad + \int_0^{jh+\check{t}} e^{\boldsymbol{A}^c(jh+\check{t}-s)}\boldsymbol{B}_1^c\tilde{\boldsymbol{w}}(s)ds + \int_0^{jh+\check{t}} e^{\boldsymbol{A}^c(jh+\check{t}-s)}\boldsymbol{B}_2^c\tilde{\boldsymbol{u}}(s)ds \\ &= e^{\boldsymbol{A}^c\check{t}}\tilde{\boldsymbol{x}}(jh) + \int_0^{\check{t}} e^{\boldsymbol{A}^c(\check{t}-\check{s})}\boldsymbol{B}_1^c\check{\boldsymbol{w}}(j,\check{s})d\check{s} + \int_0^{\check{t}} e^{\boldsymbol{A}^c(\check{t}-\check{s})}\boldsymbol{B}_2^c\check{\boldsymbol{u}}(j,\check{s})d\check{s}, \end{aligned} \tag{8.34}$$

where $j = 0, 1, \ldots$ and $\check{\boldsymbol{w}}(j, \check{s})$ is a more general way to write $\check{\boldsymbol{w}}(jh + \check{s})$ (and similarly for $\check{\boldsymbol{u}}(j, \check{s})$ and other cases that will follow). In particular, the discrete state, defined as $\check{\boldsymbol{x}}(j) \stackrel{\text{def}}{=} \tilde{\boldsymbol{x}}(jh)$, evolves by

$$\check{\boldsymbol{x}}(j+1) = e^{\boldsymbol{A}^c h}\check{\boldsymbol{x}}(jh) + \int_0^{h} e^{\boldsymbol{A}^c(h-\check{s})}\boldsymbol{B}_1^c\check{\boldsymbol{w}}(j,\check{s})d\check{s} + \int_0^{h} e^{\boldsymbol{A}^c(h-\check{s})}\boldsymbol{B}_2^c\check{\boldsymbol{u}}(j,\check{s})d\check{s}, \tag{8.35}$$

which can be written, in the operator notation[2] used in [11], as

$$\check{\boldsymbol{x}}(j+1) = \check{\boldsymbol{A}}\check{\boldsymbol{x}}(j) + \check{\boldsymbol{B}}_1\check{\boldsymbol{w}}(j) + \check{\boldsymbol{B}}_2\check{\boldsymbol{u}}(j). \tag{8.36}$$

Similarly, the output error equation, in operator notation is

$$\check{\boldsymbol{z}}(j) = \check{\boldsymbol{C}}_1\check{\boldsymbol{x}}(j) + \check{\boldsymbol{D}}_{11}\check{\boldsymbol{w}}(j) + \check{\boldsymbol{D}}_{12}\check{\boldsymbol{u}}(j). \tag{8.37}$$

where

$$\begin{aligned} \check{\boldsymbol{C}}_1\check{\boldsymbol{x}}(j) &= \boldsymbol{C}_1 e^{\boldsymbol{A}^c\check{t}}\check{\boldsymbol{x}}(j), \\ \check{\boldsymbol{D}}_{11}\check{\boldsymbol{w}}(j) &= \int_0^{\check{t}} \left[\boldsymbol{C}_1 e^{\boldsymbol{A}^c\check{t}-\check{s}}\mathbf{1}_{(\check{t}-\check{s})}\boldsymbol{B}_1^c + \boldsymbol{D}_{11}\delta(\check{t}-\check{s})\right]\boldsymbol{w}(j,\check{s})d\check{s}, \\ \check{\boldsymbol{D}}_{12}\check{\boldsymbol{u}}(j) &= \int_0^{\check{t}} \left[\boldsymbol{C}_1 e^{\boldsymbol{A}^c\check{t}-\check{s}}\mathbf{1}_{(\check{t}-\check{s})}\boldsymbol{B}_2^c + \boldsymbol{D}_{12}\delta(\check{t}-\check{s})\right]\boldsymbol{u}(j,\check{s})d\check{s}, \end{aligned} \tag{8.38}$$

[2]In this notation, the same symbol is used for an operator and its kernel.

and $\mathbf{1}_{(t)}$ is the unit step function

$$\mathbf{1}_{(t)} = \begin{cases} 1 & t > 0 \\ 0 & t \leq 0 \end{cases}. \tag{8.39}$$

The measurement output equation is

$$\breve{\boldsymbol{y}}(j) = \breve{\boldsymbol{C}}_2 \breve{\boldsymbol{x}}(j) + \breve{\boldsymbol{D}}_{21} \breve{\boldsymbol{w}}(j) + \breve{\boldsymbol{D}}_{22} \breve{\boldsymbol{u}}(j), \tag{8.40}$$

where the operators $\breve{\boldsymbol{C}}_2$, $\breve{\boldsymbol{D}}_{21}$ and $\breve{\boldsymbol{D}}_{22}$ can be easily derived by inspecting (8.38).

The system $\mathcal{P}^c$ in (8.33) has a continuous-time lifted representation given by

$$\breve{\mathcal{P}} : \quad \begin{bmatrix} \breve{\boldsymbol{x}}(j+1) \\ \breve{\boldsymbol{z}}(j) \\ \breve{\boldsymbol{y}}(j) \end{bmatrix} = \begin{bmatrix} \breve{\boldsymbol{A}} & \breve{\boldsymbol{B}}_1 & \breve{\boldsymbol{B}}_2 \\ \breve{\boldsymbol{C}}_1 & \breve{\boldsymbol{D}}_{11} & \breve{\boldsymbol{D}}_{12} \\ \breve{\boldsymbol{C}}_2 & \breve{\boldsymbol{D}}_{21} & \breve{\boldsymbol{D}}_{22} \end{bmatrix} \begin{bmatrix} \breve{\boldsymbol{x}}(j) \\ \breve{\boldsymbol{w}}(j) \\ \breve{\boldsymbol{u}}(j) \end{bmatrix}. \tag{8.41}$$

Let the sampler, $\mathfrak{S}_h$, be a map from the space of piecewise continuous vector-valued functions into the space of vector- valued sequences such that, for constant sampling time h,

$$\boldsymbol{f}(jh) = \mathfrak{S}_h \boldsymbol{f}(t) \quad \Leftrightarrow \quad \boldsymbol{f}(jh) = \lim_{t \to jh^-} \boldsymbol{f}(t) = \boldsymbol{f}(jh^-). \tag{8.42}$$

The ZOH, $\mathfrak{H}_h$, is the hold operator defined as

$$\begin{aligned} &\boldsymbol{f}(t) = \mathfrak{H}_h \boldsymbol{f}(jh) \quad \Leftrightarrow \quad \forall t \in [jh, (j+1)h) : \boldsymbol{f}(t) = \boldsymbol{f}(jh), \\ &\boldsymbol{f}(t=0) = 0. \end{aligned} \tag{8.43}$$

Now, let us add $\mathfrak{S}_h$ and $\mathfrak{H}_h$ to $\breve{\mathcal{P}}$ (at its measurement outputs and controlled inputs respectively). The equation in (8.41) becomes

$$\grave{\mathcal{P}} : \quad \begin{bmatrix} \grave{\boldsymbol{x}}(j+1) \\ \grave{\boldsymbol{z}}(j) \\ \grave{\boldsymbol{y}}(j) \end{bmatrix} = \underbrace{\begin{bmatrix} \breve{\boldsymbol{A}} & \breve{\boldsymbol{B}}_1 & \breve{\boldsymbol{B}}_2 \mathfrak{H}_h \\ \breve{\boldsymbol{C}}_1 & \breve{\boldsymbol{D}}_{11} & \breve{\boldsymbol{D}}_{12} \mathfrak{H}_h \\ \mathfrak{S}_h \breve{\boldsymbol{C}}_2 & \mathfrak{S}_h \breve{\boldsymbol{D}}_{21} & \mathfrak{S}_h \breve{\boldsymbol{D}}_{22} \mathfrak{H}_h \end{bmatrix}}_{\grave{\mathcal{A}}} \begin{bmatrix} \grave{\boldsymbol{x}}(j) \\ \grave{\boldsymbol{w}}(j) \\ \grave{\boldsymbol{u}}(j) \end{bmatrix}, \tag{8.44}$$

and it is shown in [11] that, after loop-shifting and scaling,

$$\grave{\mathcal{A}} = \begin{bmatrix} \grave{\boldsymbol{A}} & \grave{\boldsymbol{B}}_1 & \grave{\boldsymbol{B}}_2 \\ \grave{\boldsymbol{C}}_1 & \grave{\boldsymbol{D}}_{11} & \grave{\boldsymbol{D}}_{12} \\ \grave{\boldsymbol{C}}_2 & \boldsymbol{0} & \boldsymbol{0} \end{bmatrix} = \begin{bmatrix} e^{\boldsymbol{A}^c h} & e^{\boldsymbol{A}^c (h-s)} \boldsymbol{B}_1^c & \int_0^h e^{\boldsymbol{A}^c r} dr \boldsymbol{B}_2^c \\ \boldsymbol{C}_1 e^{\boldsymbol{A}^c h} & \boldsymbol{C}_1 e^{\boldsymbol{A}^c (t-s)} \mathbf{1}_{(t-s)} \boldsymbol{B}_1^c & \boldsymbol{C}_1 \int_0^t e^{\boldsymbol{A}^c r} dr \boldsymbol{B}_2^c \\ \boldsymbol{C}_2 & \boldsymbol{0} & \boldsymbol{0} \end{bmatrix}. \tag{8.45}$$

Note that $\grave{\mathcal{A}}$ is a matrix of operators written in the operator shorthand notation defined in a similar way as in (8.38). $\grave{\mathcal{P}}$ is a system with infinite-dimensional input and output vectors. The following theorem shows how $\grave{\mathcal{P}}$

can be reduced to $\mathcal{P}$, an equivalent system with finite input and output dimensions. Moreover, if $\mathcal{K}$ is the discrete feedback controller, the relationship between $\|\mathbb{F}(\grave{P},\mathcal{K})\|_{\ell_2}$ (the induced ℓ_2-norm of the closed-loop system defined in (8.21)) and $\|\mathbb{F}(\mathcal{P},\mathcal{K})\|_{\ell_2}$ (the induced ℓ_2-norm of the finite-dimensional equivalent closed-loop system) is given. (Recall that the notation $\mathbb{F}(\mathcal{G},\mathcal{H})$ is used to denote the closed-loop mapping from the exogenous input to the regulated output of the generalized plant $\mathcal{G}$ in feedback with controller $\mathcal{H}$; see Section 8.3.1.2.)

Theorem 8.1 ([11]) *Given the system matrix* $\grave{\mathcal{A}}$ *in* (8.45) *with* $\|\grave{\boldsymbol{D}}_{11}\| < 1$, *form the matrices*

$$\grave{\boldsymbol{B}}_1(\boldsymbol{I} - \grave{\boldsymbol{D}}_{11}^T\grave{\boldsymbol{D}}_{11})^{-1}\grave{\boldsymbol{B}}_1^T = \boldsymbol{T}_B^T \begin{bmatrix} \boldsymbol{\Sigma}_b & \boldsymbol{0} \\ \boldsymbol{0} & \boldsymbol{0} \end{bmatrix} \boldsymbol{T}_B,$$

$$\begin{bmatrix} \grave{\boldsymbol{C}}_1^T \\ \grave{\boldsymbol{D}}_{12}^T \end{bmatrix} (\boldsymbol{I} - \grave{\boldsymbol{D}}_{11}^T\grave{\boldsymbol{D}}_{11})^{-1} \begin{bmatrix} \grave{\boldsymbol{C}}_1 & \grave{\boldsymbol{D}}_{12} \end{bmatrix} = \boldsymbol{T}_{CD}^T \begin{bmatrix} \boldsymbol{\Sigma}_{cd} & \boldsymbol{0} \\ \boldsymbol{0} & \boldsymbol{0} \end{bmatrix} \boldsymbol{T}_{CD}, \tag{8.46}$$

where $\boldsymbol{\Sigma}_b$ *and* $\boldsymbol{\Sigma}_{cd}$ *are diagonal and nonsingular, and* $\boldsymbol{T}_B$ *and* $\boldsymbol{T}_{CD}$ *are finite dimensional matrices. Define the finite-dimensional system*

$$\mathcal{P}: \quad \begin{bmatrix} \boldsymbol{x}(j+1) \\ \boldsymbol{z}(j+1) \\ \boldsymbol{y}(j+1) \end{bmatrix} = \begin{bmatrix} \boldsymbol{A} & \boldsymbol{B}_1 & \boldsymbol{B}_2 \\ \boldsymbol{C}_1 & \boldsymbol{0} & \boldsymbol{D}_{12} \\ \boldsymbol{C}_2 & \boldsymbol{0} & \boldsymbol{0} \end{bmatrix} \begin{bmatrix} \boldsymbol{x}(j) \\ \boldsymbol{w}(j) \\ \boldsymbol{u}(j) \end{bmatrix}, \tag{8.47}$$

where

$$\begin{aligned}
&\boldsymbol{B}_1 \overset{def}{=} \boldsymbol{T}_B^T \begin{bmatrix} \boldsymbol{\Sigma}_b^{\frac{1}{2}} \\ \boldsymbol{0} \end{bmatrix}, \quad \begin{bmatrix} \boldsymbol{C}_1 & \boldsymbol{D}_{12} \end{bmatrix} \overset{def}{=} \begin{bmatrix} \boldsymbol{\Sigma}_{cd}^{\frac{1}{2}} & \boldsymbol{0} \end{bmatrix} \boldsymbol{T}_{CD}, \\
&\boldsymbol{A} \overset{def}{=} \grave{\boldsymbol{A}} + \grave{\boldsymbol{B}}_1\grave{\boldsymbol{D}}_{11}^T\left(\boldsymbol{I} - \grave{\boldsymbol{D}}_{11}\grave{\boldsymbol{D}}_{11}^T\right)^{-1}\grave{\boldsymbol{C}}_1, \\
&\boldsymbol{B}_2 \overset{def}{=} \grave{\boldsymbol{B}}_1\grave{\boldsymbol{D}}_{11}^T\left(\boldsymbol{I} - \grave{\boldsymbol{D}}_{11}\grave{\boldsymbol{D}}_{11}^T\right)^{-1}\grave{\boldsymbol{D}}_{12} + \grave{\boldsymbol{B}}_2, \\
&\boldsymbol{C}_2 \overset{def}{=} \grave{\boldsymbol{C}}_2.
\end{aligned} \tag{8.48}$$

Then, the following are equivalent:

(i) $\mathbb{F}(\grave{\mathcal{P}},\mathcal{K})$ *is internally stable and* $\|\mathbb{F}(\grave{P},\mathcal{K})\|_{\ell_2} < 1$

(ii) $\mathbb{F}(\mathcal{P},\mathcal{K})$ *is internally stable and* $\|\mathbb{F}(\mathcal{P},\mathcal{K})\|_{\ell_2} < 1$.

Remark 8.2 *Theorem 8.1 offers a solution to the calculation of the input-to-output 2-gain of a plant in feedback with any discrete controller. The necessary 'removal' of the operator* $\breve{\boldsymbol{D}}_{11}$, *described in [11] and accomplished by the result of [11, Lemma 5], is only possible for LTI controllers. However, the interest here is in periodically time-varying controllers. The approach is easily extendable to a linear periodic controller, assuming that the analytical lemma in [11, Lemma 5] is only used after the discrete-time lifting of the controller (and the plant) to remove the periodicity in p. Thus, available tools in MATLAB® can be easily modified based on that idea.* •

Note that defining the ℓ_2 problem as $\|\mathbb{F}(\mathcal{P},\mathcal{K})\|_{\ell_2} < 1$ rather than the more general $\|\mathbb{F}(\mathcal{P},\mathcal{K})\|_{\ell_2} < \gamma$, for $\gamma > 0$, is only a matter of scaling.

In summary, for a given continuous-time plant $\mathcal{P}^c$, discrete-time controller $\mathcal{K}$, ZOH element $\mathfrak{H}_h$ and a sampling element $\mathfrak{S}_h$, the continuous-time lifting in (8.45) and the result of Theorem 8.1 can be used to obtain the equivalent discrete-time time-invariant generalized plant $\mathcal{P}$ such that

$$\|\mathbb{F}(\mathcal{P}^c, \mathfrak{H}_h\mathcal{K}\mathfrak{S}_h)\|_{\mathcal{L}_2} < \gamma \quad \Leftrightarrow \quad \|\mathbb{F}(\mathcal{P},\mathcal{K})\|_{\ell_2} < \gamma, \tag{8.49}$$

where the $\mathcal{L}_2$-norm of a continuous signal is defined as

$$\|\boldsymbol{\nu}(t)\|^2_{\mathcal{L}_2} = \int_0^T \|\boldsymbol{\nu}(t)\|^2 dt. \tag{8.50}$$

$\mathcal{P}$ is an equivalent discretization of $\mathcal{P}^c$ for 2-norm problems.

Let $\mathcal{G}^c$ be a generalized continuous-time system with a ZOH used on its controlled inputs and a sampler used on its measured outputs. Let $\mathcal{G}$ be the discrete system obtained by applying the continuous-time lifting to $\mathcal{G}^c$. Then, the ℓ_2-gain of $\mathcal{G}$ has the same magnitude as the $\mathcal{L}_2$-gain of $\mathcal{G}^c$.

For shorthand notation, let us define a continuous-time lifting operator.

Definition 8.1 *The continuous-time lifting operator* $\mathbb{L}_C\{\cdot,\cdot\}$ *is defined as*

$$\mathcal{G} = \mathbb{L}_C\{\mathcal{G}^c, h\}, \tag{8.51}$$

where h is the periodicity (or sampling time). ▲

Note that an example for $\mathcal{G}^c$ is given in Figure 8.6 where $\mathcal{G}^c = \mathcal{P}^c$ and the continuous-time signals $\tilde{\boldsymbol{w}}(t)$ and $\tilde{\boldsymbol{z}}(t)$ are to be lifted. Moreover, here the controller $\hat{\mathcal{K}}$ is periodic due to the communication schedulers $\mathcal{S}_A$, $\mathcal{S}_D$ and $\mathcal{S}_S$, while the controller used by [11] is LTI. If such a controller is first discrete-time lifted then the work by [11], for ℓ_2-gain computation, can be readily applied (see Remark 8.2).

8.4.2 NCS model

Let the NCS be described by the diagram in Figure 8.6. The distributed plant $\mathcal{P}^c$ is given in (8.33) and the sampler $\mathfrak{S}_h$ and ZOH $\mathfrak{H}_h$ are defined in (8.42)-(8.43). The two-DOF discrete controller is the usual and it is given in (3.8). Since the feedback loop of the NCS is closed via a network, sensor-to-controller, demand-to-controller and controller-to-actuator communication must be scheduled. This is represented by the communication schedulers (periodic state space models) $\mathcal{S}_S$, $\mathcal{S}_D$ and $\mathcal{S}_A$ (given in (3.17), (3.16) and (3.15) respectively).

The inputs of $\mathcal{K}$ are the outputs of $\mathcal{S}_S$ and $\mathcal{S}_D$, and the output of $\mathcal{K}$ is the input of $\mathcal{S}_A$. Following the same procedure as the one in Section 8.3.1.1, these

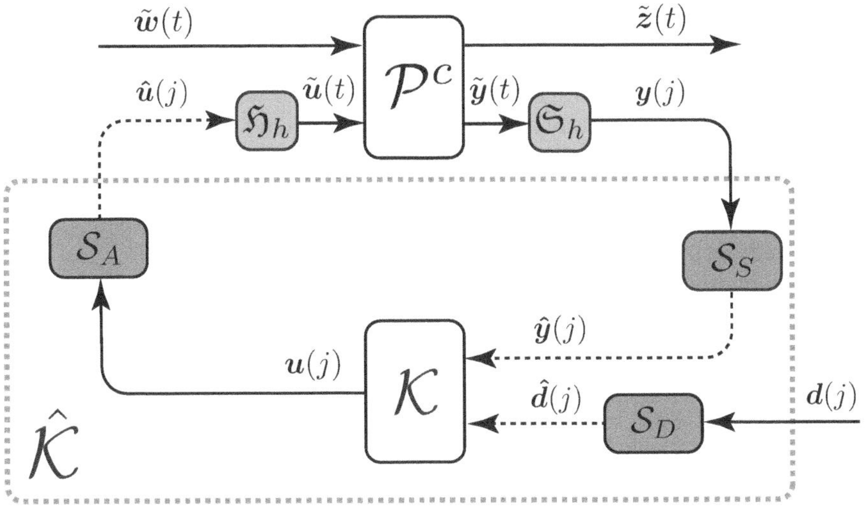

FIGURE 8.6: Continuous-time plant with ZOH, sampler, sampled-data controller and schedulers.

systems are first merged to form the augmented controller $\hat{\mathcal{K}}$ (see Figure 8.6) and then discrete-time lifted to form the discrete-time lifted controller

$$\bar{\mathcal{K}}: \quad \begin{bmatrix} \hat{x}_K(pl+p) \\ \bar{u}(pl) \end{bmatrix} = \begin{bmatrix} \tilde{A}_{K,p} & \tilde{B}_{K1,p} & \tilde{B}_{K2,p} \\ \bar{C}_K & \bar{D}_{K1} & \bar{D}_{K2} \end{bmatrix} \begin{bmatrix} \hat{x}_K(pl) \\ \bar{y}(pl) \\ \bar{d}(pl) \end{bmatrix} \tag{8.52}$$

(which is the same as the one in (8.22) but it has been rewritten in this form to be consistent with the notation of this section). Note that since the discrete-time lifting is applied to the ℓ_2-gain stable system $\hat{\mathcal{K}}$, then the resulting ℓ_2-gain of the lifted system $\bar{\mathcal{K}}$ is identical to the gain of $\hat{\mathcal{K}}$.

The generalized plant $\mathcal{P}^c$ in Figure 8.6 is in continuous-time and therefore it is not defined in the same time domain as the controller (which is discrete). The continuous-time lifting technique [11] (also, see Section 8.4.1) will be used to obtain a lifted discrete plant. This is illustrated in Figure 8.7. Since

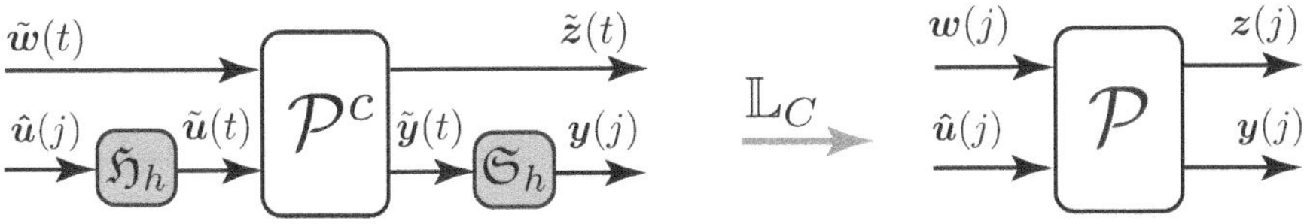

FIGURE 8.7: Continuous-time lifting applied to the continuous plant.

$\mathcal{P}^c$ is the generalized continuous plant, then the equivalent discrete plant is

$\mathcal{P} = \mathbb{L}_C\{\mathcal{P}^c, h\}$ (see Definition 8.1), where $\mathcal{P}$ is given in (8.47) and h is the sampling time. The induced ℓ_2-gain of the closed-loop system involving $\mathcal{P}$ is the same as the induced $\mathcal{L}_2$-gain as for the closed-loop with $\mathcal{P}^c$. The continuous-time lifting, together with some geometric projection and loop-shifting allows the computation of a discrete-time plant representation $\grave{\mathcal{P}}$ from the lifted plant $\mathcal{P}$ so that the closed-loop $\mathbb{F}(\grave{\mathcal{P}}, \hat{\mathcal{K}})$ has the same induced ℓ_2-gain as $\mathbb{F}(\mathcal{P}, \hat{\mathcal{K}})$ (see Section 8.4.1).

After the discrete-time lifting, the dimensions of the augmented controller input and output signals have increased by a factor of p. To match the input and output signals of the plant with the discrete-time lifted controller's inputs and outputs, the equivalent discretized plant $\mathcal{P}$ in (8.47) also needs to be discrete-time lifted. The discrete-time lifting operator is therefore applied to $\mathcal{P}$ to obtain $\bar{\mathcal{P}} = \mathbb{L}_D\{\mathcal{P}, p\}$. $\bar{\mathcal{P}}$ is the equivalent higher dimensional system described by

$$\bar{\mathcal{P}}: \quad \begin{bmatrix} \hat{\boldsymbol{x}}(pl+p) \\ \bar{\boldsymbol{z}}(pl) \\ \bar{\boldsymbol{y}}(pl) \end{bmatrix} = \begin{bmatrix} \tilde{\boldsymbol{A}} & \tilde{\boldsymbol{B}}_1 & \tilde{\boldsymbol{B}}_2 \\ \bar{\boldsymbol{C}}_1 & \bar{\boldsymbol{D}}_{11} & \bar{\boldsymbol{D}}_{12} \\ \bar{\boldsymbol{C}}_2 & \bar{\boldsymbol{D}}_{21} & \bar{\boldsymbol{D}}_{22} \end{bmatrix} \begin{bmatrix} \hat{\boldsymbol{x}}(pl) \\ \bar{\boldsymbol{w}}(pl) \\ \bar{\boldsymbol{u}}(pl) \end{bmatrix}. \tag{8.53}$$

This is shown in Figure 8.3 although $\mathcal{P}$ in Section 8.3.1 is not the same as $\mathcal{P}$ defined here (the former is the usual discretization while the latter is an equivalent discretization for 2-norm problems).

Finally, the discrete LTI closed-loop system is, as shown in Figure 8.4, $\bar{\mathcal{P}}_{cl} = \mathbb{F}(\bar{\mathcal{P}}, \bar{\mathcal{K}})$ ($\bar{\mathcal{P}}_{cl}$ is not the same as the previous one).

8.4.3 Cost function

From the previous discussion, it emerges that, for a stable $\bar{\mathcal{P}}_{cl}$ and $\mathbb{F}(\mathcal{P}^c, \mathfrak{H}_h\hat{\mathcal{K}}\mathfrak{S}_h)$,

$$\|\bar{\mathcal{P}}_{cl}\|_{\ell_2} = \|\mathbb{F}(\mathcal{P}^c, \mathfrak{H}_h\mathcal{K}\mathfrak{S}_h)\|_{\mathcal{L}_2}, \tag{8.54}$$

where $\|\cdot\|_{\ell_2}$ and $\|\cdot\|_{\mathcal{L}_2}$ are the induced ℓ_2- and $\mathcal{L}_2$-gain respectively. Hence, the induced $\mathcal{L}_2$-norm of the hybrid system with communication scheduling can be found by calculating the ℓ_2-gain of the equivalent LTI discrete system $\bar{\mathcal{P}}_{cl}$, for which a solution exists.

The optimization parameter is the communication sequence, σ. Let $\gamma_{sd} \in \mathbb{R}$ and $\gamma_{sd} \geq \|\bar{\mathcal{P}}_{cl}\|_{\ell_2} > 0$. Then, the optimization problem considered is: minimize γ_{sd} by finding a communication sequence σ such that, for a given γ_{sd}, the induced ℓ_2-norm of the mapping from $[\bar{\boldsymbol{d}}(pl)^T \ \bar{\boldsymbol{w}}(pl)^T]^T$ to $\bar{\boldsymbol{z}}(pl)$ is less than γ_{sd}. This can be calculated by appropriate modification of the MATLAB® functions `sdhinfnorm` and `sdn_step` to allow for periodic controllers as explained previously. For the feasibility issue see Remark 8.1.

8.5 Optimal design with an example

The optimal design consists of finding a communication sequence, between nodes, that minimizes a cost. The optimization procedures were presented in Chapter 5 and their applications were discussed in Sections 6.3 and 7.3 (pp. 137 and 150) for the proposed quadratic costs. In this chapter, three $\mathcal{H}_\infty$-based cost functions were presented. The optimization process here will be equivalent to the ones previously encountered with the main difference that the new $\mathcal{H}_\infty$-based cost functions are used instead.

The cost functions presented in Sections 8.3 and 8.4 allow for the treatment of model uncertainties as well as demand scheduling. Their effectiveness will be demonstrated by a practical implementation on an automotive system in the next chapter. For this reason, no simulation examples will be given for those. However, a numerical example will be used next to show the effectiveness of the proposed cost of Section 8.2.

Example 8.1 *The same setup as for Example 7.1 (p. 152) is used here for comparative reasons. In this example, two inverted pendulums with different dynamics, given by different rod lengths, have to be controlled. Since sensors and actuators communicate to the controller via a limited bandwidth communication bus, the problem is to find an optimal scheduling communication sequence. Contrary to Example 7.1, position demand signals are considered here. The demands, or reference signals, come into play in the usual way. Two errors (one for each pendulum system) are created by the position demand signals and the actual positions from the sensors. The errors are the input of the LQR controller (see Section 7.4, p. 152). The two demand signals are associated to the two demand nodes*

$$\begin{aligned}\Delta(1d) &= \{\textit{Pendulum 1 position demand}\},\\ \Delta(2d) &= \{\textit{Pendulum 2 position demand}\}.\end{aligned} \tag{8.55}$$

Hence, the full set of nodes, which includes the demands, is

$$\Delta = \{\Delta(1p), \Delta(1v), \Delta(1a), \Delta(1d), \Delta(2p), \Delta(2v), \Delta(2a), \Delta(2d)\}. \tag{8.56}$$

The problem here is to find a communication sequence that minimizes the cost in (8.20). This time (for no particular reason apart from wanting to use a different algorithm than in Example 7.1) the PS02 algorithm is used. The returned optimized sequence is

$$\begin{aligned}\sigma_{PSO2} = \{&\Delta(2a), \Delta(2p), \Delta(2v), \Delta(1a), \Delta(1d), \Delta(2a), \Delta(2p), \Delta(2v), \Delta(1p),\\ &\Delta(1d), \Delta(2a), \Delta(2p), \Delta(2v), \Delta(1v), \Delta(1d), \Delta(2a), \Delta(2p), \Delta(2v), \Delta(2d)\}.\end{aligned} \tag{8.57}$$

It was found in 35 seconds. This sequence is compared to an RR one, that, for this example, is

$$\sigma_{RR} = \{\Delta(1p), \Delta(1v), \Delta(1a), \Delta(1d), \Delta(2p), \Delta(2v), \Delta(2a), \Delta(2d)\}. \quad (8.58)$$

The unit step response for the optimized sequence in comparison to the RR and the ideal case is shown in Figure 8.8 (recall that the 'ideal' system is the one with dedicated connection between demands, sensors, actuators and controller and no need for scheduling). Although the response for the RR sequence seems to be closer to the ideal situation for pendulum 1, it completely destabilizes pendulum 2. In reality these sequence are rejected by the proposed algorithms as an unfeasible one because of Remark 8.1. The response for the PSO2-optimized sequence is stable and shows minimal deviation from the ideal situation. It is difficult to speculate the exact reason why the optimized sequence performs better. Nonetheless, it can be noticed that pendulum 2's sensors and actuators appear more often in the sequence, therefore receiving more 'attention' (which is expected since pendulum 2 is the more unstable). The demand nodes, although they do not affect the closed-loop dynamics, need to be scheduled in practical situations. This was possible for this formulation. ♦

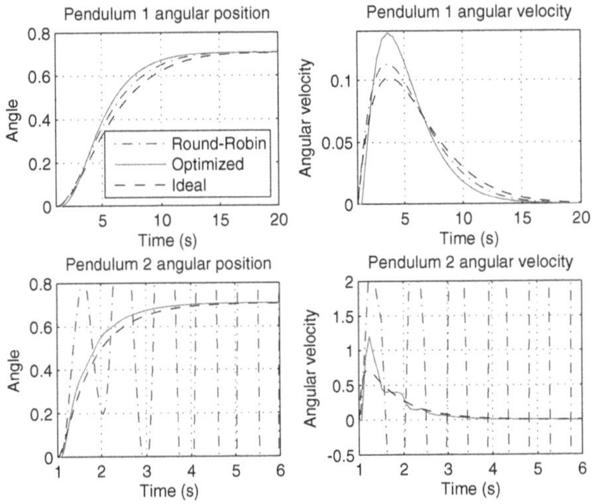

FIGURE 8.8: Comparison between ideal network, RR schedule, and optimized schedule with demands. The RR schedule is not even capable to stabilize pendulum 2.

8.6 Summary

The advantage of an $\mathcal{H}_\infty$ approach is that unmodeled dynamics (like delays) and parameter uncertainties can be considered, thus adding robustness to the overall design. In this chapter, the analysis was extended to an $\mathcal{H}_\infty$ framework. The NCS was modeled in the usual way (Chapter 3) and the optimization performed with the algorithms of Chapter 5. The difference is that the cost function for the sequence optimization was reformulated using tools from robust control theory. Three $\mathcal{H}_\infty$-based cost functions were derived. One of those was an exact ℓ_2-gain from the exogenous input to the regulated output calculated using the continuous- and discrete-time lifting techniques. With this approach, demand signals can be scheduled optimally. This formulation is the most practically relevant and it will be used in the next chapter to optimize the communication between electronic control units of an automotive control system.

9

Application to an automotive control system

CONTENTS

Control theory is a discipline which lies between applied mathematics and engineering. Research and development of new techniques in the field of control would be quite uninteresting if such techniques were not validated and implemented in real practical systems.

9.1 Introduction

In the previous chapters, it was shown how to model NCSs in order to take into account the limited time-triggered sensor-to-controller and controller-to-actuator communication. It was then possible to design, offline, a communication sequence that is optimal in the sense that it minimizes some cost function. The final aim is to implement such communication sequences in a practical system in the form of a scheduler. A scheduler is a mechanism that coordinates messages exchange according to the predefined sequence.

The reason for studying this type of NCSs is that they closely describe large spatially distributed systems. An example of such systems are vehicles where microcontrollers and devices communicate to each other via a bus. The vehicle bus standard is the Controller Area Network (CAN) that is slowly being replaced by a new, faster and more flexible protocol called FlexRay[1]. CAN is traditionally an event-triggered protocol but it can be defined as a time-triggered one (for example the TTCAN, ISO 11898-4). On the other hand, FlexRay allows a combination of time-triggered and event-triggered communication exploiting the advantages of both protocols.

Although vehicles are not the only systems that can be described by NCSs (others might be airplanes or robots), they offer a versatile testbed for control and communication optimization. For this reason, a vehicular case study will be used here to validate the theoretical results.

The next section describes the model of a realistic automotive system followed by the design of an optimal controller. The vehicle model and the controller are implemented in a custom built Hardware-In-the-Loop (HIL) system, described in Section 9.3. The HIL system is used as testbed and the results from the experiments are presented in Section 9.4. Finally, in 9.5, some experiments are performed on a FlexRay simulation tool provided by the company Vector Informatik GmbH[2].

9.2 Vehicle model and controller design

In this section, the model of a cruise-driveline-temperature automotive system is derived followed by the design of a Linear Quadratic Gaussian (LQG) controller with integral action.

9.2.1 A cruise-driveline-temperature automotive system

The plant to be controlled is represented by a vehicle moving along a highway. The model includes the dynamics of the vehicle's longitudinal motion, the driveline torsion, and the air-conditioning systems.

9.2.1.1 Linearized longitudinal dynamics

The nonlinear equation of motion and suitable parameters used for the vehicle's longitudinal dynamics are those derived in [112] and used in [110, 111].

[1] `www.flexray.com`.
[2] `www.vector.com`.

In [112], The 'Pacejka magic model' was used to capture the dynamics of the tire/road interface [10, 38]. The model has been linearized and the linearization consisted of the following.

- A first-order Taylor series approximation around an operating point (vehicle speed) of 60 mph for the quadratic drag and torque functions.
- Eliminating the nonlinear gear switching function by assuming that the operating point is 60 mph and using a fixed gear ratio appropriate for that speed.
- Assuming that braking action can be obtained by a negative throttle signal. This assumption is reasonable for electric vehicles where a negative (with respect to the sense of motion) torque can be applied by the electric motor(s).

Furthermore, a first order filter is used to model the throttle actuator dynamics [123]. Finally, the linearized longitudinal dynamics are given by the continuous-time state space model

$$\begin{aligned}\begin{bmatrix}\dot{x}_f(t)\\ \dot{x}_v(t)\end{bmatrix} &= \begin{bmatrix}-\frac{1}{\tau_s} & 0\\ \frac{\eta_f d_1}{R_i m} & \frac{-2d_2}{m}\end{bmatrix}\begin{bmatrix}x_f(t)\\ x_v(t)\end{bmatrix} + \begin{bmatrix}\frac{K_s}{\tau_s}\\ 0\end{bmatrix} u_t(t)\\ y_v(t) &= \begin{bmatrix}0 & 1\end{bmatrix}\begin{bmatrix}x_f(t)\\ x_v(t)\end{bmatrix},\end{aligned} \tag{9.1}$$

where $x_f(t)$ is the filtered throttle signal, $x_v(t)$ is the vehicle velocity, $u_t(t)$ is the throttle input, $y_v(t)$ is the output (velocity) and

$$\begin{aligned} d_1 &= (528.7 + 0.152n - 0.0000217n^2)\eta_g,\\ d_2 &= C_d \rho A_f v_{op},\\ n &= \frac{v_{op}\eta_r \eta_g \eta_f}{R_i}. \end{aligned} \tag{9.2}$$

The throttle input $u_t(t)$ is expressed as a percentage of the total pedal excursion so that if $u_t(t) = 0$ the throttle pedal is completely released, if $u_t(t) = 50$ the throttle pedal is pressed to half (or 50%), *etc.* The parameters used for the model are explained and given explicitly in Table 9.1.

9.2.1.2 Driveline dynamics

With the addition of the driveline dynamics, the vehicle is modeled as shown in Figure 9.1. The driveline is the system that transfers the torque from the engine to the wheels. Among many other components, the driveline has a drive-shaft that can be modeled as a damped torsional flexibility [65, 45] as

$$J_1\ddot{\alpha}_1(t) = T_E(t) - k(\alpha_1(t) - \alpha_2(t)) - c(\dot{\alpha}_1(t) - \dot{\alpha}_2(t)), \tag{9.3a}$$

Symbol	Value	Unit	Description
η_g	1.2		Gear ratio in between the 4th and 5th gear
η_f	2.82		Gear ratio of the differential
η_r	9.549		Ratio radian per second/revolutions per minute
R_i	0.3	m	Vehicle's wheel radius
C_d	0.29		Aerodynamic drag coefficient
ρ	1.23	$\mathrm{kg \cdot m^{-3}}$	Density of air
A_f	2	$\mathrm{m^2}$	Frontal area of the vehicle
m	1854	kg	Vehicle's mass
τ_s	0.1	s	Throttle filter time constant
K_s	1		Throttle filter gain
v_{op}	60	mph	Vehicle's velocity operating point

TABLE 9.1: Parameters for the vehicle longitudinal model from [112].

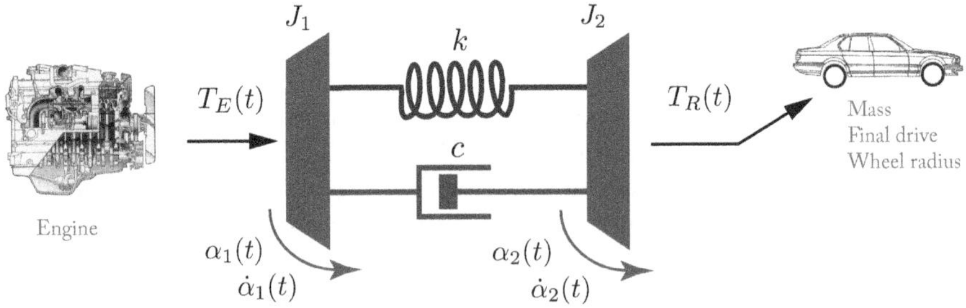

FIGURE 9.1: Schematic of the two-DOF driveline model. The model is composed by two inertias connected by a linear damper and a linear spring.

$$J_2\ddot{\alpha}_2(t) = k(\alpha_1(t) - \alpha_2(t)) + c(\dot{\alpha}_1(t) - \dot{\alpha}_2(t)) - T_R(t), \tag{9.3b}$$

where $\alpha_1(t)$ and $\alpha_2(t)$ are the flywheel and wheel angles, J_1 and J_2 are the driveline and vehicle inertias, $T_E(t)$ is the engine torque given by

$$T_E(t) = d_1 x_f(t), \tag{9.4}$$

$T_R(t)$ is the road torque and c and k are the viscous damping and stiffness coefficients respectively. $T_R(t)$ can be calculated as follows. The equation of motion of the vehicle from [112] is

$$m\dot{x}_v(t) = \frac{T_R(t)}{c_1} - d_2 x_v(t), \tag{9.5}$$

where

$$c_1 = \eta_f R_i. \tag{9.6}$$

Also, the relationship to convert angular to linear acceleration (and to account for the fact that the differential gear is after the driveline) is

$$\dot{x}_v(t) = c_1\ddot{\alpha}_2(t). \tag{9.7}$$

Substituting (9.7) into (9.5) gives the road torque as

$$m\eta_f R_i\ddot{\alpha}_2(t) = \frac{T_R(t)}{\eta_f R_i} - d_2 x_v(t) \quad \Rightarrow \quad T_R(t) = m_1\ddot{\alpha}_2(t) + d_3\dot{\alpha}_2(t), \tag{9.8}$$

where

$$m_1 = mc_1^2 \qquad \text{and} \qquad d_3 = d_2 c_1. \tag{9.9}$$

The final equations for the shaft torsion are obtained by substituting (9.4) into (9.3a) to give

$$\begin{aligned} &J_1\ddot{\alpha}_1(t) = d_1 x_f(t) - k(\alpha_1(t) - \alpha_2(t)) - c(\dot{\alpha}_1(t) - \dot{\alpha}_2(t)) \\ \Rightarrow \quad &\ddot{\alpha}_1(t) = \frac{d_1}{J_1}x_f(t) - \frac{k}{J_1}(\alpha_1(t) - \alpha_2(t)) - \frac{c}{J_1}\dot{\alpha}_1(t) - \frac{c}{J_1}\dot{\alpha}_2(t), \end{aligned} \tag{9.10}$$

and by substituting (9.8) into (9.3b) to give

$$\begin{aligned} &J_2\ddot{\alpha}_2(t) = k(\alpha_1(t) - \alpha_2(t)) + c(\dot{\alpha}_1(t) - \dot{\alpha}_2(t)) - m_1\ddot{\alpha}_2(t) + d_3\dot{\alpha}_2(t) \\ \Rightarrow \quad &\ddot{\alpha}_2(t) = \frac{k}{J_2 + m_1}(\alpha_1(t) - \alpha_2(t)) + \frac{c}{J_2 + m_1}\dot{\alpha}_1(t) + \frac{d_3 - c}{J_2 + m_1}\dot{\alpha}_2(t). \end{aligned} \tag{9.11}$$

The model of the longitudinal vehicle's motion with the driveline is, in state

Symbol	Value	Unit	Description
J_1	97	$kg \cdot m^2$	Effective driveline inertia
J_2	180	$kg \cdot m^2$	Effective vehicle inertia
k	2600	$N \cdot m \cdot rad^{-1}$	Stiffness coefficient
c	1000	$N{\cdot}m{\cdot}s{\cdot}rad^{-1}$	Linear damping coefficient

TABLE 9.2: Parameters for the driveline model from Jaguar and Ford [45].

space form,

$$\begin{bmatrix} \dot{x}_f(t) \\ \dot{\alpha}_1(t) - \dot{\alpha}_2(t) \\ \ddot{\alpha}_1(t) \\ \ddot{\alpha}_2(t) \end{bmatrix} = \begin{bmatrix} -\frac{1}{T_s} & 0 & 0 & 0 \\ 0 & 0 & 1 & -1 \\ \frac{d_1}{J_1} & -\frac{k}{J_1} & -\frac{c}{J_1} & \frac{c}{J_1} \\ 0 & \frac{k}{J_2+m_1} & \frac{c}{J_2+m_1} & \frac{d_3-c}{J_2+m_1} \end{bmatrix} \begin{bmatrix} x_f(t) \\ \alpha_1(t) - \alpha_2(t) \\ \dot{\alpha}_1(t) \\ \dot{\alpha}_2(t) \end{bmatrix} + \begin{bmatrix} \frac{K_s}{T_s} \\ 0 \\ 0 \\ 0 \end{bmatrix} u_t(t)$$

$$\begin{bmatrix} y_v(t) \\ y_s(t) \end{bmatrix} = \begin{bmatrix} 0 & 0 & 0 & c_1 \\ 0 & 1 & 0 & 0 \end{bmatrix} \begin{bmatrix} x_f(t) \\ \alpha_1(t) - \alpha_2(t) \\ \dot{\alpha}_1(t) \\ \dot{\alpha}_2(t) \end{bmatrix}. \tag{9.12}$$

The parameters for the driveline model are shown in Table 9.2. They are the ones given by Jaguar and Ford in [45].

9.2.1.3 Air-conditioning system

The final step is to add the air-conditioning system to the longitudinal motion and driveline dynamics. The model used for the air-conditioning is the one of [74] given below

$$\begin{bmatrix} \dot{T}_e(t) \\ \dot{T}_i(t) \end{bmatrix} = \begin{bmatrix} -a_1 & 0 \\ b_{31} & -a_2 \end{bmatrix} \begin{bmatrix} T_e(t) \\ T_i(t) \end{bmatrix} + \begin{bmatrix} b_{21} & b_{22} \\ 0 & 0 \end{bmatrix} \begin{bmatrix} \omega_c(t) \\ \alpha_v(t) \end{bmatrix}, \tag{9.13}$$

where $\omega_c(t)$ is the compressor speed and $\alpha_v(t)$ is the valve opening input to control the cabin temperature T_i. T_e is the evaporator's temperature that will not be controlled here. The assumption is that the compressor speed is proportional to the engine speed therefore, contrary to [74], $\omega_c(t)$ is not considered to be a control input but rather a coupling element. The model parameters are given in Table 9.3

The mechanical coupling between the engine and the compressor is represented by a belt and pulleys drive system as shown in Figure 9.2. From the diagram it follows that

$$F(t) = \frac{T_{A1}(t)}{r_1} \quad \text{and} \quad F(t) = \frac{T_{A2}(t)}{r_2} \quad \Rightarrow \quad T_{A1}(t) = \frac{r_1}{r_2} T_{A2}(t), \tag{9.14}$$

where $T_{A1}(t)$ is the torque provided by the engine, $T_{A2}(t)$ the torque given to

Symbol	Value	Unit	Description
a_1	0.7985		From system identification in [74]
a_2	0.00237		From system identification in [74]
b_{21}	−1.2613		From system identification in [74]
b_{22}	0.37163		From system identification in [74]
b_{31}	0.00237		From system identification in [74]
J_A	20	$kg \cdot m^2$	Engine inertia
c_a	10	$N \cdot m \cdot rad^{-1}$	Damping coefficient
r_1	0.02	m	Radius of the pulley's wheel connected to the engine
r_2	0.04	m	Radius of the pulley's wheel connected to the compressor

TABLE 9.3: Parameters for the air-conditioning model.

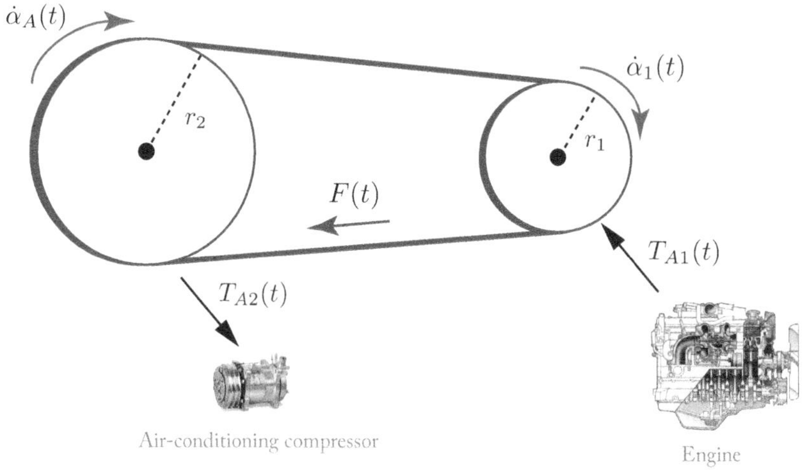

FIGURE 9.2: Schematic of the belt and pulleys drive system. The air-conditioning compressor shaft is mechanically connected to the engine shaft via a belt.

the compressor and r_1 and r_2 the radii of the pulleys connected to the engine and to the compressor shaft respectively. Also,

$$\dot{\alpha}_A(t) = \frac{r_1}{r_2}\dot{\alpha}_1(t) \quad \text{and} \quad \ddot{\alpha}_A(t) = \frac{r_1}{r_2}\ddot{\alpha}_1(t), \tag{9.15}$$

where $\dot{\alpha}_1(t)$ and $\dot{\alpha}_A(t)$ are the angular velocities of the engine and the compressor respectively. The extra load on the engine is modeled as (from (9.10))

$$J_1\ddot{\alpha}_1(t) = d_1 x_f - k(\alpha_1(t) - \alpha_2(t)) - c(\dot{\alpha}_1(t) - \dot{\alpha}_2(t)) - T_{A1}(t), \tag{9.16}$$

and the air conditioning rotational shaft dynamics are given by

$$J_A\ddot{\alpha}_A(t) = T_{A2}(t) - c_A\dot{\alpha}_A(t) \quad \Rightarrow \quad T_{A2}(t) = J_A\ddot{\alpha}_A(t) + c_A\dot{\alpha}_A(t), \tag{9.17}$$

where J_A and c_A are the inertia and damping coefficients respectively. These values, as well as r_1 and r_2 are not available. They have therefore been chosen in a way that simulation results match the results of [74, Figure 4]. The chosen values are given in Table 9.3.

Substituting (9.15) into (9.17) and the resulting equation into (9.14) gives

$$T_{A1}(t) = \left(\frac{r_1}{r_2}\right)^2 J_A\ddot{\alpha}_1(t) + \left(\frac{r_1}{r_2}\right)^2 c_A\dot{\alpha}_1(t). \tag{9.18}$$

Finally, substituting (9.18) into (9.16) gives

$$\begin{aligned} J_1\ddot{\alpha}_1(t) &= d_1 x_f(t) - k(\alpha_1(t) - \alpha_2(t)) - c(\dot{\alpha}_1(t) - \dot{\alpha}_2(t)) \\ &\quad - \frac{r_1^2}{r_2^2}J_A\ddot{\alpha}_1(t) + \frac{r_1^2}{r_2^2}c_A\dot{\alpha}_1(t) \\ \Rightarrow \quad \ddot{\alpha}_1(t) &= \frac{d_1}{J_B}x_f(t) - \frac{k}{J_B}(\alpha_1(t) - \alpha_2(t)) - \frac{\frac{r_1^2}{r_2^2}c_A + c}{J_B}\dot{\alpha}_1(t) + \frac{c}{J_B}\dot{\alpha}_2(t), \end{aligned} \tag{9.19}$$

where

$$J_B = J_1 + \frac{r_1^2}{r_2^2}J_A. \tag{9.20}$$

The model of the longitudinal motion of the vehicle with the driveline and air-conditioning is, in state space form,

$$\begin{aligned} \dot{\boldsymbol{x}}(t) &= \boldsymbol{A}^c\boldsymbol{x}(t) + \boldsymbol{B}^c\boldsymbol{u}(t) \\ \boldsymbol{y}(t) &= \boldsymbol{C}\boldsymbol{x}(t), \end{aligned} \tag{9.21}$$

where

$$\boldsymbol{x}(t) = \begin{bmatrix} x_f(t) \\ \alpha_1(t) - \alpha_2(t) \\ \dot{\alpha}_1(t) \\ \dot{\alpha}_2(t) \\ T_e(t) \\ T_i(t) \end{bmatrix}, \quad \boldsymbol{u}(t) = \begin{bmatrix} u_t(t) \\ \alpha_v(t) \end{bmatrix}, \quad \boldsymbol{y}(t) = \begin{bmatrix} y_v(t) \\ y_s(t) \\ y_t(t) \end{bmatrix},$$

$$\boldsymbol{A}^c = \begin{bmatrix} -\frac{1}{\tau_s} & 0 & 0 & 0 & 0 & 0 \\ 0 & 0 & 1 & -1 & 0 & 0 \\ \frac{d_1}{J_B} & -\frac{k}{J_B} & -\frac{\frac{r_1^2}{r_2^2}c_A + c}{J_B} & \frac{c}{J_B} & 0 & 0 \\ 0 & \frac{k}{J_2+m_1} & \frac{c}{J_2+m_1} & \frac{d_3-c}{J_2+m_1} & 0 & 0 \\ 0 & 0 & \frac{r_1}{r_2}b_{21} & 0 & -a_1 & 0 \\ 0 & 0 & 0 & 0 & b_{31} & -a_2 \end{bmatrix}, \quad \boldsymbol{B}^c = \begin{bmatrix} \frac{K_s}{\tau_s} & 0 \\ 0 & 0 \\ 0 & 0 \\ 0 & 0 \\ 0 & b_{22} \\ 0 & 0 \end{bmatrix},$$

$$\boldsymbol{C} = \begin{bmatrix} 0 & 0 & 0 & c_1 & 0 & 0 \\ 0 & 1 & 0 & 0 & 0 & 0 \\ 0 & 0 & 0 & 0 & 0 & 1 \end{bmatrix}. \tag{9.22}$$

This is a MIMO plant with inputs

$$\boldsymbol{u}(t) = \begin{bmatrix} \textit{Throttle pedal position} \\ \textit{Expansion valve opening} \end{bmatrix} \tag{9.23}$$

and outputs

$$\boldsymbol{y}(t) = \begin{bmatrix} \textit{Vehicle velocity} \\ \textit{Drive-shaft torsion} \\ \textit{Cabin temperature} \end{bmatrix}. \tag{9.24}$$

9.2.2 Design of an observer and an LQR controller with integral action

The controller for the above plant is required to track a velocity demand and a cabin temperature demand signal in an optimal way. The chosen approach is to design an optimal state observer, so that an estimate of the plant states is available, and then an LQR controller having as an input the state estimate and the two integrated error signals. The overall controller will be a two-DOF controller which takes as an input the two demand signals and the three plant outputs. The proposed controller is equivalent to an LQG controller with integral action for the vehicle velocity and cabin temperature. The controller is designed for the continuous-time plant model in (9.21)-(9.22).

The aim is to design a continuous-time controller of the form

$$\mathcal{K}^c : \quad \left\{ \begin{array}{l} \dot{\boldsymbol{x}}_K(t) = \boldsymbol{A}_K^c \boldsymbol{x}_K(t) + \boldsymbol{B}_{K1}^c \boldsymbol{y}(t) + \boldsymbol{B}_{K2}^c \boldsymbol{d}(t) \\ \boldsymbol{u}(t) = \boldsymbol{C}_K \boldsymbol{x}_K(t) + \boldsymbol{D}_{K1} \boldsymbol{y}(t) + \boldsymbol{D}_{K2} \boldsymbol{d}(t) \end{array} \right. , \tag{9.25}$$

where $\boldsymbol{x}_K(t) \in \mathbb{R}^8$ are the controller states, $\boldsymbol{y}(t) \in \mathbb{R}^3$ are the plant outputs, $\boldsymbol{d}(t) \in \mathbb{R}^2$ are the demand signals given by

$$\boldsymbol{d}(t) = \begin{bmatrix} d_v(t) \\ d_t(t) \end{bmatrix} = \begin{bmatrix} \textit{Vehicle velocity demand} \\ \textit{Cabin temperature demand} \end{bmatrix}, \tag{9.26}$$

$\boldsymbol{u}(t) \in \mathbb{R}^2$ are the control signals and $\boldsymbol{A}_K^c, \boldsymbol{B}_{K1}^c, \boldsymbol{B}_{K2}^c, \boldsymbol{C}_K, \boldsymbol{D}_{K1}$ and $\boldsymbol{D}_{K2}$ are constant matrices of appropriate dimensions.

Consider the closed-loop system in Figure 9.3 where $d_v(s)$, $d_t(s)$, $\boldsymbol{u}(s)$ and

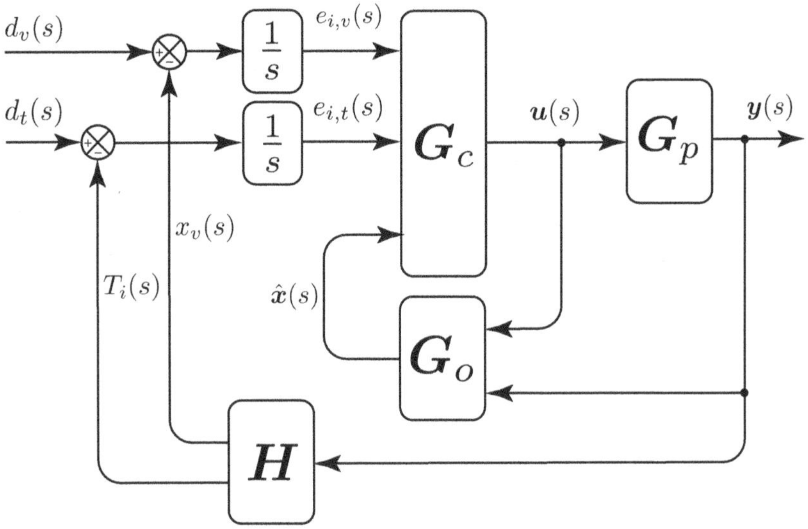

FIGURE 9.3: Schematic for LQG controller design with integral action.

$\boldsymbol{y}(s)$ are the Laplace transform of $d_v(t)$, $d_t(t)$, $\boldsymbol{u}(t)$ and $\boldsymbol{y}(t)$ respectively. $\boldsymbol{G}_c$ is a static controller to be designed, the matrix

$$\boldsymbol{H} = \begin{bmatrix} 1 & 0 & 0 \\ 0 & 0 & 1 \end{bmatrix}, \tag{9.27}$$

is used to select signals $x_v(s)$ and $T_i(s)$ from the vector $\boldsymbol{y}(s)$ and

$$\boldsymbol{G}_p = \boldsymbol{C}(s\boldsymbol{I} - \boldsymbol{A}^c)^{-1}\boldsymbol{B}^c \tag{9.28}$$

is the transfer function matrix of the vehicle model in (9.21)-(9.22). $\boldsymbol{G}_o$ is the optimal observer to be designed which has a state space realization of the form

$$\dot{\hat{\boldsymbol{x}}}(t) = \boldsymbol{A}^c\hat{\boldsymbol{x}}(t) + \boldsymbol{B}^c\boldsymbol{u}(t) + \boldsymbol{L}(\boldsymbol{y}(t) - \boldsymbol{C}\hat{\boldsymbol{x}}(t)). \tag{9.29}$$

$\boldsymbol{G}_c$ is an LQR controller designed for an augmented plant which includes two extra states, $\alpha_2^i(t)$ and $T_i^i(t)$, which are the integrated version of the states $\alpha_2(t)$ and $T_i(t)$ (see (9.22)). This augmented plant is

$$\dot{\boldsymbol{x}}_{aug}(t) = \underbrace{\begin{bmatrix} \boldsymbol{A}^c & \boldsymbol{0}_{6\times 2} \\ \boldsymbol{E} & \boldsymbol{0}_{2\times 2} \end{bmatrix}}_{\boldsymbol{A}^c_{aug}} \boldsymbol{x}_{aug}(t) + \underbrace{\begin{bmatrix} \boldsymbol{B}^c \\ \boldsymbol{0}_{2\times 2} \end{bmatrix}}_{\boldsymbol{B}^c_{aug}} \boldsymbol{u}(t) + \begin{bmatrix} \boldsymbol{0}_{6\times 2} \\ \boldsymbol{I}_2 \end{bmatrix} \boldsymbol{d}(t), \tag{9.30}$$

where

$$\boldsymbol{x}_{aug}(t) = \begin{bmatrix} \boldsymbol{x}(t)^T & \alpha_2^i(t) & T_i^i(t) \end{bmatrix}^T, \tag{9.31}$$

and $\boldsymbol{E} = \begin{bmatrix} \boldsymbol{0}_{2\times 3} & -\boldsymbol{H} \end{bmatrix}$. The demand vector $\boldsymbol{d}(t)$ is considered to be a disturbance. For the plant in (9.30) an LQR controller, $\boldsymbol{G}_c$, can be designed in the usual way as

$$\boldsymbol{G}_c = \boldsymbol{R}_{lqr}^{-1}(\boldsymbol{B}^c_{aug})^T \boldsymbol{S}, \tag{9.32}$$

where $\boldsymbol{S}$ is the solution of the Riccati equation

$$(\boldsymbol{A}^c_{aug})^T \boldsymbol{S} + \boldsymbol{S}\boldsymbol{A}^c_{aug} - \boldsymbol{S}\boldsymbol{B}^c_{aug}\boldsymbol{R}_{lqr}^{-1}(\boldsymbol{B}^c_{aug})^T\boldsymbol{S} + \boldsymbol{Q}_{lqr} = \boldsymbol{0}, \tag{9.33}$$

and $\boldsymbol{R}_{lqr}$ and $\boldsymbol{Q}_{lqr}$ are design weighting matrices.

Let

$$\boldsymbol{G}_c = \begin{bmatrix} \boldsymbol{G}_{c1} & \boldsymbol{G}_{c2} \end{bmatrix}, \tag{9.34}$$

thus, the control law, in the time domain is

$$\boldsymbol{u}(t) = \boldsymbol{G}_{c1}\hat{\boldsymbol{x}}(t) + \boldsymbol{G}_{c2}\boldsymbol{e}_i(t), \tag{9.35}$$

where $\hat{\boldsymbol{x}}(t)$ is the inverse Laplace transform of the observer's state vector and $\boldsymbol{e}_i(t)$ is the inverse Laplace transform of the signal

$$\boldsymbol{e}_i(s) = \begin{bmatrix} e_{i,v}(s) \\ e_{i,t}(s) \end{bmatrix} = \begin{bmatrix} \frac{1}{s}(d_v(s) - x_v(s)) \\ \frac{1}{s}(d_t(s) - T_i(s)) \end{bmatrix}, \tag{9.36}$$

(see Figure 9.3). Substituting (9.35) into (9.29) and rearranging the equation gives

$$\dot{\hat{\boldsymbol{x}}}(t) = \underbrace{(\boldsymbol{A}^c + \boldsymbol{B}^c\boldsymbol{G}_{c1} - \boldsymbol{L}\boldsymbol{C})}_{\boldsymbol{A}^c_o}\hat{\boldsymbol{x}}(t) + \underbrace{\begin{bmatrix} \boldsymbol{L} & \boldsymbol{B}^c\boldsymbol{G}_{c2} \end{bmatrix}}_{\boldsymbol{B}^c_o} \begin{bmatrix} \boldsymbol{y}(t) \\ \boldsymbol{e}_i(t) \end{bmatrix}. \tag{9.37}$$

The matrix $\boldsymbol{L}$ is chosen optimally to be

$$\boldsymbol{L} = (\boldsymbol{R}_o^{-1}\boldsymbol{C}\boldsymbol{S})^T, \tag{9.38}$$

where $\boldsymbol{S}$ is the solution of the Riccati equation

$$\boldsymbol{A}^c\boldsymbol{S} + \boldsymbol{S}(\boldsymbol{A}^c)^T - \boldsymbol{S}(\boldsymbol{C})^T\boldsymbol{R}_o^{-1}\boldsymbol{C}\boldsymbol{S} + \boldsymbol{Q}_o = \boldsymbol{0}, \tag{9.39}$$

and $\boldsymbol{R}_o$ and $\boldsymbol{Q}_o$ are design weighting matrices. (Notice that this is the dual problem of LQR design.)

Let

$$\boldsymbol{G}_o = (s\boldsymbol{I} - \boldsymbol{A}_o^c)\boldsymbol{B}_o^c, \tag{9.40}$$

be the transfer function matrix for the observer in (9.37) which can be partitioned into

$$\boldsymbol{G}_o = \begin{bmatrix} \boldsymbol{G}_{o1} & \boldsymbol{G}_{o2} \end{bmatrix}. \tag{9.41}$$

Hence, the state estimate, in the frequency domain, is given by

$$\hat{\boldsymbol{x}}(s) = \begin{bmatrix} \boldsymbol{G}_{o1} & \boldsymbol{G}_{o2} \end{bmatrix} \begin{bmatrix} \boldsymbol{y}(s) \\ \boldsymbol{e}_i(s) \end{bmatrix}. \tag{9.42}$$

From (9.36), it follows that

$$\begin{bmatrix} \boldsymbol{y}(s) \\ \boldsymbol{e}_i(s) \end{bmatrix} = \begin{bmatrix} \boldsymbol{I} & \boldsymbol{0} \\ -\operatorname{diag}\left(\frac{1}{s}, \frac{1}{s}\right) \boldsymbol{H} & \operatorname{diag}\left(\frac{1}{s}, \frac{1}{s}\right) \end{bmatrix} \begin{bmatrix} \boldsymbol{y}(s) \\ \boldsymbol{d}(s) \end{bmatrix}. \tag{9.43}$$

Let the Laplace transform of the control law in (9.35) be

$$\boldsymbol{u}(s) = \boldsymbol{G}_{c1}\hat{\boldsymbol{x}}(s) + \boldsymbol{G}_{c2}\boldsymbol{e}_i(s). \tag{9.44}$$

Substituting the equation for the state estimate in (9.42) into the control law in (9.44) and rearranging to gather the signals together gives

$$\boldsymbol{u}(s) = \begin{bmatrix} \boldsymbol{G}_{c1}\boldsymbol{G}_{o1} & \boldsymbol{G}_{c1}\boldsymbol{G}_{o2} + \boldsymbol{G}_{c2} \end{bmatrix} \begin{bmatrix} \boldsymbol{y}(s) \\ \boldsymbol{e}_i(s) \end{bmatrix}. \tag{9.45}$$

Finally, substituting (9.43) into (9.45) gives

$$\boldsymbol{u}(s) = \underbrace{\begin{bmatrix} \boldsymbol{G}_{c1}\boldsymbol{G}_{o1} & \boldsymbol{G}_{c1}\boldsymbol{G}_{o2} + \boldsymbol{G}_{c2} \end{bmatrix} \begin{bmatrix} \boldsymbol{I} & \boldsymbol{0} \\ -\operatorname{diag}\left(\frac{1}{s}, \frac{1}{s}\right) \boldsymbol{H} & \operatorname{diag}\left(\frac{1}{s}, \frac{1}{s}\right) \end{bmatrix}}_{\mathcal{K}_{TFM}} \begin{bmatrix} \boldsymbol{y}(s) \\ \boldsymbol{d}(s) \end{bmatrix}. \tag{9.46}$$

Equation (9.46) is the control law in terms of the plant output $\boldsymbol{y}(s)$ and the demand signal $\boldsymbol{d}(s)$ and $\mathcal{K}_{TFM}$ is the 2×5 transfer function matrix of the controller. Let $\boldsymbol{y}(t)$, $\boldsymbol{d}(t)$ and $\boldsymbol{u}(t)$ be the inverse Laplace transform of the signals $\boldsymbol{y}(s)$, $\boldsymbol{d}(s)$ and $\boldsymbol{u}(s)$, then the controller in (9.25) is a state space realization of the controller $\mathcal{K}_{TFM}$ in (9.46). The weighting matrices for the design of this controller are

$$\begin{aligned} \boldsymbol{Q}_{lqr} &= \operatorname{diag}(0.1, 10000, 0.1, 1, 1, 10, 100000, 100), \\ \boldsymbol{R}_{lqr} &= \operatorname{diag}(1, 1), \\ \boldsymbol{Q}_o &= \operatorname{diag}(1, 0.001, 1, 1, 1, 1,), \\ \boldsymbol{R}_o &= \operatorname{diag}(1, 1000, 1). \end{aligned} \tag{9.47}$$

These matrices have been chosen to give a realistic response considering the physical actuator limitations. The matrices for the controller state space representation of (9.25), calculated with MATLAB®, are given explicitly below:

$$
\boldsymbol{A}_K^c = \begin{bmatrix} -40.07 & -2207 & -329.1 & -6155 & 0.969 & 93.11 & -6319 & 8.648 \\ 0 & -0.0009873 & 1 & -0.8882 & 0 & 0.01629 & 0 & 0 \\ 1.771 & -1.115 & -0.5574 & -0.03096 & 0 & 0.01255 & 0 & 0 \\ 0 & 0.6638 & 0.06636 & -0.7697 & 0 & 0.01608 & 0 & 0 \\ 0.0003346 & -5.3 & -25.22 & 2.975 & -1.019 & -56.34 & -5.082 & -3.713 \\ 0 & 1.629 \cdot 10^{-5} & 0 & 0.01151 & 0.00237 & -1.002 & 0 & 0 \\ 0 & 0 & 0 & 0 & 0 & 0 & 0 & 0 \\ 0 & 0 & 0 & 0 & 0 & 0 & 0 & 0 \end{bmatrix},
$$

$$
\boldsymbol{B}_{K1}^c = \begin{bmatrix} 8.697 \cdot 10^{-6} & 1.069 \cdot 10^{-7} & -2.88 \cdot 10^{-7} \\ -0.1321 & 0.0009873 & -0.01629 \\ 0.1684 & -0.0001473 & -0.01255 \\ 0.816 & -0.0001562 & -0.01608 \\ -6.008 & -0.007621 & 0.7718 \\ -0.0136 & -1.629 \cdot 10^{-5} & 0.9994 \\ 1 & 0 & 0 \\ 0 & 0 & 1 \end{bmatrix}, \quad \boldsymbol{B}_{K2}^c = \begin{bmatrix} 0 & 0 \\ 0 & 0 \\ 0 & 0 \\ 0 & 0 \\ 0 & 0 \\ 0 & 0 \\ -1 & 0 \\ 0 & -1 \end{bmatrix},
$$

$$
\boldsymbol{C}_K = \begin{bmatrix} -1.004 & -110.4 & -16.45 & -307.8 & 0.04845 & 4.655 & -315.9 & 0.4324 \\ 0.0009003 & -14.28 & 0.02077 & -5.672 & -0.5939 & -149.5 & -13.67 & -9.991 \end{bmatrix},
$$

$$
\boldsymbol{D}_{K1} = \begin{bmatrix} 0 & 0 & 0 \\ 0 & 0 & 0 \end{bmatrix}, \quad \boldsymbol{D}_{K2} = \begin{bmatrix} 0 & 0 \\ 0 & 0 \end{bmatrix}. \tag{9.48}
$$

9.3 HIL from TTE systems

For the practical validation of control and scheduling of NCSs, a modified version of an HIL Adaptive Cruise-Control System (ACCS) from TTE Systems Ltd[3] is used (see Figure 9.4). A description of the original system is first given.

The ACCS is an HIL test facility supporting the development and assessment of distributed embedded vehicle control systems. The test facility is implemented over three desktop computers and ten ARM Olimex [4] LPC-2378-STK development boards and utilizes low-cost interfaces, such as standard parallel (IEEE-1284, DB-25) and serial (RS-232, DB-9) ports.

The main principle of HIL testing is the employment of real-time simulation of the control process' dynamical model to synthesize real signals for use as inputs to the system under test. The loop is closed by feeding the outputs from the system back into the simulation inputs. The three computers serve to do the following.

[3] `www.tte-systems.com.`

[4] `www.olimex.com`

FIGURE 9.4: Hardware-in-the-loop testbed: 10 ECUs adaptive cruise control system from TTE®-systems.

- *Simulation PC.* Run the vehicle and highway simulation. This computer runs a detailed dynamical model of a motor vehicle traveling along a three-lane highway, along with third-party vehicles employing accurate driver models. It generates measurement signals from virtual sensors and reads real signal to control virtual actuators.
- *Fault-injector PC.* This computer runs a statistical model of the system under test and injects actual faults into both hardware and software during the test process.
- *GUI PC.* Graphically display what is happening in the current simulation.

Communication between the physical nodes (ARM development boards) in the network is via a CAN bus. To avoid confusion with the abstract definition of 'node' adopted so far (Definition 3.1, p. 50), nodes, which are 'physical nodes' will be referred to as Electronic Control Units (ECUs). The simulation PC runs a vehicle simulation, hence providing sensor signals (*e.g.* vehicle velocity) to the ECUs. One of the ECUs, called the *master*, runs the control algorithm and the bus scheduling. The other ECUs, called *slaves*, either deal with signals coming from the simulation or send signals (*e.g.* throttle control signal) to the simulation. See Figure 9.5 for a close-up picture of the master and a slave ECU.

CAN is often viewed as an event-triggered protocol, but time-triggered behavior using other algorithms can be achieved. In fact, the protocol used for the HIL is a time-triggered protocol called Time Triggered CAN with

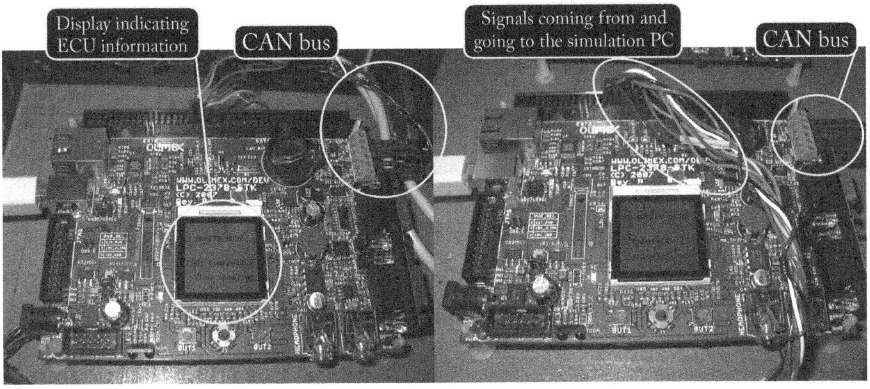

FIGURE 9.5: Close-up picture of two Olimex LPC-2378-STK development boards. The one on the left implements the master ECU while the other one a slave ECU.

Shared Clock (TTC-SC) [9] . This is a TDMA protocol where the idea is to synchronize the execution of tasks on the individual nodes by sharing a single clock. This is achieved by having one of the ECUs acting as a master. The master ECU generates periodic timer interrupts to drive its own task scheduler. When a timer interrupt occurs, the master node also generates a tick message that is broadcasted to the entire network. This message is received by each of the slave ECUs and used to generate a local and synchronized interrupt for the local task scheduler. In this way, it is easier to schedule communication between slave ECUs and master ECU and, if everything is working correctly, no contention occurs on the bus (although identifiers with priority are still assigned). In the original set up, communication between slave ECUs and the master ECU is scheduled with an RR protocol. The master ECU tick message and the acknowledgment message from the slaves contain up to 7 bytes per message that are used for data transfers. Slave-to-slave communication is not permitted; all communication is directed via the master.

9.3.1 Modifications to the original HIL

The original HIL has been modified in order to allow testing of the proposed design methods. The main changes to the original system are:

(i) replacement of the original vehicle model with the more complex one described in Section 9.2.1

(ii) replacement of the original controller (PID) with the LQG controller with integral action described in Section 9.2.2

(iii) modification in the controller ECU to include a scheduling mechanism that allows the user to implement any periodic communication sequence between nodes (thus, not restricted to the simple RR scheduling).

The changes required for (i) and (ii) involved the definition of new signals between ECUs and the simulation PC (*e.g.* the cabin temperature). For achieving (iii), the code in the master ECU has been modified. The new algorithm allows to schedule the bus deterministically according to a predefined communication sequence between slaves implemented in the master. Such a setup can emulate the static frame communication of a modern FlexRay network although bandwidths are not as high. For simplicity, task scheduling has not been considered at this stage, although it was shown in Section 3.6.2 (p. 72) that it is possible. Furthermore, no testing has been carried out with injection of faults.

The modified HIL comprises eight ECUs, where one is the master and the other seven are the slaves. The architecture is illustrated in Figure 9.6.

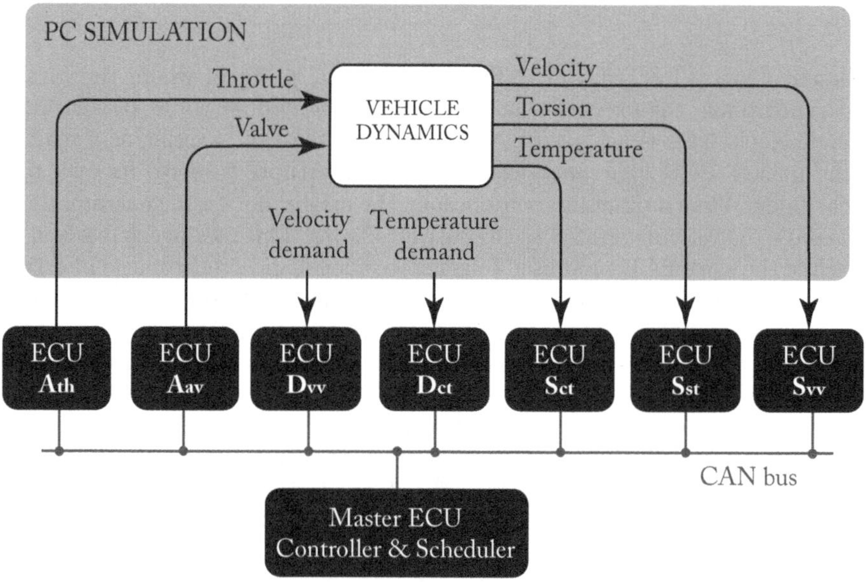

FIGURE 9.6: Architecture of the modified HIL system: seven slave ECUs receive/send information from/to the simulation PC. One ECU is the master which implements the control algorithm and the scheduling of the slave ECUs. ECU-to-ECU communication is via a CAN bus.

Figure 9.7 shows the block diagram configuration of the HIL setup that derives from the networked representation in Figure 9.6. The vehicle model is the one in (9.21)-(9.22). Note that a negative throttle in this case means braking. This applies to the case where throttle and brakes are implemented on the same ECU and have the same dynamics or to the case of some electric vehicle. This model has been implemented as an algorithm with fast sampling

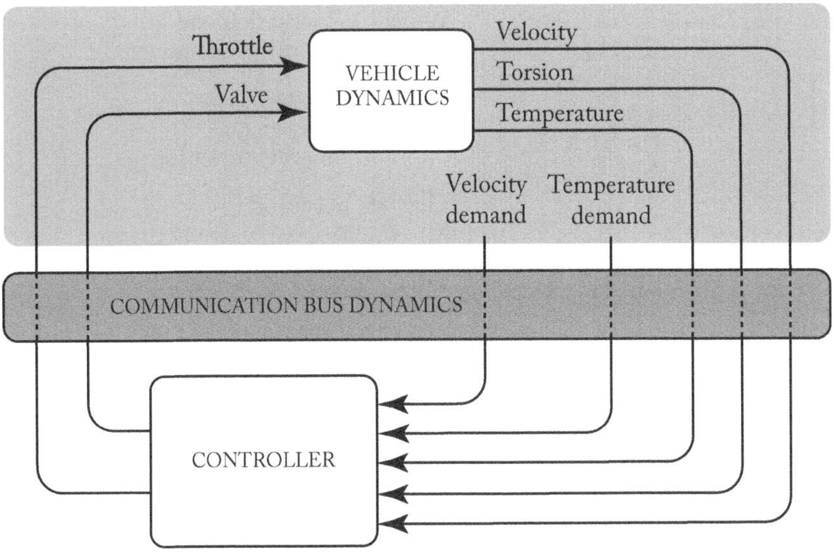

FIGURE 9.7: Closed-loop diagram of the HIL system with bus dynamics.

time (at least 20 times faster than the controller's) in the simulation PC. The previously designed continuous-time two DOF LQG controller with integral action in (9.25) (where the matrices are explicitly given in (9.48)) has been discretized with a sampling time $h = 0.025$. The controller algorithm has been implemented in the master ECU. The master has an interrupt routine that executes every h seconds. The messages transmitted via the CAN bus, to and from the controller, are the ones from the three sensors, the two demands and the two actuators. Each of these signals are dealt with by a dedicated slave ECU. The name of each slave ECU with the associated signal is given in Table 9.4 and shown in Figure 9.6. Only one ECU can be scheduled at every time tick. This scheduling mechanism is shown in Figure 9.7 by the block denoted as 'Communication bus dynamics'.

The problem is finding a sequence of communication between these nodes and the controller such that a specified cost function is minimized leading to an improved performance.

9.4 Experiments on the HIL

In this section, the optimization tools previously discussed will be used to optimize the communication between the ECUs of the HIL. First of all it should

ECU name	Associated signal
D_{vv}	Vehicle Velocity demand
D_{ct}	Cabin Temperature demand
S_{vv}	Vehicle Velocity sensor
S_{st}	Shaft Torsion sensor
S_{ct}	Cabin Temperature sensor
A_{th}	Throttle actuator
A_{av}	Air-conditioning Valve actuator

TABLE 9.4: ECUs name, with associated sensor, actuator, or demand signal.

be noticed, following Remark 5.1 (p. 113), that, if the simple RR schedule had to be used, a sequence σ of length of at least $p = 7$ (there are 7 nodes) needs to be used to guarantee feasibility. An exhaustive search to find the minimum cost for all the possible 7^7 permutations of σ takes more than 24 hours on the available desktop PC with 1.86 GHz processor and 3.24 GB of RAM. This may be acceptable if $p = 7$ is a reasonable sequence length. Unfortunately, it will be shown that the HIL system cannot be even stabilized with such a short sequence. The reason is that the same communication resources are allocated to fast and safety-critical processes (like velocity and drive-shaft torsion control) and to slow non-critical ones (like the cabin temperature control). Therefore, longer sequences have to be utilized excluding any possibility of optimizing by an exhaustive search.

9.4.1 Simulation results

Before showing the HIL experimental results, some MATLAB® simulation results of the system will be presented. The simulation consists of a velocity step response of the vehicle in feedback with the controller. The vehicle is allowed to settle to a set point velocity of 65 mph and a set point cabin temperature of 20° C. After 3 minutes, the driver demands an increase in velocity of 5 mph.

Three results are shown and illustrated in Figure 9.8. The first one is a continuous-time simulation (performed by solving ODEs). This simulation is for the continuous-time plant and controller and represents the *ideal* situation where the system is not implemented as an NCS. It can be seen from the response that the vehicle velocity reaches its new set point of 70 mph in around 8 seconds with minimal overshoot. The acceleration of the vehicle causes the drive-shaft to reach a peak torsion of almost 4.5 degrees. The throttle has a peak value of 50%. The vehicle acceleration causes a drop of temperature that is due to the increase of the compressor speed. However, in less than 2 minutes the temperature sets again to the desired point of 20 degrees due to the increase in the expansion valve opening. This simulation shows the

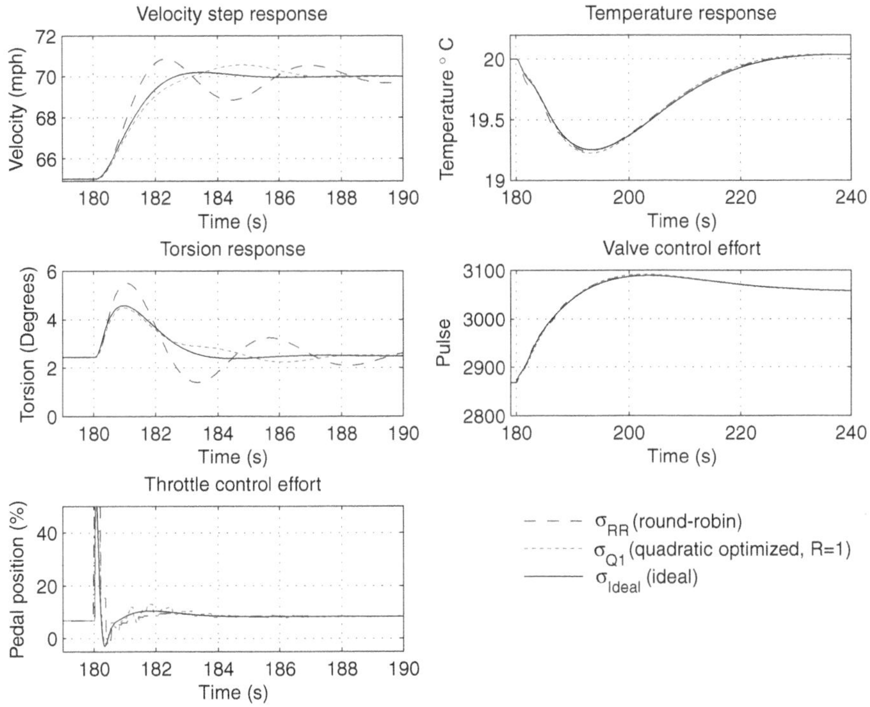

FIGURE 9.8: Vehicle's simulated step response: σ_{RR} is an RR sequence while σ_{Q1} is optimized under the quadratic cost of Section 7.2.2. The 'ideal' response is for the system not implemented as an NCS. The optimized sequence gives a response that is closer to the ideal situation when compared to the RR.

Name	Sequence	Method
σ_{RR}	$\{D_{vv}, S_{vv}, S_{st}, A_{th}, D_{ct}, S_{ct}, A_{av}\}$	RR
σ_{Q1}	$\{D_{vv}, S_{vv}, A_{av}, A_{th}, S_{vv}, A_{av}, S_{vv}, A_{th}, S_{vv}, A_{th}, A_{av}, S_{st}, A_{av}, S_{vv}, A_{av}, A_{th}, A_{av}, S_{ct}, A_{th}, D_{ct}\}$	Quadratic cost, $\boldsymbol{R} = \text{diag}(1,1)$

TABLE 9.5: Sequences used for testing of quadratic performance (see Table 9.4 and Figure 9.6 for the list of ECUs with associated signals). σ_{RR} is an RR sequence while σ_{Q1} is optimized under the quadratic cost of Section 7.2.2.

coupling between the components of the vehicle model as well as the ability of the controller to track velocity and temperature. Also, it should be noted that the air-conditioning dynamics are much slower than the rest of the dynamics. It is assumed that these were the desired characteristics and the controller was designed to meet these requirements.

In the next two tests, the system is implemented as an NCS. This means that the controller has been discretized and the communication is scheduled according to a predefined communication sequence. This setup mimics the situation of the HIL system with ECU scheduling. Two sequences are tested. The first one is the simple RR schedule. The other one is a sequence optimized using the PSO1 algorithm for a sequence length of $p = 20$. The cost function used is the quadratic cost of Section 7.2.2, p. 147. Recall that, for this cost, a weighting matrix $\boldsymbol{R}$ needs to be selected (see Section 7.2.3, p. 148). In this example, $\boldsymbol{R} = \text{diag}(1,1)$.

These two sequences, and their names, are given in Table 9.5. The responses are shown in Figure 9.8. It is clear from the figure that an RR schedule makes the dynamics deviate considerably from the ideal situation. On the other hand, the use of an optimized schedule minimizes this unavoidable deviation. Notice that the differences for the cabin temperature response and valve opening are negligible for σ_{RR} and σ_{Q1}. This is due to the fact that they have much slower dynamics compared to the rest of the system.

In the experimental results of the next sections it will be demonstrated that, because of the bus communication between physical nodes, the performance of the vehicle control system inevitably deteriorates. However, it will be shown how the use of optimized scheduling sequences can minimize this detrimental effect.

9.4.2 HIL quadratic performance results

In the practical experiments that will follow, the controller is discretized and implemented in the master ECU. Sensor, demand and actuator signal will be made available to the controller node via the slave nodes. Communication is via a shared bus with a time-triggered protocol. In other words, the vehicle control system will be implemented as a real physical NCS.

As a first experiment, the same sequences used for the simulation results in the previous section are implemented in the HIL for ECU scheduling. The sequences are σ_{RR} and σ_{Q1} (see Table 9.5). In the HIL rig, the RR scheduling of σ_{RR} is not even capable of stabilizing the system, therefore results are not shown. The results from the optimal sequence σ_{Q1} are shown in Figure 9.9. Comparing this experimental result to the analogous simulated one in

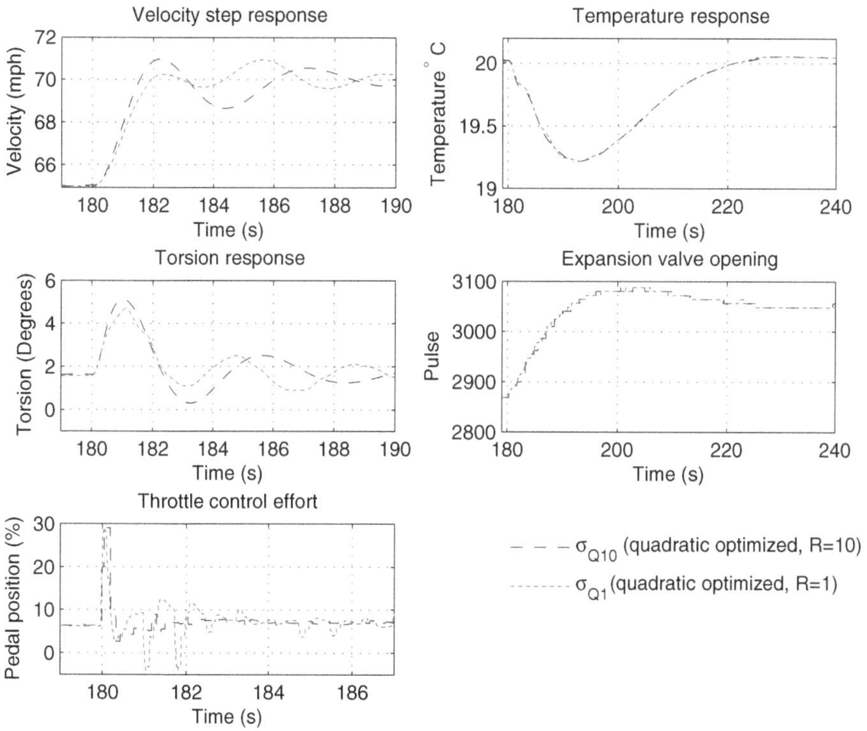

FIGURE 9.9: Experimental step response for quadratic performance: σ_{Q1} is optimized (under the quadratic cost) for $\boldsymbol{R} = \text{diag}(1, 1)$ while σ_{Q10} is optimized for $\boldsymbol{R} = \text{diag}(10, 10)$. Both sequences do not perform as well as expected from simulation because of input uncertainties.

Figure 9.8 suggests that there is a considerable discrepancy between the simulation and the experimental setup. In the experiment, the optimized sequence does not perform as well as expected.

The next experiment consists in re-optimizing the sequence but this time for an input weighting matrix of $\boldsymbol{R} = \text{diag}(10, 10)$. The optimized sequence is given in Table 9.6 and the system response, under this sequencing, is also given in Figure 9.9. As expected, because of the higher weight to the actuator signals,

Name	Sequence	Method
σ_{Q10}	$\{S_{st}, A_{av}, S_{st}, A_{th}, D_{vv}, A_{th}, S_{st}, A_{av}, A_{th}, A_{av}, D_{vv}, S_{st}, A_{th}, S_{st}, S_{vv}, A_{av}, S_{st}, A_{th}, D_{vv}, D_{ct}\}$	Quadratic cost, $\boldsymbol{R} = \text{diag}(10, 10)$

TABLE 9.6: Sequence used for further testing of quadratic performance (see Table 9.4 and Figure 9.6 for the list of ECUs with associated signals). σ_{Q10} is optimized under the quadratic cost of Section 7.2.2.

the throttle actuator response has improved. However, the velocity and torsion responses are still not satisfactory and do not match the simulation results. In general, experience in testing suggests that the experimental responses tend to be more oscillatory (hence more unstable) than the simulated ones. It should be pointed out that stability, like for the RR schedule, could be regained by faster sampling times. However, a 'critical' sampling time has been chosen here for the sake of experimentation so that small dynamical changes, due to different schedules, show large variations in the results.

Investigation of the problem of instability and oscillatory behavior suggested that lack of robustness due to input uncertainty may be the cause. Many practical systems have unmodeled delays and uncertainties; this is the case here. These uncertainties come mainly from delays but also from quantization errors and numerical precision in the microcontrollers. It will be shown in the next section that the proposed robust scheduling design methods can cater for this.

9.4.3 HIL robust performance results

In the previous sections, the quadratic cost of Chapter 7 was used to find optimal sequences. Although the results in simulation were satisfactory, practical application testing showed that supposedly 'optimal' sequences did not perform as expected. The problem was identified as a lack of robustness due to input uncertainty.

The robust scheduling problem was analyzed in Sections 8.3 and 8.4 (pp. 162 and 164) for the discrete and the sampled-data case respectively. The two methods are equivalent apart from the fact that, for the sampled-data case, a 'discretization' of the plant (in the usual meaning) is not necessary. Plant discretization is an approximation and therefore the resulting cost function will also be approximated. For the sampled-data case, it was shown that the calculated cost is exact. For this reason, the sampled-data approach will be used in the following experimental examples.

The lack of robustness due to input uncertainty can be captured by the relative multiplicative uncertainty introduced in the input path as shown in 9.10. In the diagram, $\mathcal{W}_1$ and $\mathcal{W}_2$ are two pair of filters carefully chosen to represent the input uncertainty of the system. The product $\mathcal{W}_1\mathcal{W}_2$ represents

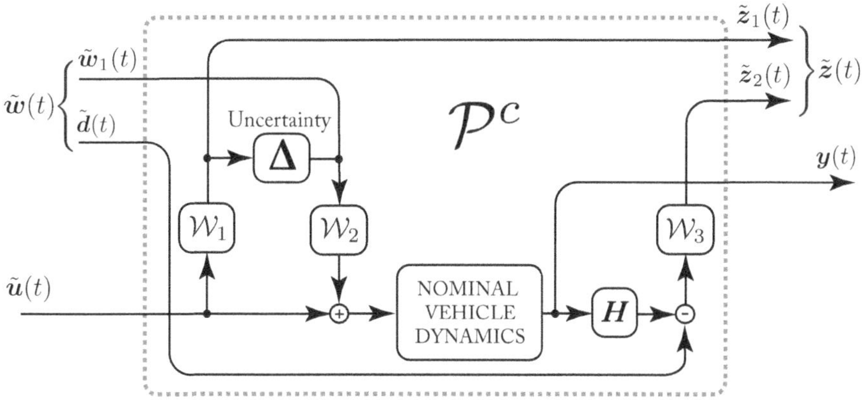

FIGURE 9.10: Vehicle dynamics with input uncertainty and performance measure.

the relative uncertainty in the frequency domain. Usually, the low frequency domain is less affected by uncertainty and $\|\mathcal{W}_1(j\omega)\mathcal{W}_2(j\omega)\| = 1$ ($j = \sqrt{-1}$) implies 100% uncertainty. The peak values in the frequency responses between $\tilde{\boldsymbol{w}}_1(t)$ and $\tilde{\boldsymbol{z}}_1(t)$ is what needs to be minimized. Moreover, to measure performance, an error signal has been created between the demand $\tilde{\boldsymbol{d}}(t)$ and the actual system output $\boldsymbol{y}(t)$ (where $\tilde{\boldsymbol{d}}(t) = \mathfrak{H}_h \hat{\boldsymbol{d}}(j)$ is the demand subject to the scheduling and the ZOH). This signal is weighted by the filter $\mathcal{W}_3$ to produce the performance output signal $\tilde{\boldsymbol{z}}_2(t)$. In Figure 9.10, the matrix $\boldsymbol{H}$ is used to select only the velocity and temperature signals from the output vector (this matrix is given in (9.27)).

The demand signal, which is an input to the discrete controller, is not directly available and it is obtained by feeding it via $\mathcal{S}_D$, $\mathcal{K}$ and $\mathcal{S}_A$ (see Figure 8.6, p. 171) by appropriate augmentation of their system matrices with identity and zero matrices. Hence, the control signal $\boldsymbol{u}(j)$ in Figure 8.6 must be augmented to include the demand signal.

Robust performance is achieved by minimizing the induced $\mathcal{L}_2$-gain between the virtual input signal $\tilde{\boldsymbol{w}}(t) = [\tilde{\boldsymbol{w}}_1(t)^T \ \tilde{\boldsymbol{d}}(t)^T]^T$ and the performance output $\tilde{\boldsymbol{z}}(t) = [\tilde{\boldsymbol{z}}_1(t)^T \ \tilde{\boldsymbol{z}}_2(t)^T]^T$. Let $\mathcal{P}_{cl}$ describe the closed-loop system shown in Figure 8.6 (p. 171) with input $\tilde{\boldsymbol{w}}(t)$ and output $\tilde{\boldsymbol{z}}(t)$ *i.e.* $\mathcal{P}_{cl} : \tilde{\boldsymbol{w}}(t) \mapsto \tilde{\boldsymbol{z}}(t)$.[5] Then, consider $\gamma_{sd} = \|\mathcal{P}_{cl}\|_{\mathcal{L}_2}$ which is the induced $\mathcal{L}_2$-norm of the system $\mathcal{P}_{cl}$. The objective is to find the communication sequence σ such that γ_{sd} is minimized. The result is a simultaneous improvement in overall robustness and performance. The trade-off between robustness and performance can be regulated by tuning the weighting filters $\mathcal{W}_1$, $\mathcal{W}_2$ and $\mathcal{W}_3$ (see [114] for a tutorial). Notice that this improvement is only due to the appropriate bus scheduling and does not involve controller design.

[5]Notice that, to be able to use Figure 8.6 for this particular case, the exogenous input $\tilde{\boldsymbol{w}}(t)$ in Figure 8.6 must correspond to $\tilde{\boldsymbol{w}}_1(t)$ in Figure 9.10.

Name	Sequence	Method
σ_I	$\{D_{vv}, S_{vv}, S_{st}, A_{th}, S_{vv}, S_{st}, A_{th}, D_{ct}, S_{ct}, A_{av}\}$	Intuitive
σ_P	$\{D_{vv}, S_{vv}, A_{th}, A_{th}, S_{vv}, D_{ct}, A_{av}, A_{th}, A_{av}, A_{th}, A_{av}, A_{th}, D_{vv}, S_{vv}, A_{th}, S_{ct}, A_{th}, S_{st}, A_{th}, S_{vv}\}$	Only performance
σ_{RP}	$\{A_{th}, S_{st}, A_{th}, S_{vv}, A_{th}, S_{ct}, A_{th}, S_{st}, A_{th}, D_{vv}, A_{th}, S_{vv}, A_{th}, S_{st}, A_{th}, D_{ct}, A_{th}, A_{av}, A_{th}, S_{st}, A_{th}, S_{vv}\}$	Robust performance

TABLE 9.7: Sequences used for testing of robust performance (see Table 9.4 and Figure 9.6 for the list of ECUs with associated signals). σ_I is an intuitively good sequence while σ_P and σ_{RP} are optimized by the proposed algorithms to guarantee performance only and robust performance respectively.

This experiment involves the comparison of the results from three different scheduling sequences, given in Table 9.7. σ_I is an intuitive sequence that has been found by common sense without using any optimization method. The reasoning behind this choice is that the velocity demand (D_{vv}) and velocity and torsion sensor signals (S_{vv}, S_{st}) are updated first and then the throttle (A_{th}) is controlled. The velocity and torsion control pattern (S_{vv}, S_{st}, A_{th}) is repeated and then the temperature (D_{ct}, S_{ct}, A_{av}) is controlled (only once since it has slower dynamics and it is not safety-critical). σ_P is an optimized sequence but only for performance. This means that, referring to the diagram in Figure 9.10, the weighting filters $\mathcal{W}_1$ and $\mathcal{W}_2$ are set to zero. $\mathcal{W}_3$ is a pair of filters, one for the velocity and one for the temperature signal. These two filters have been designed to approximate the inverse of the sensitivity transfer function of the system [114]. The frequency response of $\mathcal{W}_3$ is shown in the bottom plot of Figure 9.11.

Finally, σ_{RP} is an optimized sequence for performance and robustness to input uncertainty. The main sources of uncertainty for the rig are delays. Delays cause actuator additive uncertainty that can be modeled as $\mathcal{D} - \mathcal{I}$ where $\mathcal{D}$ is the delay model and $\mathcal{I}$ is the identity model. $\mathcal{D}$ was obtained by the first order Padé approximation of a delay equal to h. This is shown in Figure 9.10, as two (throttle and valve) diagonal filters $\mathcal{W}_2$. $\mathcal{W}_1$ are also two (diagonal) low pass filters. The product $\mathcal{W}_1\mathcal{W}_2$ expresses the relative uncertainty of the throttle and valve actuators and its frequency response is shown in the top plot of Figure 9.11. Most of the uncertainty is in high frequency although the filters' transfer function is proper. It should be mentioned that the optimization algorithms took less than 4 minutes to find the sequences σ_P and σ_{RP} in Table 9.7. This is longer than the previous optimizations because, for the sampled-data scheduling, calculating the cost is a more computationally expensive operation.

The responses are shown in Figure 9.12. Sequence σ_I, although intuitively

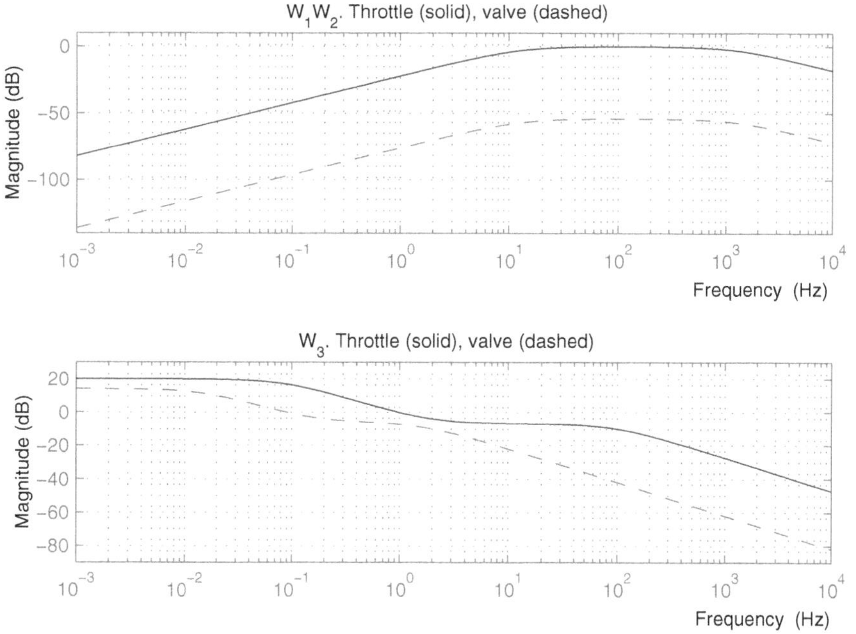

FIGURE 9.11: Weighting filters frequency response. The product $\mathcal{W}_1\mathcal{W}_2$ expresses the relative uncertainty of the throttle and valve actuators. Most of the uncertainty is in high frequency although the filters' transfer function is proper. $\mathcal{W}_3$ has been designed to approximate the inverse of the sensitivity transfer function of the system.

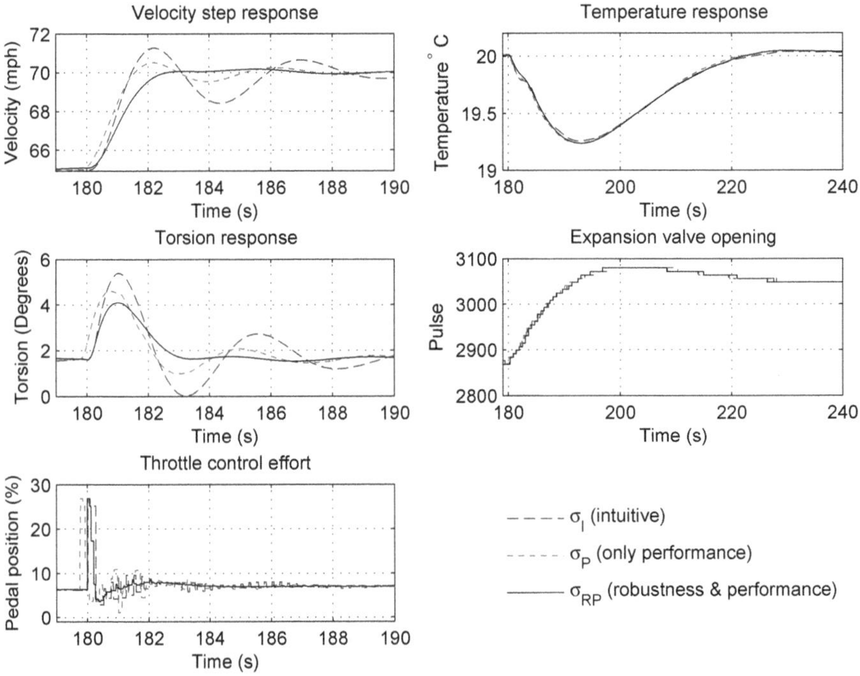

FIGURE 9.12: Experimental step response for $\mathcal{H}_\infty$ performance: σ_I is an intuitively good sequence while σ_P and σ_{RP} are optimized for performance and for robust performance respectively. The addition of robustness attenuates the oscillations.

good, shows the worst performance in terms of overshoot ($\sim$ 22%), oscillations and settling time ($\sim$ 12 seconds). Sequence σ_P, optimized for performance only, shows an improvement in terms of oscillations, settling time ($\sim$ 9 seconds) and control of shaft torsion but the overshoot is still pronounced. Sequence σ_{RP}, optimized for robust performance, shows the best response with minimal overshoot, minimal oscillation and a settling time of $\sim$ 3 seconds. As expected, since the temperature is a slow process compared to the others, the differences between the responses for the three different sequences is negligible.

This experiment has shown how robust performance can be achieved by using optimized sequences for the bus scheduling.

9.5 Experiments with FlexRay

9.5.1 Brief overview of FlexRay and its development tools

FlexRay is a new automotive communication protocol designed to substitute old protocols like CAN. FlexRay features a high data rate (up to 10 megabits per second for each channel), combined time- and event-triggered communication, redundancy and fault-tolerance. It is the determinism and the error tolerance that makes FlexRay ideal for the increasingly complex automotive performance and safety requirements which include drive-by-wire, steer-by-wire and brake-by-wire systems. For a tutorial on FlexRay see [100, 86].

The FlexRay protocol supports hybrid time- and event-triggered communication via the definition of a pre-set *communication cycle*. The FlexRay communication cycle (or simply 'cycle') is the fundamental element of the medium access. Its duration is fixed and it is assigned when the network is designed. The cycle comprises four main parts (see Figure 9.13).

- *The static segment.* This is made up of slots reserved for deterministic data that arrives at predefined time periods (time-triggered or contention-free communication).
- *The dynamic segment.* This is used for transmission of event-triggered data (its behavior is similar to the one of CAN).
- *The symbol window.* This is used for network maintenance and signaling the start of the network (not shown in Figure 9.13).
- *The network idle time.* This is used for clock synchronization.

At the design stage, the static and dynamic segments, as well as many other parameters, can be configured by using commercial software. The one used here is the Network Designer and CANoe from Vector Informatik GmbH.

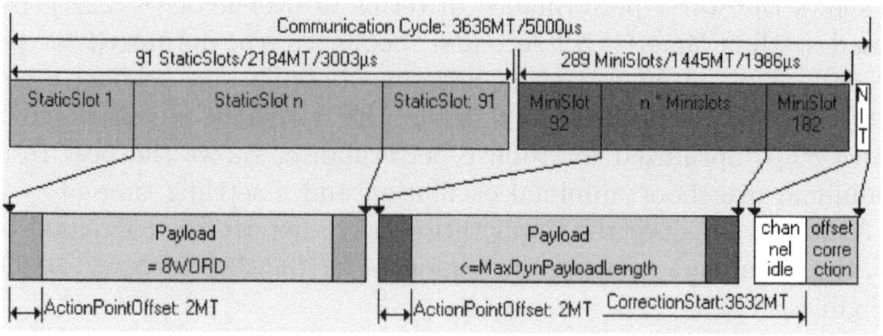

FIGURE 9.13: FlexRay communication cycle as shown in the Network Designer development tool. Slots 1 to 91 (left) represent the static segment. Slots 92 to 182 (right) represent the dynamic segment. The final slot, 'NIT' (far right), is the network idle time.

CANoe is a development tool that supports simulation, diagnostics and testing for ECUs. It can be interfaced with MATLAB/Simulink® allowing data to be passed to or received from a Simulink model. For the purpose of the following test, the two main configurations needed are the *simulation set up* and the *FlexRay schedule designer.* The simulation setup defines the physical nodes (or ECUs) that are connected to the FlexRay network (see Figure 9.14). The

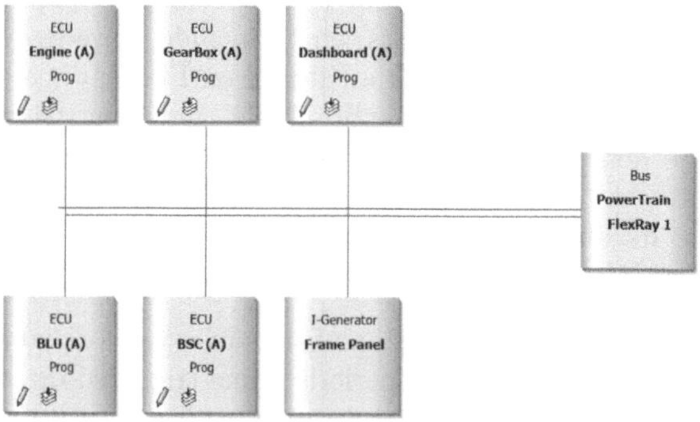

FIGURE 9.14: Configuration of FlexRay network as shown in the CANoe development tool. This configuration defines what nodes are attached to the communication bus (or buses, as FlexRay allows for redundancy).

FlexRay schedule designer is used for the actual scheduling of frames (which are signals) in the communication cycle (see Figure 9.15). Scheduling of the static segment in the cycle means assigning frames to the available (pre-set) time slots. A frame is a set of data that, for the control case, represent one or more signals. The scheduling of frames within the cycle is similar to the implementation of the scheduler in the master ECU of the HIL rig previously discussed. Hence, the length of the cycle corresponds to the periodicity of the

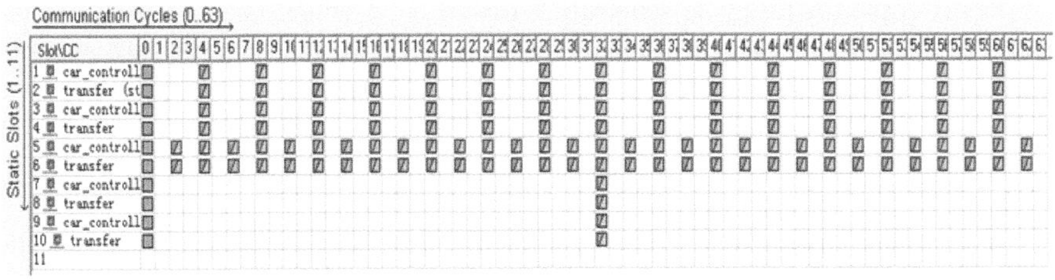

FIGURE 9.15: FlexRay scheduler designer as shown in the Network Designer development tool.

communication sequence defined in this book as ph where p is the number of time slots available and h is the sampling time or the duration of each slot. The practical limitation imposed by FlexRay is a maximum and minimum bound on ph that also depends on the size and the number of slots in the static and dynamic segment.

Another interesting characteristic of FlexRay is the possibility to perform different frame scheduling for different cycles as well as frame scheduling within a cycle. This means that cycles do not need to be identical. The constraints[6] are that cyclic scheduling of frames must be periodic with a maximum periodicity of 64 and that frames can only be scheduled every 2^n, $n = 0, 1, \ldots, 6$, cycles. For example, if frame X has a cyclic periodicity of one, its associated signal(s) will be transmitted every cycle, if it has a cyclic periodicity of two, its associated signal(s) will be transmitted every two cycles, and so on for a maximum periodicity of 64 at steps of integer powers of 2. Scheduling of cycles is illustrated in Figure 9.15 (this is a snapshot of the Network Designer's 'frame scheduling designer'). The vertical axis of the grid ('Static Slots (1:11)') are frames scheduled within the cycle and the horizontal axis ('Communication Cycles (0..63)') shows the scheduling of frames for cycle 0 to cycle 63.

9.5.2 Optimal cycle scheduling for FlexRay

From the previous section it emerges that scheduling of frames in cycles corresponds to assigning sampling rates to individual signals in a multirate system (see Section 5.6, p. 124, for a description of multirate system's optimization). If a signal is only scheduled once in the cycle and if the length of a cycle is ph then the sampling rate to be assigned to each signal can be chosen to be from the set

$$\left\{ \frac{1}{ph}, \frac{1}{2ph}, \frac{1}{4ph}, \frac{1}{8ph}, \frac{1}{16ph}, \frac{1}{32ph}, \frac{1}{64ph} \right\}. \tag{9.49}$$

[6]Some of the scheduling constraints could be relaxed by using so-called Protocol Data Units (PDUs).

Hence, optimizing cycle scheduling for an NCS which uses a FlexRay protocol is equivalent to optimizing an NCS which is in fact a multirate system. This is the scenario used for the experiment that will follow. The problem is to find an optimal scheduling of frames for the cycles that satisfies the constraint that frames can only have transmission rates that are elements of the set in (9.49).

If no other optimization constraints are introduced to this problem, the optimal solution would be to schedule every frame for every cycle, *i.e.* with a periodicity of 1. This is the solution that allows the maximum exchange of data at the expense of 'bus occupancy' . Let the bus occupancy O be defined as in (5.14). It can be easily calculated that the maximum bus occupancy for FlexRay is equal to the number of slots in the cycle or is equal to the number of frames to be scheduled if a frame is only scheduled once in the cycle (as in the case considered here).

A limit, O_{max}, on the maximum allowed bus occupancy will be used as a constraint for the frame scheduling optimization of FlexRay as explained in Section 5.6.1, p. 125.

9.5.3 Results for the FlexRay setup

In this experiment, it will be shown that it is possible to maintain a desired performance (or improve it) and reduce the bus occupancy by optimizing the scheduling of cycles for FlexRay using the proposed methods.

The automotive cruise-driveline-torsion control system example introduced in Section 9.2 will be reconsidered. The continuous-time model of the vehicle in (9.21)-(9.22) and the sampled data controller (a discretized version of the one in (9.25) with matrices given in (9.48)) are implemented in Simulink as shown in Figure 9.16 by the blocks denoted as 'Vehicle' and 'Controller' respectively. The Simulink model is interfaced to CANoe, in order to perform the FlexRay bus simulation, via the blocks denoted as 'CANoe'. Each signal (three sensors and two actuators) is assigned to a unique frame for a total of 5 frames that need to be scheduled. In a sense, a frame here could be thought of an ECU as depicted in Figure 9.6 (scheduling of the two demand signals is not considered in this case for simplicity). The communication cycle is configured with 11 available slots for the static part of the cycle (see Figure 9.15) and no dynamic part is used. This configuration results in a cycle length of 0.024 seconds. The cycle length is in fact the base sampling time h and it was chosen to be close to the one used in the HIL setup which was 0.025 seconds.

The experiment consists of testing a non-optimal cycle schedule that gives a satisfactory performance and then trying to find an optimal schedule that can maintain the performance but reduce the bus occupancy. Let the tuple of sampling times (Definition 5.6, p. 125) be defined as

$$\zeta \stackrel{\text{def}}{=} (h_{velocity}, h_{torsion}, h_{temperature}, h_{throttle}, h_{valve}), \tag{9.50}$$

where the subscripts indicate, referring to Figure 9.7, to which signal (or

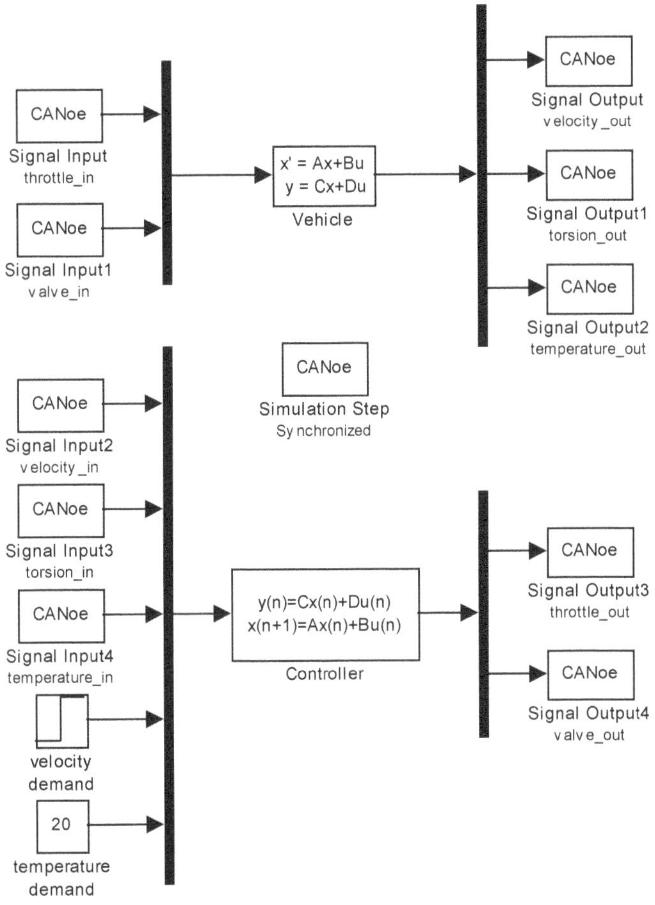

FIGURE 9.16: Simulink model for the vehicle and the cruise-driveline-torsion controller. The vehicle and controller models are interfaced to the FlexRay simulation via the blocks denoted as 'CANoe'.

frame) the sampling time is associated. The non-optimal cycle schedule is the one that schedules each frame every 4 cycles, *i.e.*

$$\zeta_n = (4h, 4h, 4h, 4h, 4h), \tag{9.51}$$

where $h = 0.024$ seconds is the cycle length (or the base sampling time). According to (5.14), ζ_n gives a bus occupancy $O_n = 1.25$.

The aim is to find an optimal tuple of sampling times ζ_{opt} with bus occupancy $O_{opt} < O_n$ that is able to preserve the performance given by ζ_n in a velocity step change test. In other words, if γ_n is the cost given by ζ_n and γ_{opt} the cost given by ζ_{opt}, then the optimization process will be considered successful if $O_{opt} < O_n$ and $\gamma_{opt} \leq \gamma_n$. The cost function used here is the one described in Section 8.4 and used to generate σ_{RP} in Table 9.7. This cost function, with the model described in Section 9.4.3, proved to give good robust performance results for the experiments on the HIL. For the non-optimal case the cost is $\gamma_n = 11.83$. The optimal tuple of sampling times, found by the GA is

$$\zeta_{opt} = (4h, 4h, 32h, 2h, 32h), \tag{9.52}$$

which has a bus occupancy $O_{opt} = 1.0625$ and a cost $\gamma_{opt} = 10.47$. Clearly $O_{opt} < O_n$ and $\gamma_{opt} \leq \gamma_n$, which was the aim of the optimization. The velocity step response for ζ_n and ζ_{opt} is given in Figure 9.17. The figure shows that, although the bus occupancy for ζ_{opt} is smaller than the one for ζ_n, the step transient behavior is very similar. In fact, for the velocity and the torsion, the response for ζ_{opt} seems to be slightly better (also suggested by the lower cost). This result can be justified intuitively by noticing that slow processes, like the 'temperature' and the 'valve', are sampled less often and fast processes, like the 'throttle', are sampled more often with the optimized scheduling.

A reduced bus occupancy means that less information is sent via the bus. This example has shown that optimizing the scheduling of frames in the FlexRay communication cycle can significantly reduce the amount of data sent via the bus without sacrificing the transient behavior of the control system. In general, the sampling rates of any multirate system could be optimized using the proposed tools.

9.6 Summary

In this chapter, the previously proposed tools for communication optimization in a linear robust framework have been applied practically. A hardware-in-the-loop system, implementing a realistic automotive control problem, was set up as a testbed. This consists of a PC simulation of a vehicle controlled by real microprocessor boards (the electronic control units) communicating to each

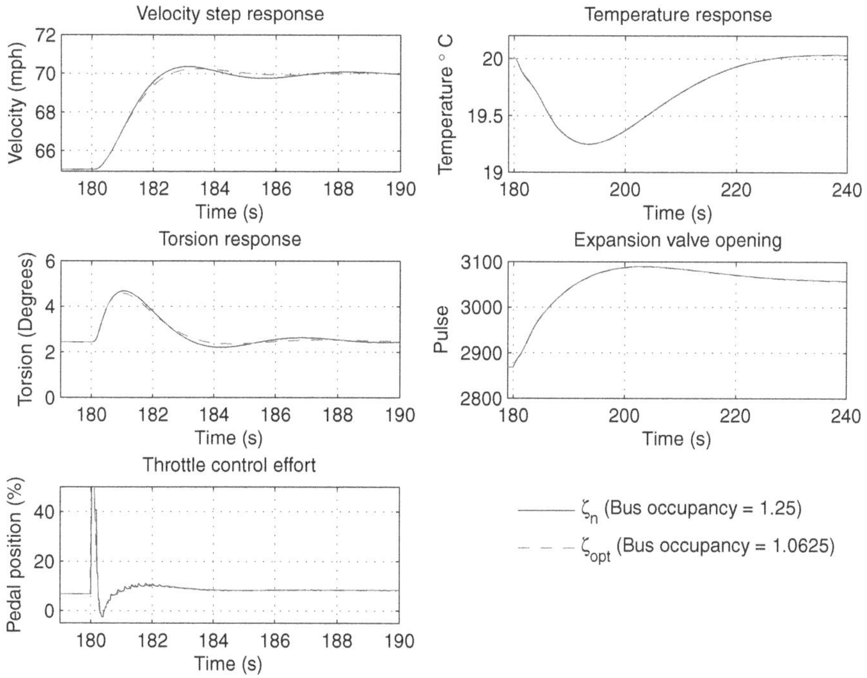

FIGURE 9.17: FlexRay simulated step response for optimal (ζ_{opt}) and non-optimal (ζ_n) cycle scheduling. Although the responses of the two schedules are very similar, optimal scheduling has reduced the bus occupancy from 1.25 to 1.0625, satisfying all the scheduling constraints.

other via a controller area network. Vehicle modeling and controller design were illustrated. The electronic control unit bus access was scheduled using optimized sequences. It was noticed that, due to delays in the actuator signals, the inclusion of robustness was a necessity. Hence, the proposed $\mathcal{H}_\infty$-based formulation proved to be the most practically relevant. In general, optimized schedules performed better than intuitively good ones.

The same setup was also used for experimenting with the new automotive communication protocol, FlexRay. Results showed that the flexibility of the proposed optimization tools can reduce the bus occupancy without sacrificing the performance and controller robustness.

The framework of $\mathcal{H}_\infty$-robustness is often an ideal tool to deal with uncertainty and also nonlinearities. However, robustness tools alone may not always be sufficient to account for nonlinearities, as they can introduce significant levels of conservatism. The next chapter deals with a computational framework for NCS integration in the context of nonlinear systems.

10

Schedule design for nonlinear NCSs

CONTENTS

Nonlinearity of practical dynamical systems always introduces complexity in the analysis for stability and performance. The methodology to deal with such nonlinearity is not trivial, in particular for networked control systems. For this purpose, an appropriate approximation method and an effective optimization framework are developed and described here.

10.1 Introduction

In previous chapters, linear NCSs have been discussed in detail. It has been shown that the introduction of the network provides an explicit effect on the dynamic behavior of the control system. It will be now presented how this framework can be extended to nonlinear systems.

Although a variety of techniques have been developed for the design of optimal communication sequences for linear NCSs, the literature for nonlinear NCSs seems to concentrate more on the stability problem (although $\mathcal{L}_2$

gain stability as in [91] may lead to suitable designs and the work of [98] is an exception for mildly nonlinear systems), and methods for optimal communication design have been somehow ignored [7, 13]. However, computational tools to verify performance and stability of nonlinear systems are nowadays readily available. The semidefinite programming approach has been extended from the linear matrix inequality framework to the idea of sum-of-squares (SOS).

The SOS method was first presented by P.A. Parrilo [96]. The existence of an SOS representation for a polynomial is sufficient for its non-negativity [96], which provides a computationally tractable way to obtain the Lyapunov function for stability analysis [3, 13]. In [83], the stability analysis of switched hybrid systems via a Lyapunov analysis is carried out. Equally, nonlinear hybrid systems are analyzed for their stability by [95]. An optimal control design for a continuous-time problem has been recently proposed in [15]. In [13], an NCS using hybrid systems formulations was investigated for stability based on LMI computations. Stability for nonlinear systems with delays are analyzed in [94, 97] using Lyapunov-Krasovskii functions and an SOS-computational approach. There have been some applications of SOS for discrete-time system control. The authors of [119] propose a design condition in terms of SOS for the guaranteed cost control of a discrete fuzzy system, while a condition for robust stability of an LTI discrete-time system is addressed in [143] and solved by the SOS method.

In this chapter, performance for the integration of the nonlinear NCSs is considered in particular for the actuator of the plant. Thus, a system is investigated where only the actuator nodes share a common bus. The communication is again assumed to be time-triggered and all the non-updated control signals will remain the same until updated. The analysis process extends over ideas from Chapters 6 and 7. The plant, a nonlinear affine system, is approximated in terms of its nonlinearities using a Taylor series approach. This allows the use of the SOS framework. Considering the limitation of the communication medium, a scheduler that defines the way that data is transmitted on the channel is introduced, and a similar process of augmentation and lifting as that in Chapter 7 is carried out. The relevant optimization for the nonlinear NCSs with the SOS framework is described concretely for a fixed communication schedule. The Lyapunov equation for the closed-loop system is solved and the relevant Lyapunov function is computed. This permits to compute a performance cost of the integrated NCS in a local sense using the SOS approach (also under the assumption that the initial value of the state variables obeys the standard Gaussian distribution). This cost computation will allow (as for the linear NCS problem in Chapter 7) for a search of the optimal communication schedule which creates the best performing integrated NCS.

10.2 Discretization of nonlinear affine systems

Since the network is naturally a discrete-time process and the controller is a sampled-data one, the plant dynamics are first discretized. In general, a delay-free nonlinear plant system in state-space form can be represented as follows [62]:

$$\frac{d\boldsymbol{x}(t)}{dt} = \boldsymbol{f}(\boldsymbol{x}(t)) + \mathbf{g}(\boldsymbol{x}(t))\boldsymbol{u}(t), \tag{10.1}$$

where $\boldsymbol{x} \in X \subset \mathbb{R}^{n_x}$ (X an open and connected set) is the vector of the states and $\boldsymbol{u} \in \mathbb{R}^{n_u}$ is the controlled input. It is assumed that $\boldsymbol{f}(\boldsymbol{x})$ and $\mathbf{g}(\boldsymbol{x})$ are real analytic vector fields on X. Here, the sample and hold scheme is introduced, and an equidistant grid on the time axis with a sampling period h is considered. The states $\boldsymbol{x}(t)$ are sampled in time with the sampling time h, where the control input $\boldsymbol{u}$ is created by a sample-and-hold element for the same sampling time as the states (see Section 2.3, p. 16).

Under the ZOH assumption, an approximate sampled-data representation of system (10.1) within one sampling interval can be obtained via taking up-to the Nth-order terms of the Taylor series expansion in h and the resulting coefficients would be calculated by taking successive partial time derivatives of the right-hand side of (10.1) [62]:

$$\boldsymbol{x}(j+1) = \boldsymbol{x}(j) + \sum_{l=1}^{N} \frac{d^l\boldsymbol{x}}{dt^l}\frac{h^l}{l!}\bigg|_{t=t_j} + \underbrace{\sum_{l=N+1}^{\infty} \frac{d^l\boldsymbol{x}}{dt^l}\frac{h^l}{l!}\bigg|_{t=t_j}}_{remainder}, \tag{10.2}$$

where $\boldsymbol{x}(j)$ is the value of state vector $\boldsymbol{x}$ at time $t = t_j = jh$.

Note that this discrete description of the nonlinear system (10.1) is exact for $N \to \infty$.

10.3 Sampled-data model of nonlinear NCS

Having provided the discrete representation for a single input system (10.2) (from [62]), it is now necessary to obtain a compact nonlinear representation of a multi-input system subject to a ZOH-input vector with n_u elements, u_i. Considering the fact that these actuators are obtaining control signal information via a communication bus, a model for the network with time triggered communication (see Chapter 3 and 7) is to be introduced.

10.3.1 Time discretization of a multi-input nonlinear affine system

The state-space form expression for a multi-input nonlinear affine NCS is

$$\begin{aligned} \frac{d\boldsymbol{x}(t)}{dt} &= \boldsymbol{f}(\boldsymbol{x}(t)) + \sum_{i=1}^{n_u} \boldsymbol{g}_i(\boldsymbol{x}(t))u_i(t) \\ &= \boldsymbol{f}(\boldsymbol{x}(t)) + \boldsymbol{g}_1(\boldsymbol{x}(t))u_1(t) + \ldots + \boldsymbol{g}_{n_u}(\boldsymbol{x}(t))u_{n_u}(t), \end{aligned} \tag{10.3}$$

where $\mathbf{g} = \begin{bmatrix} \boldsymbol{g}_1 & \boldsymbol{g}_2 & \cdots & \boldsymbol{g}_{n_u} \end{bmatrix}$.

Here, $\tilde{\mathfrak{A}}^{[l]}(\boldsymbol{x}, \boldsymbol{u})$ expresses the lth order derivative of the state variable $\boldsymbol{x}$, then $\tilde{\mathfrak{A}}^{[1]}(\boldsymbol{x}, \boldsymbol{u})$ for system (10.3) is

$$\tilde{\mathfrak{A}}^{[1]}(\boldsymbol{x}, \boldsymbol{u}) = \boldsymbol{f}(\boldsymbol{x}) + \sum_{i=1}^{n_u} \boldsymbol{g}_i(\boldsymbol{x})u_i. \tag{10.4}$$

This allows to introduce $\tilde{\mathfrak{A}}^{[l]}(\boldsymbol{x}, \boldsymbol{u})$ which is an lth degree polynomial in $\boldsymbol{u}$:

$$\begin{aligned} \tilde{\mathfrak{A}}^{[l]}(\boldsymbol{x}, \boldsymbol{u}) &= \frac{\partial \tilde{\mathfrak{A}}^{[l-1]}(\boldsymbol{x}, \boldsymbol{u})}{\partial \boldsymbol{x}}(\boldsymbol{f}(\boldsymbol{x}) + \sum_{i=1}^{n_u} \boldsymbol{g}_i(\boldsymbol{x})u_i) \\ &= \tilde{\mathfrak{A}}_0^{[l]} + \tilde{\mathfrak{A}}_1^{[l]}\tilde{\boldsymbol{U}}_1 + \cdots + \tilde{\mathfrak{A}}_l^{[l]}\tilde{\boldsymbol{U}}_l, \end{aligned} \tag{10.5}$$

which can be proven by induction from (10.3). In the expression above, $\tilde{\mathfrak{A}}_i^{[l]}$ and $\tilde{\boldsymbol{U}}_i$ are coefficient matrices in combination with the control signal of i th degree, respectively, where $i = 1, \ldots, l$, $l = 1, \ldots, N$. Moreover, the matrix $\tilde{\boldsymbol{U}}_i$ is defined in the following format:

$$\begin{aligned} \tilde{\boldsymbol{U}}_1 &= [u_1, \cdots, u_{n_u}]^T, \\ \tilde{\boldsymbol{U}}_i &= [\tilde{\boldsymbol{U}}_{i-1,1}\tilde{\boldsymbol{U}}_1^T, \cdots, \tilde{\boldsymbol{U}}_{i-1,n_u{}^{i-1}}\tilde{\boldsymbol{U}}_1^T]^T, i = 2, \ldots, l. \end{aligned} \tag{10.6}$$

Therefore, equation (10.5) can be expressed as

$$\begin{aligned} \tilde{\mathfrak{A}}^{[l]} =& \tilde{\mathfrak{A}}_0^{[l]} + \tilde{\mathfrak{a}}_{1,1\cdots n_u}^{[l]}\tilde{\boldsymbol{U}}_1 + \cdots \\ &+ \tilde{\mathfrak{a}}_{l,1\cdots n_u}^{[l]}\tilde{\boldsymbol{U}}_{l-1,1}\tilde{\boldsymbol{U}}_1 + \cdots + \tilde{\mathfrak{a}}_{l,((n_u{}^{l-1}-1)n_u+1)\cdots n_u{}^l}^{[l]}\tilde{\boldsymbol{U}}_{l-1,n_u{}^{l-1}}\tilde{\boldsymbol{U}}_1, \end{aligned} \tag{10.7}$$

where $\tilde{\mathfrak{a}}_{i,k_1\cdots k_2}^{[l]}$ is the matrix created from the k_1^{th} to k_2^{th} columns of matrix $\tilde{\mathfrak{A}}_i^{[l]}$, *i.e.*

$$\tilde{\mathfrak{a}}_{i,k_1\cdots k_2}^{[l]} = \begin{bmatrix} \tilde{\mathfrak{a}}_{i,k_1}^{[l]} & \tilde{\mathfrak{a}}_{i,k_1+1}^{[l]} & \cdots & \tilde{\mathfrak{a}}_{i,k_2}^{[l]} \end{bmatrix}.$$

Using the discretization scheme, the multi-input nonlinear affine system can now be computed in a more compact form by

$$\begin{aligned} \boldsymbol{x}(j+1) =& \boldsymbol{x}(j) + \boldsymbol{H}(h)|_{\boldsymbol{x}=\boldsymbol{x}(j),u_i=u_i(j)}\mathfrak{A}_0|_{\boldsymbol{x}=\boldsymbol{x}(j),u_i=u_i(j)} \\ &+ \boldsymbol{H}(h)|_{\boldsymbol{x}=\boldsymbol{x}(j),u_i=u_i(j)}\mathfrak{A}|_{\boldsymbol{x}=\boldsymbol{x}(j),u_i=u_i(j)}\boldsymbol{E}\tilde{\boldsymbol{U}}_1(j), \end{aligned} \tag{10.8}$$

where

$$H(\tau) = \begin{bmatrix} \tau I_{n_x} & \frac{\tau^2}{2!} I_{n_x} & \cdots & \frac{\tau^N}{N!} I_{n_x} \end{bmatrix}, \tag{10.9}$$

$$\mathfrak{A}_0 = \begin{bmatrix} (\tilde{\mathfrak{A}}_0^{[1]})^T & (\tilde{\mathfrak{A}}_0^{[2]})^T & \cdots & (\tilde{\mathfrak{A}}_0^{[N]})^T \end{bmatrix}^T, \tag{10.10}$$

$$E = \begin{bmatrix} I_{n_u} & I_{n_u} & \cdots & I_{n_u} \end{bmatrix}^T, E \in \mathbb{R}^{((1-{n_u}^{N-1})/(1-n_u))\times n_u}, \tag{10.11}$$

$$\tilde{U}_1(j) = \begin{bmatrix} u_1(j) & u_2(j) & \ldots & u_{n_u}(j) \end{bmatrix}^T, \tag{10.12}$$

$$\mathfrak{A} = \begin{bmatrix} \tilde{\mathfrak{a}}^{[1]}_{1,1\cdots n_u} & 0 & \ldots & 0 \\ \tilde{\mathfrak{a}}^{[2]}_{1,1\cdots n_u} & & \ldots & 0 \\ \vdots & \vdots & & \vdots \\ \tilde{\mathfrak{a}}^{[N]}_{1,1\cdots n_u} & & \ldots & \tilde{\mathfrak{a}}^{[N]}_{N,(({n_u}^{N-1}-1)n_u+1)\cdots {n_u}^N} \tilde{U}_{N-1,{n_u}^{N-1}}(j) \end{bmatrix}. \tag{10.13}$$

In the expressions above, n_x is the number of states, n_u the number of inputs and N is the highest order of terms taken from the Taylor series expansion in terms of the sampling time, h. Let $\tilde{x}(j+1) = \begin{bmatrix} x(j+1)^T & 1 \end{bmatrix}^T$, then the more compact form is obtained

$$\tilde{x}(j+1) = F|_{x=x(j),u_i=u_i(j)}\tilde{x}(j) + G|_{x=x(j),u_i=u_i(j)}\tilde{U}_1(j) \tag{10.14}$$

where

$$F = \tilde{F}(h), \quad \tilde{F}(\tau) = \begin{bmatrix} I_{n_x} & H(\tau)\mathfrak{A}_0 \\ 0_{1\times n_x} & 1 \end{bmatrix}, \quad G = \tilde{G}(h), \quad \tilde{G}(\tau) = \begin{bmatrix} H(\tau)\mathfrak{A}E \\ 0_{1\times n_u} \end{bmatrix}.$$

Note that nonlinearities now lie in the matrices $\mathfrak{A}_0 \in \mathbb{R}^{n_x N}$ and $\mathfrak{A} \in \mathbb{R}^{n_x N\times((1-{n_u}^{(N-1)})/(1-n_u))}$, which are included in matrices F and G respectively, and either of them is a nonlinear function of the states x and the step-wise constant input u, *i.e.* $\mathfrak{A}_0 = \mathfrak{A}_0(x,u)$ and $\mathfrak{A} = \mathfrak{A}(x,u)$.

10.3.2 Scheduling of actuator information

Consider the general architecture of NCS shown in Figure 6.1 (p. 132). The discrete multi-input system is now subjected to the communication bus, which feeds control signals u_i to the actuators. Here the situation that the actuator nodes share a common bus is considered and only a limited number of actuator signals can be updated at every time tick. From Definition 3.5 and Eq.(3.15), the communication sequence σ is fully characterised by the bus scheduler for the actuator, $\mathcal{S}_A$. It can be represented by

$$\mathcal{S}_A : \begin{cases} x_A(j+1) = \bar{S}_A(j)x_A(j) + S_A(j)\hat{U}_1(j) \\ \hat{u}(j) = \bar{S}_A(j)x_A(j) + S_A(j)\hat{U}_1(j) \end{cases}, \tag{10.15}$$

where $\hat{U}_1(\cdot)$ is the full control vector produced by the controllers, $v(\cdot)$ is the actual control input applied to the plant consisting of signals updated within the sampling interval and retained by the ZOH. The plant model with the limited control communication channel can now be expressed as

$$\tilde{\boldsymbol{x}}(j+1) = \boldsymbol{F}\tilde{\boldsymbol{x}}(j)+\boldsymbol{G}\boldsymbol{v}(j) = \boldsymbol{F}\tilde{\boldsymbol{x}}(j)+\boldsymbol{G}(\bar{\boldsymbol{S}}_A(k)\boldsymbol{v}(j-1)+\boldsymbol{S}_A(k)\hat{\boldsymbol{U}}_1(j)), \tag{10.16}$$

where $\boldsymbol{F} := \boldsymbol{F}(\boldsymbol{x}(j), \boldsymbol{v}(j))$ and $\boldsymbol{G} := \boldsymbol{G}(\boldsymbol{x}(j), \boldsymbol{v}(j))$ (for brevity, the arguments in the nonlinear function are omitted). For $j = pl + k$, the following dynamics after merging of the two system equations (10.15) and (10.16) are obtained

$$\hat{\boldsymbol{x}}(pl + k + 1) = \hat{\boldsymbol{F}}(k)\hat{\boldsymbol{x}}(pl + k) + \hat{\boldsymbol{G}}(k)\hat{\boldsymbol{U}}_1(pl + k) \tag{10.17}$$

where

$$\hat{\boldsymbol{x}}(pl+k) = \begin{bmatrix} \tilde{\boldsymbol{x}}(pl+k) \\ \boldsymbol{v}(pl+k-1) \end{bmatrix}, \quad \hat{\boldsymbol{F}}(k) = \begin{bmatrix} \boldsymbol{F} & \boldsymbol{G}\bar{\boldsymbol{S}}_A(k) \\ 0 & \bar{\boldsymbol{S}}_A(k) \end{bmatrix}, \quad \hat{\boldsymbol{G}}(k) = \begin{bmatrix} \boldsymbol{G}\boldsymbol{S}_A(k) \\ \boldsymbol{S}_A(k) \end{bmatrix}.$$

This is the augmented model of the nonlinear plant which includes the network dynamics. These dynamics can only be computed for a fixed, given communication sequence.

10.4 Quadratic cost function for NCS performance

Now that a model of the plant with the communication dynamics is available, the problem is to find a cost function that measures the optimality of a particular communication sequence for the given plant and controller. This cost is based on the continuous-time plant and an asymptotically stabilizing discrete controller which might have been designed for the plant without knowledge of the communication system.

10.4.1 Cost function for the sampled-data system

Consider the given continuous time, infinite horizon quadratic control cost

$$J \underset{\substack{\boldsymbol{u} \\ s.t.(10.3)}}{=} \int_0^\infty \left(\boldsymbol{x}(t)^T \boldsymbol{Q}_{c1}\boldsymbol{x}(t) + \boldsymbol{u}(t)^T \boldsymbol{Q}_{c2}\boldsymbol{u}(t)\right) dt, \tag{10.18}$$

where $\boldsymbol{u}(t) = \begin{bmatrix} u_1(t) & u_2(t) & \cdots & u_{n_u}(t) \end{bmatrix}^T$, and assume that $\boldsymbol{Q}_{c1} \geq 0$ and $\boldsymbol{Q}_{c2} > 0$ are both weighting matrices for a desirable ideal closed-loop response. The control signals $\boldsymbol{u}(t)$ (piecewise constant functions) are in this case suitably chosen, differentiable functions of the sampled states $\boldsymbol{x}(jh)$.

Note that, alternatively, a stochastic framework is easily obtained by evaluating the expectation $E\{J\}$ of J assuming a Gaussian distribution for the initial states $\boldsymbol{x}(0)$. A suitable straightforward approach will be presented later, while, for simplicity, much of the initial discussion will focus on the deterministic cost.

For the sampled-data system (10.14), the equivalent discrete cost function is

$$J \underset{\substack{\tilde{U}_1 \\ s.t.(10.14)}}{=} \sum_{j=0}^{\infty} \begin{bmatrix} \tilde{x}(jh) \\ \tilde{U}_1(jh) \end{bmatrix}^T \begin{bmatrix} Q_1 & Q_{12} \\ Q_{12}^T & Q_2 \end{bmatrix}_{\tilde{U}} \begin{bmatrix} \tilde{x}(jh) \\ \tilde{U}_1(jh) \end{bmatrix} \tag{10.19}$$

where

$$\begin{bmatrix} Q_1 & Q_{12} \\ Q_{12}^T & Q_2 \end{bmatrix}_{\tilde{U}} = \int_0^h \begin{bmatrix} \tilde{F}(\tau)^T & 0 \\ \tilde{G}(\tau)^T & I \end{bmatrix} \left[\begin{array}{cc|c} Q_{c1} & 0 & 0 \\ 0 & 0 & 0 \\ \hline 0 & 0 & Q_{c2} \end{array}\right] \begin{bmatrix} \tilde{F}(\tau)^T & \tilde{G}(\tau)^T \\ 0 & I \end{bmatrix} d\tau. \tag{10.20}$$

Note for $N \to \infty$ in (10.8)-(10.14), there is no approximation between (10.18) and (10.19). After the augmentation with the scheduler of (10.15), the cost function becomes

$$J \underset{\substack{\hat{U}_1 \\ s.t.(10.14)}}{=} \sum_{j=0}^{\infty} \begin{bmatrix} \tilde{x}(j) \\ v(j) \end{bmatrix}^T \begin{bmatrix} Q_1 & Q_{12} \\ Q_{12}^T & Q_2 \end{bmatrix} \begin{bmatrix} \tilde{x}(j) \\ v(j) \end{bmatrix}. \tag{10.21}$$

Thus, for the augmented system obtained above, the associated cost is

$$J \underset{\substack{\hat{U}_1 \\ s.t.(10.17)}}{=} \sum_{l=0}^{\infty} \sum_{k=0}^{p-1} \begin{bmatrix} \hat{x}(pl+k) \\ \hat{U}_1(pl+k) \end{bmatrix}^T \begin{bmatrix} \hat{Q}_1 & \hat{Q}_{12} \\ \hat{Q}_{12}^T & \hat{Q}_2 \end{bmatrix} \begin{bmatrix} \hat{x}(pl+k) \\ \hat{U}_1(pl+k) \end{bmatrix}, \tag{10.22}$$

where the $\hat{Q}(k)$ is derived widely following the principal ideas of (6.14)-(6.17) in Chapter 6. However, since the plant here is nonlinear and the relationships in Section 10.3.2 have to be used, a higher complexity than that in Chapter 6 is to be expected. The results in Chapter 6 (and Chapter 7) consider the linear framework.

10.4.2 Removal of periodicity and cost for optimization

In order to remove the time-dependence in k of the scheduling matrices, the lifting technique as for (7.10)-(7.15) in Chapter 7 is used. The lifted system from (10.17) is:

$$\bar{x}(pl+1) = \bar{F}\hat{x}(pl) + \bar{G}\bar{U}_1(pl), \tag{10.23}$$

where

$$\begin{aligned} \bar{x}(pl) &= \begin{bmatrix} \hat{x}(pl)^T & \hat{x}(pl+1)^T & \cdots & \hat{x}(pl+p-1)^T \end{bmatrix}^T, \\ \bar{U}(pl) &= \begin{bmatrix} \hat{U}_1(pl)^T & \hat{U}_1(pl+1)^T & \cdots & \hat{U}_1(pl+p-1)^T \end{bmatrix}^T, \end{aligned}$$

and $\bar{\boldsymbol{F}}$ and $\bar{\boldsymbol{G}}$ can be easily derived. It follows that the augmented lifted plant equation is given by the last row of the matrices $\bar{\boldsymbol{F}}$ and $\bar{\boldsymbol{G}}$, therefore

$$\hat{\boldsymbol{x}}(pl+p) = \bar{\boldsymbol{F}}_p \hat{\boldsymbol{x}}(pl) + \bar{\boldsymbol{G}}_p \bar{\boldsymbol{U}}_1(pl), \tag{10.24}$$

where

$$\bar{\boldsymbol{F}}_p = \prod_{i=1}^{p} \hat{\boldsymbol{F}}(p-i),\ \bar{\boldsymbol{G}}_p = \begin{bmatrix} \boldsymbol{M}(0) & \boldsymbol{M}(1) & \cdots & \boldsymbol{M}(p-1) \end{bmatrix},$$

$$\boldsymbol{M}(k) = \left(\prod_{i=1}^{p-k-1} \hat{\boldsymbol{F}}(p-i) \right) \hat{\boldsymbol{G}}(k), k = 0, 1, \ldots, p-1.$$

Having defined the equation for the lifted system in (10.23), the closed-loop system equation with scheduler is rewritten

$$\begin{bmatrix} \hat{\boldsymbol{x}}(pl+k) \\ \hat{\boldsymbol{U}}_1(pl+k) \end{bmatrix} = \begin{bmatrix} \bar{\boldsymbol{F}}_k \hat{\boldsymbol{x}}(pl) + \bar{\boldsymbol{G}}_k \bar{\boldsymbol{U}}_1(pl) \\ \boldsymbol{E}_k \bar{\boldsymbol{U}}_1(pl) \end{bmatrix}, \tag{10.25}$$

where

$$\boldsymbol{E}_k = \begin{bmatrix} \boldsymbol{E}_k(0) & \boldsymbol{E}_k(1) & \cdots & \boldsymbol{E}_k(p-1) \end{bmatrix},\ \boldsymbol{E}_k(i) = \begin{cases} \boldsymbol{I} & \text{if } i = k \\ \boldsymbol{0} & \text{if } i \neq k \end{cases},$$

which is used to extract the particular control signal $\hat{\boldsymbol{U}}_1(pl+k)$. Equation (10.25) will be now used in (10.22) to obtain the equivalent cost function

$$J \underset{\substack{\bar{\boldsymbol{U}}_1 \\ s.t.(10.24)}}{=} \sum_{l=0}^{\infty} \sum_{k=0}^{p-1} \begin{bmatrix} \hat{\boldsymbol{x}}(pl) \\ \bar{\boldsymbol{U}}_1(pl) \end{bmatrix}^T \begin{bmatrix} \bar{\boldsymbol{Q}}_1(k) & \bar{\boldsymbol{Q}}_{12}(k) \\ \bar{\boldsymbol{Q}}_{12}^T(k) & \bar{\boldsymbol{Q}}_2(k) \end{bmatrix} \begin{bmatrix} \hat{\boldsymbol{x}}(pl) \\ \bar{\boldsymbol{U}}_1(pl) \end{bmatrix}, \tag{10.26}$$

where $\bar{\boldsymbol{Q}}(k)$ can be computed in a similar fashion as outlined in [78]. Since neither $\hat{\boldsymbol{x}}(pl), \bar{\boldsymbol{U}}_1(pl)$ is dependent on k, the non-periodic cost function for the lifted augmented system is:

$$J \underset{\substack{\bar{\boldsymbol{U}}_1 \\ s.t.(10.24)}}{=} \sum_{l=0}^{\infty} \begin{bmatrix} \hat{\boldsymbol{x}}(pl) \\ \bar{\boldsymbol{U}}_1(pl) \end{bmatrix}^T \underbrace{\left(\sum_{k=0}^{p-1} \begin{bmatrix} \bar{\boldsymbol{Q}}_1(k) & \bar{\boldsymbol{Q}}_{12}(k) \\ \bar{\boldsymbol{Q}}_{12}^T(k) & \bar{\boldsymbol{Q}}_2(k) \end{bmatrix} \right)}_{= \begin{bmatrix} \tilde{\boldsymbol{Q}}_1 & \tilde{\boldsymbol{Q}}_{12} \\ \tilde{\boldsymbol{Q}}_{12}^T & \tilde{\boldsymbol{Q}}_2 \end{bmatrix} = \tilde{\boldsymbol{Q}}} \begin{bmatrix} \hat{\boldsymbol{x}}(pl) \\ \bar{\boldsymbol{U}}_1(pl) \end{bmatrix}. \tag{10.27}$$

The cost in (10.27) gives a measure of the performance of the system for a particular communication sequence. Note that the matrix $\tilde{\boldsymbol{Q}}$ is indeed a rather complex *nonlinear* function in $\left(\hat{\boldsymbol{x}}(pl), \bar{\boldsymbol{U}}_1(\hat{\boldsymbol{x}}(pl))\right)$ only. This allows for an optimization problem to be set up where the aim is to find communication sequences that minimize the cost in (10.27).

10.5 Optimization problem

As in Chapter 7, it is now the target to determine the quadratic control cost, *i.e.* (10.18) and (10.27), in a computational framework for a given communication sequence σ. This is again under the assumption that the controller has been designed for the original plant (10.1) or the discretized system (10.14) without knowledge of the control communication system. Thus, the control vector $\tilde{\boldsymbol{U}}_1 = \begin{bmatrix} u_1 & \cdots & u_{n_u} \end{bmatrix}^T$ is a function of the state $\boldsymbol{x}(j)$. This controller is integrated across the control communication network, where a communication sequence σ needs to be chosen. For this, the cost J is established to determine the best sequence σ in a generic framework first. In the second step, an SOS approach is derived to allow for cost computation in this highly nonlinear environment.

10.5.1 Generic optimization problem

Given a fixed communication sequence σ, the cost J in (10.27) can be now computed in the following way: find a Lyapunov function $V(\cdot)$ to compute

$$J \leq \min_V V(\hat{\boldsymbol{x}}(0)), \quad \hat{\boldsymbol{x}}(0) = \begin{bmatrix} \tilde{\boldsymbol{x}}_0^T & \boldsymbol{v}_0^T \end{bmatrix}^T, \tag{10.28}$$

such that for a given initial value $\hat{\boldsymbol{x}}(0)$ and the following generic inequalities in the variable $\hat{\boldsymbol{x}} \neq 0$

$$V(\hat{\boldsymbol{x}}(p(l+1))) - V(\hat{\boldsymbol{x}}(pl)) \leq - \begin{bmatrix} \hat{\boldsymbol{x}}(pl) \\ \bar{\boldsymbol{U}}_1(pl) \end{bmatrix}^T \begin{bmatrix} \tilde{\boldsymbol{Q}}_1 & \tilde{\boldsymbol{Q}}_{12} \\ \tilde{\boldsymbol{Q}}_{12}^T & \tilde{\boldsymbol{Q}}_2 \end{bmatrix} \begin{bmatrix} \hat{\boldsymbol{x}}(pl) \\ \bar{\boldsymbol{U}}_1(pl) \end{bmatrix}, \tag{10.29}$$

$V(\hat{\boldsymbol{x}}) \geq 0$, *i.e.*, $V(\hat{\boldsymbol{x}})$ is positive definite, $\bar{\boldsymbol{U}}_1$ is a (control) function of $\hat{\boldsymbol{x}}$. Note that for an asymptotically stabilizing controller $\lim_{l \to \infty} V(\hat{\boldsymbol{x}}(pl)) = 0$ and

$$\sum_{l=0}^{\infty} (V(\hat{\boldsymbol{x}}(p(l+1))) - V(\hat{\boldsymbol{x}}(pl))) = -V(\hat{\boldsymbol{x}}(0))$$

$$\leq - \sum_{l=0}^{\infty} \begin{bmatrix} \hat{\boldsymbol{x}}(pl) \\ \bar{\boldsymbol{U}}_1(pl) \end{bmatrix}^T \begin{bmatrix} \tilde{\boldsymbol{Q}}_1 & \tilde{\boldsymbol{Q}}_{12} \\ \tilde{\boldsymbol{Q}}_{12}^T & \tilde{\boldsymbol{Q}}_2 \end{bmatrix} \begin{bmatrix} \hat{\boldsymbol{x}}(pl) \\ \bar{\boldsymbol{U}}_1(pl) \end{bmatrix} = -J.$$

Alternatively, a stochastic framework can be used where $\hat{\boldsymbol{x}}(0)$ in this case follows a normal Gaussian distribution. Hence, in this case

$$E\{J\} \leq \min_V E\{V(\hat{\boldsymbol{x}}(0))\}. \tag{10.30}$$

It is evident that the computation above is not easily carried out and a computation of the cost is in general not tractable. Note also that global stability and performance are usually not easily achieved. Thus, the cost analysis of (10.28) or (10.30) subject to (10.29) has to be carried out in a local sense for some region of attraction $\mathbb{D}$, *i.e.* $\hat{\boldsymbol{x}}(pl) \in \mathbb{D}$.

10.5.2 Lyapunov function for the SOS-approach

It is now assumed that the drift $\boldsymbol{f}(\boldsymbol{x}(t))$ and the input gain $\mathbf{g}(\boldsymbol{x}(t))$ are represented in polynomial form. This can be achieved via a Taylor series expansion approach. Hence, the dynamics of the augmented system (10.24) can be equally expressed as a Taylor series. This implies that the matrices $\tilde{\boldsymbol{Q}}_1$, $\tilde{\boldsymbol{Q}}_{12}$ and $\tilde{\boldsymbol{Q}}_2$ are also polynomial expressions in $\hat{\boldsymbol{x}}$ and $\bar{\boldsymbol{U}}_1$. In particular, if $\bar{\boldsymbol{U}}_1$ is a polynomial expression in $\hat{\boldsymbol{x}}$, a sum-of-squares approach can be used to compute J. This implies that the Lyapunov function can be SOS and is best represented as

$$V(\hat{\boldsymbol{x}}) = \begin{bmatrix} \hat{x}_1 \\ \vdots \\ \hat{x}_1^m \\ \vdots \\ \hat{x}_{n_x}^m \end{bmatrix}^T \begin{bmatrix} \boldsymbol{P}_{(1,1)} & \cdots & \boldsymbol{P}_{(1,mn_x)} \\ \vdots & & \\ \boldsymbol{P}_{(m,1)} & \cdots & \vdots \\ \vdots & & \\ \boldsymbol{P}_{(mn_x,1)} & \cdots & \boldsymbol{P}_{(mn_x,mn_x)} \end{bmatrix} \begin{bmatrix} \hat{x}_1 \\ \vdots \\ \hat{x}_1^m \\ \vdots \\ \hat{x}_{n_x}^m \end{bmatrix}, \tag{10.31}$$

where the Lyapunov matrix $\boldsymbol{P} = [\boldsymbol{P}_{(i,j)}|_{i=1,\cdots,mn_x;\ j=1,\cdots,mn_x}]$ is positive semi-definite. (In the computational framework, it will be guaranteed that $V(\cdot)$ as a function of $\hat{\boldsymbol{x}}$ is positive definite.)

As for Chapter 7, it is now possible to compute the closed loop cost in a stochastic and deterministic framework (see (7.18)). In a stochastic framework, the expected value $E\{J\}$ for a Gaussian distributed initial value $\boldsymbol{x}(0)$ is to be computed

$$\begin{aligned} E\{J\} =& E\{V(\hat{\boldsymbol{x}}(0))\} \\ =& \boldsymbol{P}_{(1,1)}E\{\hat{x}_1^2(0)\} + \boldsymbol{P}_{(2,2)}E\{\hat{x}_2^2(0)\} + \cdots \\ &+ \boldsymbol{P}_{(mn_x,mn_x)}E\{\hat{x}_{n_x}^{2m}(0)\} \\ &+ \boldsymbol{P}_{(i_1j_1,i_2j_2)}E\{\hat{x}_{i_1}^{j_1}(0)\}E\{\hat{x}_{i_2}^{j_2}(0)\}|_{i_1\neq i_2} + \cdots \end{aligned} \tag{10.32}$$

where $E\{x^{2i}\} = \frac{1}{\sqrt{2\pi}}\Gamma\left(i+\frac{1}{2}\right)2^{\left(i+\frac{1}{2}\right)}$ and $\Gamma\left(i+\frac{1}{2}\right)$ is defined by [5]:

$$\Gamma(x) = \int_0^\infty e^{-t}t^{(x-1)}dt. \tag{10.33}$$

The function $\Gamma(x)$ can be easily computed as a numerical value of high accuracy. Often $V(\cdot)$ is quadratic $V = \hat{\boldsymbol{x}}^T\boldsymbol{P}\hat{\boldsymbol{x}}$ for some positive definite $\boldsymbol{P}$ and

$$E\{J\} = tr\{\boldsymbol{P}\},\ \text{for } E\{x_i^2(0)\} = 1. \tag{10.34}$$

Ultimately, these results for the control cost (10.28) and (10.34) allow for cost computation for a given controller and a given communication sequence, assessing performance in a sum-of-squares framework within the region of attraction $\mathbb{D}$. For this, it must be emphasized that the Taylor series expansion in the state variable $\boldsymbol{x}$ for certain nonlinearities is only valid in a local sense, *i.e.* $\hat{\boldsymbol{x}}(pl) \in \mathbb{D}$.

Remark 10.1 *The models of (10.17) and (10.25) are now a representation of the nonlinear system together with the model for time-triggered communication. The performance of the closed loop system can be evaluated via (10.27). Clearly, this is only possible in an accurate sense for $N \to \infty$ (10.17). Since finite values of $N \in \mathbb{N}$ are usually used, practical guidelines derived from nonlinear sampled-data theory [90, 148] are as follows:*

i Choose $N > 0$ to be at least of the relative degree r of the nonlinear affine system (10.3) considering the state vector $\boldsymbol{x}$ as the output of the system; i.e. the system order of (10.3) is a sufficient upper bound for N.

ii Choose an asymptotically stabilizing non-linear control law and the weight matrices $\boldsymbol{Q}_{\text{c}1}$ and $\boldsymbol{Q}_{\text{c}2}$ (10.18) so that the right hand of (10.29) is negative definite at least in a local sense for $\mathbb{D}$.

Assuming that the Lyapunov function exists for (10.29), item (ii) implies that the nonlinear system (10.25) is asymptotically stable. Moreover, item (i) guarantees using [148, Theorem 1] (see also [71, Section II]) that the local single-step truncation error derived from (10.3) in relation to (10.17) is of order h^{r+1}, which is equivalent to local one-step consistency [90, Definition 1]. This implies that the controlled continuous-time plant [90, Theorem 2] is at least ultimately bounded (practically) stable in a local sense for sufficiently small h, where the ultimate bound decreases to 0 for $h \to 0$ (see also the overview of [69]).

Performance guarantees for (10.3) follow from (10.28-10.30) and the fact that the single-step truncation error is of order h^{r+1}. A later example will provide further specific evidence that the choice $N = r$ is indeed performance (and not only stability) oriented. •

10.6 An SOS-framework for local cost computation

The aim of the SOS-analysis is to compute the cost (10.27) of the integrated control system for a guaranteed region of attraction. This is also inspired by recent work in [121]. This SOS-analysis is structured into three steps:

Step 1: At first the existence of a region of attraction $\mathbb{D}$ for the closed-loop nonlinear NCS has to be established. For this, a convex region in $\hat{\boldsymbol{x}}(pl) \in \mathbb{A}_\beta$ defined using a suitably chosen polynomial $\omega(\hat{\boldsymbol{x}}(pl))$ is given via the set

$$\mathbb{A}_\beta := \left\{ \hat{\boldsymbol{x}} \middle| \beta - \omega(\hat{\boldsymbol{x}}) \geq 0 \text{ and } \omega(\hat{\boldsymbol{x}}) \geq 0 \right\}, \tag{10.35}$$

where $\beta \geq 0$ has to be found in a first optimization step for a (positive definite) polynomial $\omega(\cdot) \geq 0$. The existence of a convex set $\mathbb{A}_\beta$ will establish the existence of a region of attraction $\mathbb{D} \subset \mathbb{A}_\beta$. The constraint, $\beta - \omega(\hat{\boldsymbol{x}}) \geq 0$

from (10.35), is introduced into the optimization of (10.28)-(10.33) by using the S-procedure

$$-(V(\hat{\boldsymbol{x}}(p(l+1))) - V(\hat{\boldsymbol{x}}(pl))) - s(\hat{\boldsymbol{x}}(pl))(\beta - \omega(\hat{\boldsymbol{x}}(pl))) \\ - \begin{bmatrix} \hat{\boldsymbol{x}}(pl) \\ \bar{\boldsymbol{U}}_1(\hat{\boldsymbol{x}}(pl)) \end{bmatrix}^T \begin{bmatrix} \tilde{\boldsymbol{Q}}_1 & \tilde{\boldsymbol{Q}}_{12} \\ \tilde{\boldsymbol{Q}}_{12}^T & \tilde{\boldsymbol{Q}}_2 \end{bmatrix} \begin{bmatrix} \hat{\boldsymbol{x}}(pl) \\ \bar{\boldsymbol{U}}_1(\hat{\boldsymbol{x}}(pl)) \end{bmatrix} \geq 0, \tag{10.36}$$

where $s(\hat{\boldsymbol{x}}(pl))$ is an SOS variable and $V(\hat{\boldsymbol{x}}(pl))$ is a polynomial variable. In particular, $s(\hat{\boldsymbol{x}}(pl))$ and $V(\hat{\boldsymbol{x}}(pl))$ are positive semi-definite and positive definite polynomials respectively satisfying:

$$s(\hat{\boldsymbol{x}}(pl)) \geq 0, \ V(\hat{\boldsymbol{x}}(pl)) - \gamma\, \omega(\hat{\boldsymbol{x}}(pl)) \geq 0, \tag{10.37}$$

while $\beta \geq 0$ and $\gamma \geq 0$ are scalars. Note that β cannot be an SOS variable. The scalar γ is introduced to guarantee that $V(\cdot)$ is positive definite in $\hat{\boldsymbol{x}}$. Thus, the first step is to find the maximal value $\beta = \beta_{max} \geq 0$ subject to the SOS optimized polynomials $s(\hat{\boldsymbol{x}}(pl))$ and $V(\hat{\boldsymbol{x}}(pl))$. This can be achieved via a bisection algorithm. Note that the dynamics of $\hat{\boldsymbol{x}}(p(l+1))$ in equation (10.25) are used to create an SOS-optimization problem in the state variable $\hat{\boldsymbol{x}}(pl)$ at time step pl. Therefore, once equations (10.36) and (10.37) are satisfied, there would be a region of attraction $\mathbb{D}$, ($\mathbb{D} \subset \mathbb{A}_\beta$), for $\hat{\boldsymbol{x}}(pl)$ and the control cost of the overall NCS will be given by (10.28) or (10.30).

Step 2: Having established the existence of a region of attraction $\mathbb{D}$, this region of attraction $\mathbb{D}$ can be more explicitly determined. This is obtained by an additional constraint on the Lyapunov function:

$$\gamma\, \omega(\hat{\boldsymbol{x}}(pl)) \leq V(\hat{\boldsymbol{x}}(pl)) \leq \gamma\, c\, \omega(\hat{\boldsymbol{x}}(pl)), \tag{10.38}$$

where $c \geq 1$ is some well chosen positive constant based on the knowledge on $V(\cdot)$ in Step 1 and the singular values of the matrix $\boldsymbol{P}$ (10.31). Clearly, this guarantees that a definite region of attraction in $\hat{\boldsymbol{x}}(pl)$ for the discrete optimal problem exists which is defined by:

$$\mathbb{A}_V := \left\{ \hat{\boldsymbol{x}}(pl) \middle| V(\hat{\boldsymbol{x}}(pl)) \leq \gamma\, \beta \right\}, \tag{10.39}$$

since this implies $\omega(\hat{\boldsymbol{x}}(pl)) \leq \beta$ for $\hat{\boldsymbol{x}}(pl) \in \mathbb{A}_V$. Thus, the set

$$\mathbb{A}_{\beta/c} := \left\{ \hat{\boldsymbol{x}}(pl) \middle| \omega(\hat{\boldsymbol{x}}(pl)) \leq \frac{\beta}{c} \right\} \tag{10.40}$$

is also a region of attraction, $\mathbb{A}_{\beta/c} \subset \mathbb{A}_V \subset \mathbb{D} \subset \mathbb{A}_\beta$. Considering a fixed scalar β_{fix} within the interval $\begin{bmatrix} 0 & \beta_{max} \end{bmatrix}$, it is therefore possible to investigate the following set of SOS-inequalities

$$\begin{aligned} -(V(\hat{\boldsymbol{x}}(p(l+1))) - V(\hat{\boldsymbol{x}}(pl))) - s(\hat{\boldsymbol{x}}(pl))(\beta_{fix} - \omega(\hat{\boldsymbol{x}}(pl))) \\ - \begin{bmatrix} \hat{\boldsymbol{x}}(pl) \\ \bar{\boldsymbol{U}}_1(\hat{\boldsymbol{x}}(pl)) \end{bmatrix}^T \begin{bmatrix} \tilde{\boldsymbol{Q}}_1 & \tilde{\boldsymbol{Q}}_{12} \\ \tilde{\boldsymbol{Q}}_{12}^T & \tilde{\boldsymbol{Q}}_2 \end{bmatrix} \begin{bmatrix} \hat{\boldsymbol{x}}(pl) \\ \bar{\boldsymbol{U}}_1(\hat{\boldsymbol{x}}(pl)) \end{bmatrix} \geq 0, \\ V(\hat{\boldsymbol{x}}(pl)) - \gamma\, \omega(\hat{\boldsymbol{x}}(pl)) \geq 0, \ \gamma\, c\, \omega(\hat{\boldsymbol{x}}(pl)) - V(\hat{\boldsymbol{x}}(pl)) \geq 0, \\ s(\hat{\boldsymbol{x}}(pl)) \geq 0, \ V(\hat{\boldsymbol{x}}(pl)) \geq 0, \ \gamma \geq 0, \end{aligned} \tag{10.41}$$

to find the smallest scalar $c \geq 1$ for the variables $\gamma \geq 0$, $s(\hat{\boldsymbol{x}}(pl))$ and $V(\hat{\boldsymbol{x}}(pl))$. The smallest value $c \geq 1$ is easily obtained via a bisection algorithm for fixed β_{fix}. This characterizes a well-defined region of attraction defined by the polynomial $\omega(\hat{\boldsymbol{x}}(pl)) \geq 0$.

Step 3: For a given region of attraction $\mathbb{A}_{\beta_{fix}/c} \subset \mathbb{D}$

$$\mathbb{A}_{\beta_{fix}/c} := \left\{ \hat{\boldsymbol{x}}(pl) \middle| \omega(\hat{\boldsymbol{x}}(pl)) \leq \frac{\beta_{fix}}{c} \right\}, \tag{10.42}$$

the control cost of the control integration problem is to be found in this step, given a fixed communication sequence and constant β_{fix}. This can be obtained by considering the SOS-inequalities of (10.41) with the added optimization constraint:

$$E\{J\} \leq \min_{\gamma,V,s} E\{V(\hat{\boldsymbol{x}}(0))\}. \tag{10.43}$$

Note that in case the SOS-Lyapunov function $V(\hat{\boldsymbol{x}}(0))$ is quadratic, the optimization simplifies to $E\{J\} = \min_{\gamma,V,s} tr\{\boldsymbol{P}\}$ and the choice

$$\omega(\hat{\boldsymbol{x}}(pl)) = \hat{\boldsymbol{x}}(pl)^T \hat{\boldsymbol{x}}(pl)$$

as a quadratic polynomial is required.

Alternatively, the deterministic cost would be

$$J \leq \min_{\gamma,V,s} V(\hat{\boldsymbol{x}}(0)) \tag{10.44}$$

for a given initial state $\hat{\boldsymbol{x}}(0)$ and a fixed schedule σ. This cost computed for a fixed schedule can be now used to conduct an exhaustive search for the best performing schedule. However, it must be noted that the framework allows for faster stochastic algorithms as discussed in previous chapters.

Remark 10.2 *Scalability of the suggested NCS integration analysis approach to large order systems and long sequence lengths p appears to be limited by the need for an exhaustive search ($\sim {n_u}^p$ cost computations) and the well-known curse of dimensionality of large scale SOS-computations (e.g. [4]). Large scale SOS-problems require the exploitation of the SOS-problem structure [59, 77],* i.e. *sparsity [77] or splitting into subproblems [59] (see also later example).*

For the number of cost-computations, the equivalent linear NCS problem formulation of Chapters 6 and 7 provides evidence that an exhaustive search ($\sim {n_u}^p$ cost computations) is practically not necessary. •

10.7 Example

In order to verify the scheduling approach via this cost computation method, a particular NCS is considered, which consists of two separate inverted pendulums. The continuous-time dynamics of the system can be described by the following equation, with $\boldsymbol{x} := \begin{bmatrix}\phi_1 & \dot{\phi}_1 & \phi_2 & \dot{\phi}_2\end{bmatrix}^T$:

$$\frac{d\boldsymbol{x}}{dt} = \begin{bmatrix} \dot{\phi}_1 \\ g\sin\phi_1/l_1 \\ \dot{\phi}_2 \\ g\sin\phi_2/l_2 \end{bmatrix} + \begin{bmatrix} 0 & 0 \\ 1/M_1 l_1 & 0 \\ 0 & 0 \\ 0 & 1/M_2 l_2 \end{bmatrix} \begin{bmatrix} u_1 \\ u_2 \end{bmatrix}. \tag{10.45}$$

The two pendulums have been chosen to be different in their parameters. For simplicity, the masses are $M_1 = 2/g$ (where $g = 9.81$), $M_2 = 1/g$ while the length of both poles is the same. For simplicity, $l_1 = l_2 = g$ is chosen. In the analysis, the angular range of $[-\pi/2, \pi/2]$ is considered for both pendulums. Note the similarity of the two pendulum systems, *i.e.* the system drift is identical while the input gain differs by the constant gain of 2.

There are four state variables: the angle between each pendulum's pole and the vertical axis and the angle velocity for each pendulum. Hence, there are four sensors which are directly fed into the controller. The controller computes the control signals u_1 and u_2 for which the most up-to-date information is sent through the communication bus when scheduled, while a zero-order hold operation is applied once there is no communication packet sent.

An approximation of the sin-function using the Taylor expansion is required:

$$\sin\phi = \phi - \frac{\phi^3}{6} + \frac{\phi^5}{120} - \cdots + \frac{(-1)^{(n-1)}\phi^{2n-1}}{(2n-1)!} + \cdots. \tag{10.46}$$

Considering the angular range of interest, the sin-function is approximated to the order of 5. Using the time discretization approach detailed in Section 10.3.1 (see also [62]), the following second order approximation ($N = 2$) of each pendulum is obtained:

$$\begin{aligned} \boldsymbol{x}_i(k+1) &= \boldsymbol{x}_i(k) + T\dot{\boldsymbol{x}}_i(t)\Big|_{t=t_k} + \frac{T^2}{2}\ddot{\boldsymbol{x}}_i(t)\Big|_{t=t_k} \\ &= \begin{bmatrix} \phi_i + T\dot{\phi}_i + \frac{T^2 g}{2l_i}\Delta_{i1} \\ \dot{\phi}_i + \frac{Tg}{l_i}\Delta_{i1} + \frac{T^2 g}{2l_i}\dot{\phi}_i\Delta_{i2} \end{bmatrix} + \begin{bmatrix} \frac{T^2}{2m_i l_i} \\ \frac{T}{m_i l_i} \end{bmatrix} u_i, \end{aligned} \tag{10.47}$$

where

$$\Delta_{i1} = \phi_i - \frac{\phi_i^3}{6} + \frac{\phi_i^5}{120}, \Delta_{i2} = 1 - \frac{\phi_i^2}{2} + \frac{\phi_i^4}{24}, i = 1, 2.$$

This choice of N is also sufficient with respect to Remark 10.1, considering that each of the pendulums has a relative degree of two. Note that even the choice $N = 1$, *i.e.* the Euler approximation, is sufficient to design and prove stability for this example, as it has been theoretically and numerically validated in [69, Section 3.8.2] for the inverted pendulum. In contrast, in the case here, performance, *i.e.* the assurance of a single step truncation error of h^{r+1} for $r = N = 2$, is essential (Remark 10.1).

In this control system, the basic sampling time of $T = 0.3s$ is chosen. For this fixed sampling time, separate controllers are designed for each pendulum. Here, a discrete feedback-linearizing controller with an outer loop PD controller is used:

$$u_i = -\frac{2m_i l_i}{T^2}\left(\phi_i + T\dot{\phi}_i + \frac{T^2 g}{2l_i}\Delta_{i1}\right) + k_{di}\dot{\phi}_i + k_{pi}\phi_i. \tag{10.48}$$

This controller choice guarantees stability (according to [69] and performance Remark 10.1) of the continuous-time system. In fact, the controller stabilizes the discrete models for $N = 1$, $N = 2$ and $N = 3$, as numerically validated in Fig. 10.1. It is evident that performance for the continuous-time model and the discrete models $N = 2$ and $N = 3$ are almost identical. (Note an identical sinusoidal disturbance has been introduced for all simulations for better visualization in Fig. 10.1.)

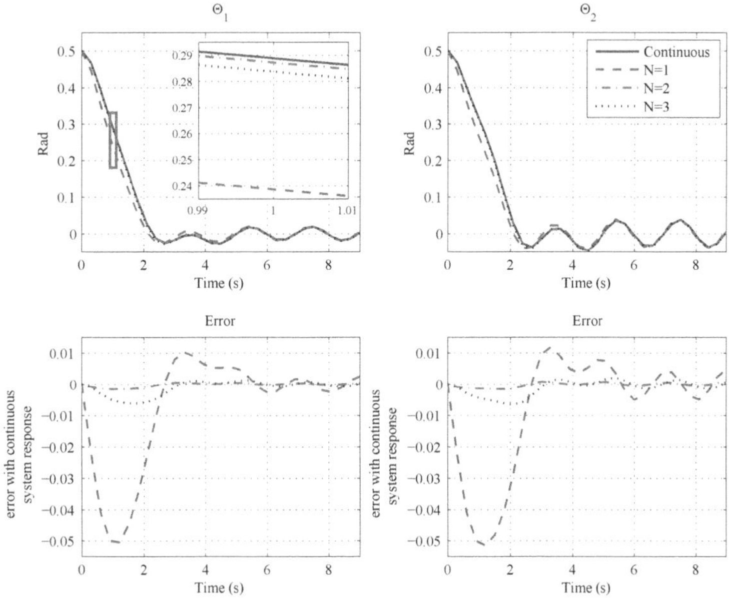

FIGURE 10.1: Continuous system and discretized system response for different N; controlled by (10.48) with sinusoidal disturbance.

For assessment of the scheduling approach, 6 different sequences σ (Table 10.1) of period $p = 3$ are applied to the system to compare the cost for achieving a desired performance, which are divided into two groups, $\sigma_1, \sigma_2, \sigma_3$ and $\sigma_4, \sigma_5, \sigma_6$. Each group focuses more on one of the pendulums, with different control communication order. Here, the two computation methods (deterministic and stochastic) for the cost, stated in (10.28) and (10.30) respectively, are of interest. Assuming the matrices of the discrete system and the expres-

TABLE 10.1: Cost (10.27) and condition number $\kappa(\boldsymbol{P})$ of Lyapunov matrix (10.31) for a fixed value c (10.40) for all six sequences obtained by two different methods: (10.28) and (10.30)

$No.$	$Sequence\ \sigma$ for $actuator$ $schedule$	c	$Cost = \min\limits_{\gamma,V,s} tr(\boldsymbol{P})$		$Cost = \min\limits_{\gamma,V,s} \hat{\boldsymbol{x}}(0)^T \boldsymbol{P} \hat{\boldsymbol{x}}(0)$		
			$\kappa(\boldsymbol{P})$	$Cost$	$\kappa(\boldsymbol{P})$	$Cost$	
						$\hat{\boldsymbol{x}}_1(0)$	$\hat{\boldsymbol{x}}_2(0)$
σ_1	{1,1,2}	2512	815.3	16.38	741.1	0.74	0.07
σ_2	{1,2,1}	305	120.0	3.19	162.5	0.26	0.02
σ_3	{2,1,1}	305	173.9	4.43	242.8	0.29	0.03
σ_4	{2,2,1}	2512	2511.1	46.77	2507.7	23.59	2.12
σ_5	{2,1,2}	305	301.9	5.86	212.3	0.47	0.04
σ_6	{1,2,2}	305	151.4	4.6	177.1	0.30	0.03

sions for the cost have been obtained, the SOS computation requires about 5 minutes for the 6 sequences in an adhoc approach.[1]

For the continuous-time two-pendulum plant with discrete control input, the following matrices of the continuous-time cost (10.20) are used:

$$\boldsymbol{Q}_{c1} = diag(0.05, 0.05, 0.05, 0.05), \boldsymbol{Q}_{c2} = diag(0.05, 0.05)$$

assuming $n_u = 2$, and only one actuator signal is scheduled at every time step. The computation of the cost function in (10.27) and the matrix $\tilde{\boldsymbol{Q}}$ requires some care. This is best done using the algebraic computational toolbox of MATLAB®, Maple®, or Mathematica®, to be used for the SOS-approach. Note that $\tilde{\boldsymbol{Q}}$ in itself is a nonlinear function of $\hat{\boldsymbol{x}}(pl)$. Thus, in this example, it has been computed as a 6th order polynomial, deleting all higher order polynomials. Similarly, the vector $\left[\hat{\boldsymbol{x}}(pl)^T \quad \bar{\boldsymbol{U}}_1(\hat{\boldsymbol{x}}(pl))^T\right]^T$ is highly nonlinear in $\hat{\boldsymbol{x}}(pl)$. The polynomial order of this vector has been limited to 3 here, which has been found sufficient in accuracy. Moreover, it is interesting to note that for the scalar function

$$\begin{bmatrix} \hat{\boldsymbol{x}}(pl) \\ \bar{\boldsymbol{U}}_1(\hat{\boldsymbol{x}}(pl)) \end{bmatrix}^T \begin{bmatrix} \tilde{\boldsymbol{Q}}_1 & \tilde{\boldsymbol{Q}}_{12} \\ \tilde{\boldsymbol{Q}}_{12}^T & \tilde{\boldsymbol{Q}}_2 \end{bmatrix} \begin{bmatrix} \hat{\boldsymbol{x}}(pl) \\ \bar{\boldsymbol{U}}_1(\hat{\boldsymbol{x}}(pl)) \end{bmatrix} \tag{10.49}$$

in $\hat{\boldsymbol{x}}(pl)$ an overall order of 6 is obtained, although an order of 12 was expected. This (naturally occurring) reduction approach is in fact necessary to obtain reliable computational results using SOS.

[1]This can be further reduced by exploiting the sparsity in the Lyapunov matrix $\boldsymbol{P}$ (10.31) and the SOS-polynomial $s(\cdot)$ (10.41). The matrix $\boldsymbol{P}$ is for each of the six sequences block-diagonal, while the $s(\cdot)$-polynomials have a widely recurring sparse structure. These facts reduce the number of variables used to less than 50%, also cutting down the computation time for all six sequences to 2.5 minutes.

The steps above now allow for efficient SOS-computation. The method discussed in Section 10.6 is used to obtain the Lyapunov function, as well as the cost of the closed-loop system with control communication. A quadratic Lyapunov function and $\omega(\hat{\boldsymbol{x}}(pl)) = \hat{\boldsymbol{x}}(pl)^T\hat{\boldsymbol{x}}(pl)$ is employed. In the first step of the SOS-analysis,

$$\beta_{max} = \{21.72, 44.40, 34.66, 21.89, 48.28, 34.73\}$$

is obtained for sequences $\sigma_1, \sigma_2, .., \sigma_6$. This permits the choice of $\beta_{fix} = 15$ for all six cases in Steps 2 and 3. However, it has not been possible to choose a consistent value for c (10.40) for all sequences. For sequences σ_1 and σ_4, it has been necessary to choose $c = 2512$ while for all other sequences, $\sigma_2, \sigma_3, \sigma_5, \sigma_6$, the value $c = 305$ was sufficient (Table 1). This is in particular due to the condition number $\kappa(\boldsymbol{P})$ of the Lyapunov matrix $\boldsymbol{P}$ (10.31) which was obtained when computing the control cost (10.27). The condition number $\kappa(\boldsymbol{P})$ should be smaller than c. Thus, a ball-shaped region of attraction with a radius of at least 0.222 is guaranteed for the controllers using $\sigma_2, \sigma_3, \sigma_5, \sigma_6$ while a significantly smaller region of attraction is guaranteed for σ_1 and σ_4. This shows that sequences σ_1 and σ_4 are not easily analysed to guarantee stable control performance. Since (10.28) and (10.30) are used as the optimization criterion, the objective functions used to compute the Lyapunov function are not the same under these two methods. Therefore, two different Lyapunov functions are obtained for one specific sequence respectively, and the Lyapunov matrix $\boldsymbol{P}$ (10.31) is different for the two methods. Note that the objective function of the method in (10.30) depends on the initial value of the state variable, the cost computed using this method varies with different initial values. The last two columns of the table show the costs based on the specific initial states $\hat{\boldsymbol{x}}_1(0) = \begin{bmatrix} 0.5 & 0 & 0.5 & 0 & 0 & 0 \end{bmatrix}^T$ and $\hat{\boldsymbol{x}}_2(0) = \begin{bmatrix} 0.15 & 0 & 0.15 & 0 & 0 & 0 \end{bmatrix}^T$. They verify what has been achieved for the stochastic approach. Note that the second set of initial values is indeed within the computed region of attraction with radius of 0.222.

The reason for the greater instability for σ_1 and σ_4 is that sequences σ_1 and σ_4 allow control for pendulum 2 and 1, respectively, only very late, which causes the pendulums to turn almost unstable. These sequences also have the highest control cost (Table 1).

Fig. 10.2 and Fig. 10.3 show the simulation results for the initial value $\hat{\boldsymbol{x}}_1(0)$ of the control performance of the controller with different communication sequences for the two pendulums. The transient response of the velocity and the position of pendulum 1 for sequences σ_1, σ_2 and σ_3 is equal to the transient response of the states of pendulum 2 for sequences σ_4, σ_5 and σ_6, while the actuator effort differs by a factor of 2. Thus, the overall control performance costs for communication sequence σ_4, σ_5 and σ_6 for this particular trajectory has to be higher than for the matching sequences σ_1, σ_2 and σ_3. Hence, simple adhoc analysis shows that sequences σ_2 and σ_3 are the best performing sequences. This is also confirmed via the SOS-analysis showing that the controller using sequence σ_2 indeed delivers the best performance.

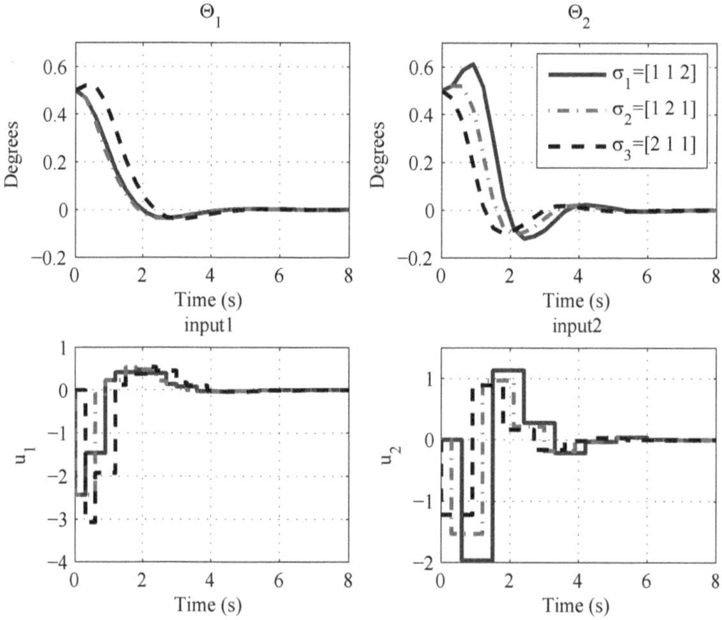

FIGURE 10.2: Control performance of control input under the first three specified communication sequences

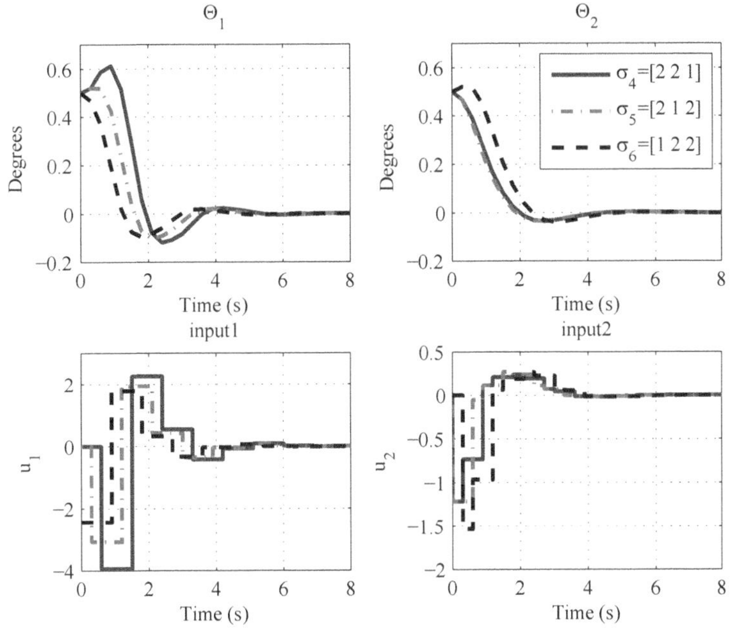

FIGURE 10.3: Control performance of control input under the last three specified communication sequences

Clearly, this appears to be a trivial result for this simple problem, but which is not easily understood for large scale systems.

10.8 Summary

This final chapter has presented a nonlinear NCS design for control optimal communication between the nodes. The system considered is nonlinear and periodically time-varying. Nonlinearities were approximated using a Taylor series which led to a sum-of-squares approach. A numerical example has shown the effectiveness and feasibility of the approach. It is evident that the computational framework easily extends to an $\mathcal{L}_2$-gain framework. Considering small gain theory (and if needed suitable Lyapunov-Krasovskii functions) design for robust stability also subject to delays is possible. The main issue is not to theoretically provide feasibility of the computational framework but to retain the computational complexity of SOS low. This can be achieved by exploiting the structure of the SOS problem, *e.g.* sparsity.

Bibliography

[1] K. Åström and B. Wittenmark. *Computer-Controlled Systems*. Prentice-Hall, 1997.

[2] J. Ahrens, X. Tan, and H. Khalil. Multirate sampled-data output feedback control with application to smart material actuated systems. *IEEE Trans. on Automatic Control*, 54(11):2518–2529, 2009.

[3] J. Anderson and A. Papachristodoulou. Input-Output Systems Analysis Using Sum of Squares Programming: A Decomposition Approach. In *MTNS, 2010*, pages 107–113.

[4] J. Anderson and A. Papachristodoulou. A network decomposition approach for efficient sum of squares programming based analysis. In *American Control Conference (ACC), 2010*, pages 4492 –4497, 2010.

[5] E Artin. *The Gamma Function.* Holt, Rinehart and Winston, 1964.

[6] K. Ärzén, B. Bernhardsson, J. Eker, A. Cervin, P. Persson, K. Nilsson, and L. Sha. Integrated control and scheduling. Technical report, Lund Institute of Technology, 1999.

[7] A. Astolfi, D. Nešić, and A. Teel. Trends in nonlinear control. In *IEEE Conference on Decision and Control, 2008*, pages 1870–1882, 2008.

[8] C. Aubrun, D. Simon, and Y.-Q. Song, editors. *Co-design Approaches for Dependable Networked Control Systems.* ISTE and Wiley, 2010.

[9] D. Ayavoo, M. Pont, M. Short, and S. Parker. Two novel shared-clock scheduling algorithms for use with 'controller area network' and related protocols. *Microprocessors and Microsystems*, 31(5):326–334, 2007.

[10] E. Bakker, H. Pacejka, and L. Lidner. A new tyre model with application in vehicle dynamic studies. *SAE Paper No 890087*, pages 101–113, 1989.

[11] B. Bamieh and J. Pearson. A general framework for linear periodic systems with applications to H_∞ sampled-data control. *IEEE Tran. on Automatic Control*, 37(4):418–435, 1992.

[12] N. Bauer, M. Donkers, W. Heemels, and N. van de Wouw. An approach to observer-based decentralized control under periodic protocols. In *American Control Conference*, number WeC16.3, 2010.

[13] N. Bauer, P.. Maas, and booktitle=IEEE Conference on Decision and Control, 2010 year=2010 Heemels, W. Stability Analysis of Networked Control Systems: A Sum of Squares Approach.

[14] M. Ben Gaid, A. Çela, and Y. Hamam. Optimal integrated control and scheduling of networked control systems with communication constraints: application to a car suspension system. *IEEE Trans. on Control Systems Technology*, 14(4):776–787, 2006.

[15] N. Boonnithivorakul and F. Pourboghrat. Optimal control design for polynomial nonlinear systems using sum of squares technique with guaranteed local optimality. In *1st virtual control conference, 2010*, 2010.

[16] R. Borges, R. Oliveira, C. Abdallah, and P. Peres. Robust $\mathcal{H}_\infty$ networked control for systems with uncertain sampling rates. *IET Control Theory & Applications*, 4(1):50–60, 2010.

[17] M. Branicky, S. Phillips, and W. Zhang. Stability of networked control systems: explicit analysis on delays. In *Proc. of the American Control Conference*, volume 4, pages 2352–2357, 2000.

[18] R. Brockett. Minimum attention control. In *Proc. of the* 36th *IEEE Conf. on Decision and Control*, volume 3, pages 2628–2632, 1997.

[19] D. Carnevale, A. Teel, and D. Nešić. A Lyapunov proof of an improved maximum allowable transfer interval for networked control systems. *IEEE Trans. on Automatic Control*, 52(5):892–897, 2007.

[20] T. Chen and B. Francis. *Optimal Sampled-data Control Systems*. Springer, London, 1995.

[21] T. Chen and L. Qiu. $\mathcal{H}_\infty$ design of general multirate sampled-data control systems. *Automatica*, 30(7):1139–1152, 1992.

[22] M. Chow and T. Yodyium. Network-based control systems: a tutorial. In *IECON The* 27th *Annual Conf. of the IEEE*, pages 1593–1602, 2001.

[23] M. Cloosterman, L. Hetel, N. van de Wouw, W. Heemels, J. Daafouz, and Nijmeijer H. Controller synthesis for networked control systems. To appear in Automatica (in press), 2010.

[24] M. Cloosterman, N. van de Wouw, W. Heemels, and H. Nijmeijer. Stability of networked control systems with large delays. In *Proc. of the* 46th *IEEE Conf. on Decision and Control*, pages 5017–5022, 2007.

[25] M. Cloosterman, N. van de Wouw, W. Heemels, and H. Nijmeijer. Stability of networked control systems with uncertain time-varying delays. *IEEE Trans. on Automatic Control*, 54(7):1575–1580, 2009.

[26] P. Colaneri, R. Scattolini, and N. Schiavoni. LQG optimal control of multirate sampled-data systems. *IEEE Trans. on Automatic Control*, 37(5):675–682, 1992.

[27] H. Dan and N. Sing Kiong. *Robust Control for Uncertain Networked Control Systems with Random Delays.* Springer, 2009.

[28] D. Dačić and D. Nešić. Observer design for linear networked control systems using matrix inequalities. In *Proc. of the* 46th *IEEE Conf. on Decision and Control*, pages 3315–3320, 2007.

[29] D. Dačić and D. Nešić. Quadratic stabilization of linear networked control systems via simultaneous protocol and controller design. *Automatica*, 43(7):1145–1155, 2007.

[30] K. Deb and M. Goyal. Optimizing engineering designs using a combined genetic search. *7th International Conf. on Genetic Algorithms*, pages 512–528, 1997.

[31] R. Diestel. *Graph theory.* Springer, 3rd edition, 2005.

[32] M. Donkers, W. Heemels, D. Bernardini, A. Bemporad, and V. Shneer. Stability analysis of stochastic networked control systems. In *American Control Conference*, number ThB16.3, 2010.

[33] M. Donkers, L. Hetel, W. Heemels, N. van de Wouw, and M. Steinbuch. Stability analysis of networked control systems using a switched linear systems approach. In *Proc. of the* 12th *Int. Conf. on Hybrid Systems: Computation and Control*, pages 150–164, 2009.

[34] A. Eqtami, D. Dimarogonas, and K. Kyriakopoulos. Event-triggered control for discrete-time systems. In *American Control Conference*, number FrA01.3, 2010.

[35] B. Friendland. *Control system design: an introduction to state-space methods.* McGraw-Hill, 1986.

[36] H. Gao, T. Chen, and J. Lam. A new delay system approach to network-based control. *Automatica*, 44(1):39–52, 2008.

[37] M. Garey and D. Johnson. *Computers and intractability: a guide to the theory of NP-completeness.* W. H. Freeman, 1979.

[38] T. Gillespie. *Fundamentals of Vehicle Dynamics.* SAE International, 1 edition, 1992.

[39] F. Göktas, J. Smith, and R. Bajcsy. Telerobotics over communication networks. In *Proc. of the IEEE Conf. on Decision and Control*, volume 3, pages 2399–2404, 1997.

[40] G. Goodwin, D. Quevedo, and E. Silva. Architectures and coder design for networked control systems. *Automatica*, 44(1):248–257, 2008.

[41] D. Görges, M. Izák, and S Liu. Optimal control of systems with resource constraints. In *Proc. of the 46th IEEE Conf. on Decision and Control*, pages 1070–1075, 2007.

[42] L. Grover. Local search and the local structure of NP-complete problems. *Operations Research Letters*, 12:235–243, 1992.

[43] D. Guan. Generalized gray codes with applications. *Proc. National Science Council ROC(A)*, 22(6):841–848, 1998.

[44] G. Gwaltney and J. Briscoe. Comparison of communication architectures for spacecraft modular avionics systems. Technical report, NASA/TM-2006-214431, 2006.

[45] C. Halse. *Nonlinear dynamics of the automotive driveline.* PhD thesis, University of Bristol, 2004.

[46] S. He, E. Prempain, and Q. Wu. An improved particle swarm optimizer for mechanical design optimization problems. *Engineering Optimization*, 36:584–605, 2004.

[47] W. Heemels, A. Teel, N. van de Wouw, and D. Nešić. Networked control systems with communication constraints: tradeoffs between transmission intervals, delays and performance. *IEEE Trans. on Automatic Control*, 55(8):1781–1796, 2010.

[48] J. Hespanha, P. Naghshtabrizi, and Y. Xu. A survey of recent results in networked control systems. In *Proc. of the IEEE*, volume 95, pages 138–162, 2007.

[49] P. Hokayem and C. Abdallah. Inherent issues in networked control systems: a survey. In *Proc. of the American Control conference*, volume 6, pages 4897–4902, 2004.

[50] D. Hristu-Varsakelis. Stabilization of networked control systems with access constraints and delays. In *Proc. of the* 45th *IEEE Conf. on Decision and Control*, pages 1123–1128, 2006.

[51] D. Hristu-Varsakelis. On the period of communication policies for networked control systems, and the question of zero-order holding. In *Proc. of the* 46th *IEEE Conf. on Decision and Control*, pages 38–43, 2007.

[52] D. Hristu-Varsakelis. Short-period communication and the role of zero-order holding in networked control systems. *IEEE Trans. on Automatic Control*, 53(15):1285–1290, 2008.

[53] D. Hristu-Varsakelis and K. Morgansen. Limited communication control. *Systems and Control Letters*, 37(4):193–205, 1999.

[54] D. Hristu-Varsakelis and L. Zhang. LQG control of networked control systems with access constraints and delays. *Int. J. of Control*, 81(8):1266–1280, 2008.

[55] O. Imer, S. Yüksel, and T. Başar. Optimal control of LTI systems over unreliable communication links. *Automatica*, 42(9):1429–1439, 2006.

[56] C. Ionete and A. Çela. Structural properties and stabilization of NCS with medium access constraints. In *Proc. of the* 45th *IEEE Conf. on Decision and Control*, pages 1141–1146, 2006.

[57] C. Ionete, A. Çela, M. Ben Gaid, and A. Reama. Controllability and observability of linear discrete-time systems with network induced variable delay. In *Proc. of the* 17th *World Congress The Int. Federation of Automatic Control*, pages 4216–4221, 2008.

[58] H. Ishii. H_∞ control with limited communication and message losses. *Systems & Control Letters*, (57):322–331, 2007.

[59] T. Jennawasin and Y. Oishi. A region-dividing technique for constructing the sum-of-squares approximations to robust semidefinite programs. *IEEE Transactions on Automatic Control*, 54(5):1029 –1035, 2009.

[60] T. Jia, Y. Niu, and X. Wang. $\mathcal{H}_\infty$ control for networked systems with data packet dropout. *Int. J. of Control, Automation and Systems*, 8(2):198–203, 2010.

[61] T. Kailath. *Linear Systems*. Prentice-Hall, 1980.

[62] N. Kazantzis and C. Kravaris. Time-discretization of nonlinear control systems via Taylor methods. *Computers & chemical engineering*, 23(6):763–784, 1999.

[63] H. Kellerer, U. Pferschy, and D. Pisinger. *Knapsack problems*. Springer, 2004.

[64] J. Kennedy and R. Eberhart. Particle swarm optimization. *IEEE International Conf. on Neural Networks*, 4:1942–1948, 1995.

[65] U. Kiencke and L. Nielsen. *Automotive Control Systems*. Springer, 2000.

[66] D. Kim, D. Choi, and P. Mohapatra. Real-time scheduling method for networked discrete control systems. *Control Engineering Practice*, 17(5):564–570, 2009.

[67] M. Kimura. Preservation of stabilizability of a continuous time-invariant linear system after discretization. *Int. J. Systems Sci.*, 21(1):65–92, 1990.

[68] H. Kopetz. Event-triggered versus time-triggered real time system. Technical Report 8/91, Inst. für Technische Informatik, Technische Universität Wien, 1991.

[69] D. Laila, D. Nešić, and A. Astolfi. Sampled-data control of nonlinear systems. In Antonio Loria, Francoise Lamnabhi-Lagarrigue, and Elena Panteley, editors, *Advanced Topics in Control Systems Theory*, volume 328 of *Lecture Notes in Control and Information Sciences*, pages 91–137. Springer Berlin / Heidelberg, 2006.

[70] G. Leen and D. Heffernan. Expanding automotive electronic systems. *IEEE Computer*, 35(1):88–93, 2002.

[71] G. Leen and D. Heffernan. Advanced Tools for Nonlinear Sampled-Data Systems' Analysis and Control. *European Journal of Control*, 13/2-3:221–241, 2007.

[72] F. Lewis. *Applied optimal control and estimation.* Prentice Hall, 1992.

[73] F. Lian, J. Moyne, and D. Tilbury. Implementation of networked machine tools in reconfigurable manufacturing systems. In *Proc. of the 2000 Japan-USA Symposium on Flexible Automation, Ann Arbor, MI*, 2000.

[74] J.-L. Lin and T.-J. Yeh. Modeling, identification and control of air-conditioning systems. *Int. J. of Refrigeration*, 30:209–220, 2007.

[75] B. Lincoln and B. Bernhardsson. LQR optimization of linear system switching. *IEEE Trans. on Automatic Control*, 47(10):1701–1705, 2002.

[76] M. Livani, J. Kaiser, and W. Jia. Scheduling hard and soft real-time communication in a controller area network. *Control Engineering Practice*, 7:1515–1523, 1999.

[77] J. Löfberg. Pre- and post-processing sum-of-squares programs in practice. *IEEE Transactions on Automatic Control*, 54(5):1007 -1011, 2009.

[78] S. Longo, G. Herrmann, and P. Barber. Optimization approaches for controller and schedule codesign in networked control. In *6th IFAC Symposium on Robust Control Design, 2009*, 2009.

[79] H. Lonn and J. Axelsson. A comparison of fixed priority and static cyclic scheduling for distributed automotive control applications. In 11th *Euromicro Conf. on Real-Time Systems*, pages 142–149, 1999.

[80] L. Lu, L. Xie, and M. Fu. Optimal control of networked systems with limited communication: a combined heuristic and convex optimization approach. In *Proc. of the* 42nd *IEEE Conf. on Decision and Control*, volume 2, pages 1194–1199, 2003.

[81] R. Luck and A. Ray. An observer-based compensator for distributed delays. *Automatica*, 26(5):903–908, 1990.

[82] D. Ma, G. Dimirovski, and J. Zhao. Hybrid state feedback controller design of networked switched control systems with packet dropout. In *American Control Conference*, number WeB16.6, 2010.

[83] E. Mojica-Nava, N. Quijano, N. Rakoto-Ravalontsalama, and A. Gauthier. A polynomial approach for stability analysis of switched systems. *Systems & Control Letters*, 59(2):98–104, 2010.

[84] L. Montestruque and P. Antsaklis. On the model-based control of networked systems. *Automatica*, 39(10):1837–1843, 2003.

[85] L. Montestruque and P. Antsaklis. Stability of model-based networked control systems with time-varying transmission times. *IEEE Trans. on Automatic Control*, 49(9):1562–1572, 2004.

[86] N. Navet and F. Simonot-Lion. *Automotive Embedded Systems Handbook*. CRC Press, 2009.

[87] D. Nešić. A unified approach to analysis and design of networked and quantized control systems. In *Chinese Control and Decision Conference*, pages 43–52, 2008.

[88] D. Nešić and A. Teel. Input-output stability properties of networked control systems. *IEEE Trans. on Automatic Control*, 49(10):1650–1667, 2004.

[89] D. Nešić and A. Teel. Input-to-state stability of networked control systems. *Automatica*, 40(12):2121–2128, 2004.

[90] D. Nešić, A. R. Teel, and P. V. Kokotovic. Sufficient conditions for stabilization of sampled-data nonlinear systems via discrete-time approximations. *Systems and Control Letters*, 38(4/5):259–270, 1999.

[91] D. Nešić and AR Teel. Input-output stability properties of networked control systems. *IEEE Transactions on Automatic Control*, 49(10):1650–1667, 2004.

[92] J. Nilsson. *Real-time control systems with delays*. PhD thesis, Lund Institute of Technology, 1998.

[93] M. Oda, T. Doi, and K. Wakata. Tele-manipulation of a satellite mounted robot by an on-ground astronaut. In *Proc. of the IEEE Int. Conf. on Robotics & Automation*, pages 1891–1896, 2001.

[94] A. Papachristodoulou, M. Peet, and S. Lall. Analysis of polynomial systems with time delays via the sum of squares decomposition. *IEEE Transactions on Automatic Control*, 54(5):1058–1064, 2009.

[95] A. Papachristodoulou and S. Prajna. Robust stability analysis of nonlinear hybrid systems. *IEEE Transactions on Automatic Control*, 54(5):1035–1041, 2009.

[96] P. Parrilo. *Structured semidefinite programs and semialgebraic geometry methods in robustness and optimization.* PhD thesis, California Institute of Technology, 2000.

[97] M. Peet, A. Papachristodoulou, and S. Lall. Positive forms and stability of linear time-delay systems. *SIAM J. on Control and Optimization*, 47(6):3237–3258, 2009.

[98] C. Peng, Y.C. Tian, and M.O. Tade. State feedback controller design of networked control systems with interval time-varying delay and nonlinearity. *International Journal of Robust and Nonlinear Control*, 18(12):1285–1301, 2008.

[99] I.G. Polushin, P.X. Liu, and C.H. Lung. On the model-based approach to nonlinear networked control systems. *Automatica*, 44(9):2409–2414, 2008.

[100] T. Pop, P. Pop, P. Eles, Z. Peng, and A. Andrei. Timing analysis of the FlexRay communication protocol. *Real-Time Systems*, 39(1-3):205–235, 2008.

[101] R. Postoyan and D. Nešić. A framework for the observer design for networked control systems. In *American Control Conference*, number ThB16.2, 2010.

[102] D. Quevedo, E. Silva, and D. Nešić. Design of multiple actuator-link control systems with packet dropouts. In *Proc. of the* 17th *World Congress The Int. Federation of Automatic Control*, pages 6642–6647, 2008.

[103] A. Ray. Introduction to networking for integrated control systems. *IEEE Control Systems Magazine*, 9(1):76–79, 1989.

[104] C. Reeves and J. Rowe. *Genetic Algorithms: Principles and Perspectives.* Kluwer Academic Publishers, 2003.

[105] H. Rehbinder and M. Sanfridson. Scheduling of a limited communication channel for optimal control. *Automatica*, 40(3):491–500, 2004.

[106] C. Robinson and P. Kumar. Sending the most recent observation is not optimal in networked control: linear temporal coding and towards the design of a control specific transport protocol. In *Proc. of the* 46th *IEEE Conf. on Decision and Control*, pages 334–339, 2007.

[107] P. Seiler and R. Sengupta. Analysis of communication losses in vehicle control problems. In *Proc. of the American Control Conference*, volume 2, pages 1491–1496, 2001.

[108] P. Seiler and R. Sengupta. An H_∞ approach to networked control. *IEEE Trans. on Automatic Control*, 50(3):356–364, 2005.

[109] L. Sha, T. Abdelzaher, K. Årzén, A. Cervin, T. Baker, A. Burns, G. Buttazzo, M. Caccamo, J. Lehoczky, and A. Mok. Real-time scheduling theory. *Real-Time Systems*, 28(2-3):101–155, 2004.

[110] M. Short and M. Pont. Hardware in the loop simulation of embedded automotive control systems. In *Proc. 8th Int. IEEE Conf. on Intelligent Transportation Systems*, pages 426–431, 2005.

[111] M. Short, M. Pont, and J. Fang. Assessment of performance and dependability in embedded control systems: methodology and case study. *Control Engineering Practice*, 16(11):1293–1307, 2008.

[112] M. Short, M. Pont, and Q. Huang. Simulation of vehicle longitudinal dynamics. Technical report, Embedded Systems Laboratory, University of Leicester, 2004.

[113] E. Silva. *A unified framework for the analysis and design of networked control systems.* PhD thesis, The University of Newcastle Callaghan, NSW 2308, Australia, 2009.

[114] S. Skogestad and I. Postlethwaite. *Multivariable feedback control: analysis & design.* Wiley, 2nd edition, 2005.

[115] J. Stoer and R. Bulirsch. *Introduction to numerical analysis.* Springer, 3rd edition, 2002.

[116] H. Sun. Stability of network-based feedback interconnection. *Int. J. of Control*, 82(8):1377–1388, 2009.

[117] Z. Sun and S. Ge. Analysis and synthesis of switched linear control systems. *Automatica*, 41(2):181–195, 2005.

[118] Z. Sun and S. Ge. Stability analysis of switched systems with stable and unstable subsystems: an average dwell time approach. *Int. J. of Systems Science*, 32(8):1055–1061, 2005.

[119] K. Tanaka, H. Ohtake, and H. Wang. A Sum of Squares Approach to Guaranteed Cost Control of Polynomial Discrete Fuzzy Systems. In *17th IFAC World Congress, 2008*, pages 6861–6866, 2008.

[120] S. Tatikonda. *Control under communication constraints.* PhD thesis, Massachusetts Institute of Technology, 2000.

[121] U. Topcu, A. Packard, P. Seiler, and G. Balas. Robust region-of-attraction estimation. *IEEE Transactions on Automatic Control*, 55(1):137–142, 2010.

[122] M. Törngren. Fundamentals on implementing real-time control applications in distributed computer systems. *Real-Time Systems*, 14(3):219–250, 1998.

[123] J. Turner and L. Austin. A review of current sensor technologies and applications within automotive and traffic control systems. In *Proc. of the Institution of Mechanical Engineers*, volume 214, Part D, No D6, pages 589–615, 2000.

[124] N. van de Wouw, D. Nešić, and W. Heemels. A discrete-time framework for stability analysis of nonlinear networked control systems. To appear in Automatica (in press), 2010.

[125] C. van Loan. Computing integrals involving the matrix exponential. *IEEE Trans. on Automatic Control*, AC-23(3):395–404, 1978.

[126] M. Velasco, J. Fuertes, C. Lin, P. Martí, and S. Brandt. A control approach to bandwidth management in networked control systems. In 30th *Annual Conf. of the IEEE Industrial Electronics Society*, volume 3, pages 2343–2348, 2004.

[127] M. Vosa. *The Simple Genetic Algorithm. Foundations and Theory.* The MIT Press, 1999.

[128] G. Walsh and H. Ye. Scheduling of networked control systems. *IEEE Control Systems Magazine*, 21(1):57–65, 2001.

[129] G. Walsh, H. Ye, and L. Bushnell. Stability analysis of networked control systems. *IEEE Trans. on Control Systems Technology*, 10(3):438–446, 2002.

[130] F. Wang and D. Liu. *Networked Control Systems.* Springer, 2008.

[131] J. Wang, T. Chen, and B. Huang. Multirate sampled-data systems: computing fast-rate models. *J. of Process Control*, 14(1):79–88, 2004.

[132] X. Wang and N. Hovakimyan. $\mathcal{L}_1$ adaptive control of event-triggered networked systems. In *American Control Conference*, number ThA03.2, 2010.

[133] X. Wang and M. Lemmon. Asymptotic stability in distributed event-triggered networked control systems with delays. In *American Control Conference*, number WeB16.5, 2010.

[134] Y. Wang and G. Yang. H-infinity performance optimization for networked control systems with limited communication channels. *J. of Control Theory and Applications*, 8(1):99–104, 2010.

[135] Z. Wang, F. Yang, D. Ho, and X. Liu. Robust $\mathcal{H}_\infty$ control for networked systems with random packet losses. *IEEE Trans. on Systems, Man and Cybernetics*, 37(4):1083–4419, 2006.

[136] J. Wu and T. Chen. Design of networked control systems with packet dropouts. *IEEE Trans. on Automatic Control*, 52(7):1314–1319, 2007.

[137] F. Xia, X. Dai, Z. Wang, and Sun Y. Feedback based network scheduling of networked control systems. *International Conf. on Control and Automation*, 2:1231–1236, 2005.

[138] F. Xia and Y. Sun. *Control and Scheduling Codesign*. Springer, 2008.

[139] J. Xiong and J. Lam. Stabilization of networked control systems with a logic ZOH. *IEEE Trans. on Automatic Control*, 54(2):358–363, 2009.

[140] J. Xu and D. Parnas. Scheduling processes with release times, deadlines, precedence and exclusion relations. *IEEE Trans. on Software Engineering*, 16(3):360–369, 1990.

[141] Y. Xu and J. Hespanha. Optimal communication logics in networked control systems. In *Proc. of the* 43th *IEEE Conf. on Decision and Control*, volume 4, pages 3527–3532, 2004.

[142] Y. Xu and J. Hespanha. Estimation under uncontrolled and controlled communications in networked control systems. In *Proc. of the* 44th *Conf. on Control Applications*, pages 842–847, 2005.

[143] J. Yanesi and A. Aghdam. Robust stability of LTI discrete-time systems using sum-of-squares matrix polynomials. In *American Control Conference, 2006*, pages 3828–3830, 2006.

[144] T. Yang. Networked control systems: a brief survey. In *IEE Proc. Control Theory and Applications*, volume 153, pages 403–412, 2006.

[145] J. Yépez, P. Martí, and J. Fuertes. Control loop scheduling paradigm in distributed control systems. In 29th *Annual Conf. of the IEEE Industrial Electronics Society*, pages 1441–1446, 2003.

[146] J. Yook, D. Tilbury, and N. Soparkar. Trading computation for bandwidth: reducing communication in distributed control systems using state estimators. *IEEE Trans. on Control Systems Technology*, 10(2):503–518, 2002.

[147] D. Yue, Q. Han, and C. Peng. State feedback controller design of networked control systems. In *Proc. of the IEEE Int. Conf. on Control Applications*, volume 1, pages 242–247, 2004.

[148] J. Yuz and G. Goodwin. On sampled-data models for nonlinear systems. *IEEE Transactions on Automatic Control*, 50(10):1477 – 1489, 2005.

[149] L. Zhang and D. Hristu-Varsakelis. LQG control under limited communication. In *Proc. of the* 44th *IEEE Conf. on Decision and Control*, pages 185–190, 2005.

[150] L. Zhang and D. Hristu-Varsakelis. Communication and control co-design for networked control systems. *Automatica*, 42(6):953–958, 2006.

[151] W. Zhang, M. Branicky, and S. Phillips. Stability of networked control systems. *IEEE Control Systems Magazine*, 21(1):84–99, 2001.

[152] W. Zhang and L. Yu. Modelling and control of networked control systems with both network-induced delay and packet-dropout. *Automatica*, 44(12):3206–3210, 2008.

[153] W. Zhang and L. Yu. A robust control approach to stabilization of networked control systems with time-varying delays. *Automatica*, 45(10):2440–2445, 2009.

[154] X. Zhao, W. Zhang, H. Zhang, and Q. Chen. Modeling of multirate input and output networked control system. In *Global Congress on Intelligent Systems*, pages 258–263, 2009.

Index